Student Solutions Manual
FOR
OXTOBY, FREEMAN
AND
BLOCK'S

CHEMISTRY
SCIENCE OF CHANGE
FOURTH EDITION

Donald J. Wink
The University of Illinois Chicago

THOMSON
™
BROOKS/COLE

Australia • Canada • Mexico • Singapore • Spain • United Kingdom • United States

Printed in the United States of America
4 5 6 7 8 9 10   11 10 09 08 07

Printer: Globus

ISBN-13: 978-0-030-33238-8
ISBN-10: 0-030-33238-9

For more information about our products,
contact us at:
Thomson Learning Academic Resource Center
1-800-423-0563

For permission to use material from this text,
contact us by:
Phone: 1-800-730-2214
Fax: 1-800-731-2215
Web: http://www.thomsonrights.com

Asia
Thomson Learning
5 Shenton Way #01-01
UIC Building
Singapore 068808

Australia
Nelson Thomson Learning
102 Dodds Street
South Street
South Melbourne, Victoria 3205
Australia

Canada
Nelson Thomson Learning
1120 Birchmount Road
Toronto, Ontario M1K 5G4
Canada

Europe/Middle East/South Africa
Thomson Learning
High Holborn House
50/51 Bedford Row
London WC1R 4LR
United Kingdom

Latin America
Thomson Learning
Seneca, 53
Colonia Polanco
11560 Mexico D.F.
Mexico

Spain
Paraninfo Thomson Learning
Calle/Magallanes, 25
28015 Madrid, Spain

TABLE OF CONTENTS

CHAPTER 1 - THE ATOMIC NATURE OF MATTER

1-1.

Table salt - <u>Heterogeneous</u> mixture of mostly sodium chloride and small amounts of other substances.

Wood - <u>Heterogeneous</u> mixture of tree cells, themselves mixtures of water and many thousands of substances.

Mercury - A <u>substance</u> and an <u>element</u>.

Air - <u>Homogeneous</u> mixture of gases.

Water - A <u>substance</u> that is a <u>compound</u> with the molecular formula H_2O.

Lemonade (fresh squeezed)- <u>Heterogeneous</u> (because of the pulp from the lemon) mixture of water and dissolved substances.

Sodium chloride - A <u>substance</u> that is a <u>compound</u> with the formula NaCl.

Ketchup - <u>Heterogeneous</u> mixture of oil, tomato sauce, vinegar (itself a mixture of water and acetic acid) and seasoning.

1-3. Some examples are: Warm water and a small amount of ice

Liquid in a seltzer bottle right after the bottle is opened

1-5. If the sample is 99.89% Si by mass, then it is 100.00 - 99.89 = 0.11% C by mass. This is equal to a mass fraction of 0.0011. The mass of C in the 10.0 g sample is then:

$$\text{mass C} = 10.0 \text{ g sample} \times \frac{0.11 \text{ g C}}{100 \text{ g sample}} = \textbf{0.011 g C}$$

1-7. The burning of paper yields many products. A few form the solid ash that remains. But most, in particular carbon dioxide and water vapor, are gases. They comprise the rest of the mass.

1-9. The law of the conservation of mass says that the mass of the system before reaction must equal the mass after the reaction. Therefore:

$$\text{mass}_{\text{potassium sulfite}} + \text{mass}_{\text{hydrochloric acid}} = \text{mass}_{\text{flask contents}} + \text{mass}_{\text{sulfur dioxide}}$$
$$1.75 \text{ g} + 19.75 \text{ g} = 21.09 \text{ g} + \text{mass}_{\text{sulfur dioxide}}$$
$$\textbf{0.41 g} = \text{mass}_{\text{sulfur dioxide}}$$

1-11. Any sample of pure ascorbic acid will have the same chemical composition, whatever the source. In the laboratory sample the ratio is: $\dfrac{\text{mass O}}{\text{mass C}} = \dfrac{40.0\text{g}}{30.0\text{g}} = \dfrac{4}{3} \Rightarrow \text{mass O} = \dfrac{4}{3} \times \text{mass C}$

Therefore, in the natural sample: $\text{mass O} = 21.3 \text{ g C} \times \dfrac{40.0 \text{ g O}}{30.0 \text{ g C}} = 28.4 \text{ g O}$

1-13

Substance	O : Ca ratio		Substance	O : Ca ratio
(a) CaO	1 : 1		(d) $CaCr_2O_7$	7 : 1
(b) $Ca_3(PO_4)_2$	8 : 3		(e) $Ca_5(PO_4)_3OH$	13 : 5
(c) $Ca(OH)_2$	2 : 1			

1-15. (a) Following Example 1–1 in the book:

$$\frac{72.64 \text{ g}}{27.36 \text{ g}} = \frac{\text{mass of 1 Fe}}{\text{mass of 1 O}}$$

$$\text{mass of 1 Fe} = \left(\frac{72.64}{27.36}\right) \times \text{mass of 1 O}$$

$$\text{mass of 1 Fe} = 2.618 \times \text{mass of 1 O}$$

The atomic mass of Fe in this case would be 2.618 times the atomic mass of O.

(b) In this case:

$$\frac{72.64 \text{ g}}{27.36 \text{ g}} = \frac{\text{mass of 3 Fe}}{\text{mass of 4 O}}$$

$$\text{mass of 3 Fe} = \left(\frac{72.64}{27.36}\right) \times \text{mass of 4 O}$$

$$\text{mass of 1 Fe} = \left(\frac{72.64}{27.36}\right)\left(\frac{4}{3}\right) \text{mass of 1 O}$$

$$\text{mass of 1 Fe} = 3.491 \text{ mass of 1 O}$$

The atomic mass of Fe in this case would be 3.491 times the atomic mass of O.

1-17. In this case, we are best getting the relative masses of the two metals by comparing them to the relative masses O. We know from Problem 15(a) that the atomic mass of Fe is 3.491 times the mass of O. We know from Problem 16(a) that the atomic mass of Zn is 4.086 times the mass of O. This means the mass of Zn to the mass of Fe is 4.086 / 3.491 = 1.171.

1-19.

Compound	Mass of Si combined with 1.00 g of N	Ratio of masses of Si in each compound
A	2.005 g	1.333 g
B	1.504 g	1.000 g

We must compare the masses of Si combined with 1.00 g of N in each compound. These are found to stand as 1.333 : 1.000, or 4 : 3. Therefore, these data are consistent with the law of multiple proportions. The first compound is a multiple of SiN if the second is Si_3N_4.

1-21. The chemical equation is: $2\,N_2O + 3\,O_2 \rightarrow 4\,NO_2$. At contstant temperature and pressure the coefficients on the substances indicate the volumes of gases in the reaction. So, for 4.0 L of NO_2 one needs **2.0 L of N_2O** and **3.0 L of O_2**.

1-23.

Symbol	Z	N	A	number of electrons
$^{12}C^-$	**6**	**6**	**12**	**7**
$^{31}P^{2+}$	15	**16**	31	13
^{11}Be	**5**	6	11	5
$^{211}Bi^{3+}$	**83**	126	**211**	**80**

1-25. We are told that the atom is about 10^4 to 10^5 times the size of the nucleus, so the atom would be about 10^5–10^6 cm, or 10^3 to 10^4 m, or 1 to 10 km. This is about the size of **Manhattan island**.

1-27. (a) Neutral ^{239}Pu has 94 protons and 239–94 = 145 neutrons, so the ratio is 145 / 94 = **1.54**.
(b) There are 94 + 1 = **95 electrons** in Pu⁻.

1-29. The value of the relative atomic masses of the combined oxygen isotopes was larger in the pre-1961 scale, by a factor of $\dfrac{16.0000}{15.9994} = 1.0000375$. *All* atomic masses were larger by that factor on the pre-1961 scale. So ^{12}C then had a relative atomic mass of $12 \times 1.0000375 = $ **12.0004**.

1-31. The average mass of any two marbles does not reflect the true distribution of marble types in the sack. The weighted average mass considers the mass of the marbles and their frequency in the sample.
In this case: average marble mass $= m_r f_r + m_g f_g$
Where: m_r = mass of a red marble f_r = fraction of marbles that are red
         m_g = mass of a green marble f_g = fraction of marbles that are green
The fractions are given by the relative numbers of each type:
$$f_r = \frac{19,490}{19,490+26,278} = 0.4258 \qquad f_g = \frac{26,278}{19,490+26,278} = 0.5742$$
average marble mass = (10.67 g)(0.4258) + (13.53 g)(0.5742) = **12.31g**.
The value (10.67 + 13.53) / 2 = 12.10 is the simple average of the mass of one green and one red marble. This does not take in into account that there are more green marbles than red ones.

1-33. For the relative atomic mass, one calculates a weighted average of the masses of the isotopes:

$$\text{relative mass copper} = A_{63}p_{63} + A_{65}p_{65}$$

Where: A_{63} = atomic mass of ^{63}Cu p_{63} = fractional abundance of ^{63}Cu

A_{65} = atomic mass of ^{65}Cu p_{65} = fractional abundance of ^{65}Cu

In this problem, the unknown is A_{65}:

$$63.546 = (62.929599)(0.6917) + A_{65}(0.3083)$$
$$63.546 = 43.52840363 + 0.3083\ A_{65}$$
$$20.01759637 = 0.3083\ A_{65}$$
$$\mathbf{64.93} = A_{11}$$

1-35. For this problem we use the atomic masses inside the back of the book. The answer is precise to the same number of places after the decimal as for the least precise atomic mass.

(a) $2(14.0067) + 5(15.9994) = \mathbf{108.0104}$

(b) $126.90447 + 18.9984032 = \mathbf{145.90287}$

(c) $2(22.98977) + 32.066 + 3(15.9994) = \mathbf{126.044}$

(d) $39.0983 + 54.90849 + 4(15.9994) = \mathbf{158.0267}$

(e) $2[14.0067 + 4(1.00794)] + 32.066 + 4(15.9994) = \mathbf{132.141}$

1-37. (a) $12.000\ \text{g}\ SF_2 \times \dfrac{1\ \text{mol}\ SF_2}{70.0628\ \text{g}\ SF_2} \times \dfrac{6.0221 \times 10^{23}\ \text{molecules}\ SF_2}{1\ \text{mol}\ SF_2}$

$$= \mathbf{1.0314 \times 10^{23}\ \text{molecules}\ SF_2}$$

(b)   1.000 g of fluorine equals: $1.000\ \text{g}\ F \times \dfrac{1\ \text{mol}\ F}{18.998\ \text{g}\ F} = 0.052637\ \text{mol}\ F$

The chemical formula indicates: $\dfrac{n_S}{n_F} = \dfrac{1}{2}$ so $n_S = 0.50000 \times n_F = 0.02632\ \text{mol}\ S$

We can now convert to mass: $0.2632\ \text{mol}\ S \times \dfrac{32.066\ \text{g}\ S}{1\ \text{mol}\ S} = \mathbf{0.8439\ g\ S.}$

(c) $12.000\ \text{g}\ S_2F_4 \times \dfrac{1\ \text{mol}\ S_2F_4}{140.124\ \text{g}\ S_2F_4} \times \dfrac{6.0221 \times 10^{23}\ \text{molecules}\ S_2F_4}{1\ \text{mol}\ S_2F_4}$

$$= \mathbf{5.157 \times 10^{22}\ \text{molecules}\ S_2F_4}$$

1.000 g of F (again) corresponds to 0.052637 mol F.

The chemical formula indicates: $\dfrac{n_S}{n_F} = \dfrac{2}{4} = \dfrac{1}{2}$

so: $n_S = 0.50000 \times n_F = 0.02632\ \text{mol}\ S$

This is the same chemical amount of S as in part (b), so mass S = **0.8440 g S.**

1-37. (continued)

(d) The number of SF_2 and S_2F_4 molecules in 12.000 g of each of these compounds is different because the molar masses are different. Because the mass of an S_2F_4 molecule is twice the mass of an SF_2 molecule, a given mass of SF_2 will contain twice as many molecules as the same mass of S_2F_4. But since the ratio $\dfrac{n_S}{n_F} = \dfrac{1}{2}$ is the same in each molecule, the relative masses contributed by S and by F are the same in each case.

1-39. Iodine has only one naturally occurring isotope, so dividing the mass of one mole of iodine atoms by Avogadro's number gives us the actual mass of one I atom.

$$\frac{126.90447 \text{ g of I}}{N_0 \text{ atoms of I}} = \frac{126.90447 \text{ g of I}}{6.022142 \times 10^{23} \text{atoms of I}} = 2.1072997 \times 10^{-22} \frac{\text{g}}{\text{atom I}}$$

For the mass in u, we only need to use the atomic mass, on the scale of the atom:

Mass of one ^{127}I atom = 126.90447 u

1-41. For the molar mass of ferrocene:

$$M = (55.847 \frac{\text{g Fe}}{\text{mol Fe}}) + 10 \times (12.011 \frac{\text{g C}}{\text{mol C}}) + 10 \times (1.00794 \frac{\text{g H}}{\text{mol H}})$$

$$= \frac{186.036 \text{ g Fe}(C_5H_5)_2}{1 \text{ mol Fe }(C_5H_5)_2}$$

(a) $$n \text{ Fe}(C_5H_5)_2 = 100.0 \text{ g Fe}(C_5H_5)_2 \times \left(\frac{1 \text{ mol Fe}(C_5H_5)_2}{186.04 \text{ g Fe}(C_5H_5)_2} \right)$$

$$= \mathbf{0.5375 \text{ mol Fe}(C_5H_5)_2}$$

(b) $$N \text{ Fe}(C_5H_5)_2 = 0.5375 \text{ mol Fe}(C_5H_5)_2 \times \left(\frac{6.022 \times 10^{23} \text{ molecules Fe}(C_5H_5)_2}{1 \text{ mol Fe}(C_5H_5)_2} \right)$$

$$= \mathbf{3.237 \times 10^{23} \text{ molecules Fe}(C_5H_5)_2}$$

1-43. (a) The mass of one mole of Li_2CO_3 is 73.894 g. The mass of Li in one mole of Li_2CO_3 is 13.882 g. We use this to get the conversion factor needed for this case:

$$65.4 \text{ g Li}_2CO_3 \times \frac{13.882 \text{ g Li}}{73.891 \text{ g Li}_2CO_3} = \mathbf{12.3 \text{ g Li}}$$

(b) The only difference in the two parts of this problem is the units for mass—grams and pounds. The relative weights of Li in $LiCO_3$ is independent of the units for mass. Therefore, if there are 12.3 g Li in 65.4 g of $LiCO_3$, then there are **12.3 lb Li** in 65.4 lb of $LiCO_3$.

1-45. If we convert 100 lb to g, we get 45,359 g of nails per barrel, or 4,536 nails per barrel. We will need to order 450,000 / 4536 = **99.2 kegs** of nails.

1-47. Some unit conversions is in order first: There are 10 deciliters (10 dL) in 1 liter and there are 1000 milliliters (1000 mL) in 1 liter. Therefore the sample size is:

$$500 \text{ mL} \times \left(\frac{1 \text{ L}}{1000 \text{ mL}} \right) \times \left(\frac{10 \text{ dL}}{1 \text{ L}} \right) = 5.00 \text{ dL}.$$

For this patient: $5.00 \text{ dL} \times \left(\frac{3.2 \times 10^{-9} \text{g Hg}}{\text{dL}} \right) = 1.6 \times 10^{-8} \text{ g} = \textbf{16 ng Hg}$

1-49. This is a good point to introduce a tabular layout for mass % problems.

	MOLAR MASS ($g\ mol^{-1}$)	MASS OF H IN ONE MOLE OF COMPOUND	MASS % H
H_2O	18.0153	$\left(\frac{2 \text{ mol H}}{1 \text{ mol } H_2O} \right) \times \left(\frac{1.00794 \text{ g H}}{1 \text{ mol H}} \right)$ $= \frac{2.0159 \text{ g H}}{1 \text{ mol } H_2O}$	$\left(\frac{2.0159 \text{ g H}}{1 \text{ mol } H_2O} \right) \times \left(\frac{1 \text{ mol } H_2O}{18.0153 \text{ g } H_2O} \right)$ $= \frac{0.1119 \text{ g H}}{1 \text{ g } H_2O} = 11.19\% \text{ H}$
$C_{12}H_{26}$	170.34	$\left(\frac{26 \text{ mol H}}{1 \text{ mol } C_{12}H_{26}} \right) \times \left(\frac{1.00794 \text{ g H}}{1 \text{ mol H}} \right)$ $= \frac{26.2064 \text{ g H}}{1 \text{ mol } C_{12}H_{26}}$	$\left(\frac{26.206 \text{ g H}}{1 \text{ mol } C_{12}H_{26}} \right) \times \left(\frac{1 \text{ mol } C_{12}H_{26}}{170.34 \text{ g } C_{12}H_{26}} \right)$ $= \frac{0.1538 \text{ g H}}{1 \text{ g } C_{12}H_{26}} = 15.38\% \text{ H}$
N_4H_6	62.0746	$\left(\frac{6 \text{ mol H}}{1 \text{ mol } N_4H_6} \right) \times \left(\frac{1.00794 \text{ g H}}{1 \text{ mol H}} \right)$ $= \frac{6.0476 \text{ g H}}{1 \text{ mol } N_4H_6}$	$\left(\frac{6.0476 \text{ g H}}{1 \text{ mol } N_4H_6} \right) \times \left(\frac{1 \text{ mol } N_4H_6}{62.0746 \text{ g } N_4H_6} \right)$ $= \frac{0.09743 \text{ g H}}{1 \text{ g } N_4H_6} = 9.743\% \text{ H}$
LiH	7.949	$\left(\frac{1 \text{ mol H}}{1 \text{ mol LiH}} \right) \times \left(\frac{1.00794 \text{ g H}}{1 \text{ mol H}} \right)$ $= \frac{1.00794 \text{ g H}}{1 \text{ mol LiH}}$	$\left(\frac{1.00794 \text{ g H}}{1 \text{ mol LiH}} \right) \times \left(\frac{1 \text{ mol LiH}}{7.949 \text{ g LiH}} \right)$ $= \frac{0.1268 \text{ g H}}{1 \text{ g LiH}} = 12.68 \% \text{ H}$

In order of increasing mass % H: $\textbf{N}_4\textbf{H}_6 < \textbf{H}_2\textbf{O} < \textbf{LiH} < \textbf{C}_{12}\textbf{H}_{26}$.

1-51. Molar mass = 19 × at. mass C + 28 × at. mass H + 2 × at. mass O

$$= 19 \times 12.011 + 28 \times 1.00794 + 2 \times 15.9994 = 288.43 \text{ g mol}^{-1}$$

$$\left(\frac{\text{g C in 1 mol}}{\text{molar mass}} \right) = \left(\frac{19 \times 12.011}{288.43} \right) = 0.7912 = \textbf{79.12\% C}$$

1-53. The masses of each element in one mole of the compound are:

$$\text{mass of Cl} = 1 \text{ mol Cl} \times \left(\frac{35.4527 \text{ g Cl}}{1 \text{ mol Cl}}\right) = 35.4527 \text{ g Cl}$$

$$\text{mass of F} = 8 \text{ mol F} \times \left(\frac{18.9984 \text{ g F}}{1 \text{ mol F}}\right) = 151.987 \text{ g F}$$

$$\text{mass of O} = 2 \text{ mol O} \times \left(\frac{15.9994 \text{ g O}}{1 \text{ mol O}}\right) = 31.9988 \text{ g O}$$

$$\text{mass of Pt} = 1 \text{ mol Pt} \times \left(\frac{195.08 \text{ g Pt}}{1 \text{ mol Pt}}\right) = 195.08 \text{ g Pt}$$

Total mass of one mole of $ClF_2O_2PtF_6$ = 414.5167 g.

The mass percentages are:

$$\left(\frac{35.4527 \text{ g Cl}}{414.5167 \text{ g compound}}\right) = 0.08553 \quad = \quad \textbf{8.553\% Cl}$$

$$\left(\frac{151.987 \text{ g F}}{414.5167 \text{ g compound}}\right) = 0.3667 \quad = \quad \textbf{36.6661\% F}$$

$$\left(\frac{31.9988 \text{ g O}}{414.5167 \text{ g compound}}\right) = 0.07720 \quad = \quad \textbf{7.71954\% O}$$

$$\left(\frac{195.08 \text{ g Pt}}{414.5167 \text{ g compound}}\right) = 0.4706 \quad = \quad \textbf{47.06\% Pt}$$

1-55. The molar mass of $PbCrO_4$ is 323.2 g mol^{-1}. The molar mass of $PbCrO_4 \bullet PbO$ is 546.4 g mol^{-1}. The mass percentages are:

$$\text{for } PbCrO_4: \quad \left(\frac{207.2 \text{ g Pb}}{323.2 \text{ g compound}}\right) = 0.6411 \quad = \textbf{64.11\% Pb}$$

$$\text{for } PbCrO_4 \bullet PbO: \quad \left(\frac{414.4 \text{ g Pb}}{546.4 \text{ g compound}}\right) = 0.7584 \quad = \textbf{75.84 \% Pb}$$

1-57. We determine the number of moles of each of the elements, assuming a 100.0 g sample. In a such a sample, the mass of each element in grams equals its mass percentage in the compound. Hence:

Element	Grams in 100.00 g	Moles in 100 g	Relative # of moles
Si	28.37 g	$28.37 \text{ g Si} \times \left(\dfrac{1 \text{ mol Si}}{28.0855 \text{ g Si}}\right) = 1.010 \text{ mol Si}$	1.000
Cl	100.00-28.37= 71.63 g	$71.63 \text{ g Cl} \times \left(\dfrac{1 \text{ mol Cl}}{35.4527 \text{ g Cl}}\right) = 2.020 \text{ mol Cl}$	2.000

The empirical formula of the compound is **SiCl₂**.

1-59.

Element	Grams in 100.00 g	Moles in 100 g	Relative # of moles
Si	53.99 g	$53.99\ g\ Si \times \left(\dfrac{1\ mol\ Si}{28.0855\ g\ Si}\right) = 1.922\ mol\ Si$	2.333
Fe	100.00 - 53.99 = 46.01 g	$46.01\ g\ Fe \times \left(\dfrac{1\ mol\ Fe}{55.847\ g\ Fe}\right) = 0.8239\ mol\ Fe$	1.000

The empirical formula must have whole number subscripts. We note that $2.333 \times 3 = 7.00$ and $1.00 \times 3 = 3.00$ (see text Appendix C-4). The empirical formula is **Fe₃Si₇**.

1-61. For compound #1:

Element	Grams in 100 g	Moles in 100 g	Relative # of moles
Ba	90.745 g	$90.745\ g\ Ba \times \left(\dfrac{1\ mol\ Ba}{137.327\ g\ Ba}\right) = 0.66080\ mol\ Ba$	1.000
N	100.000 g - 90.745 g = 9.255 g	$9.255\ g\ N \times \left(\dfrac{1\ mol\ N}{14.00674\ g\ N}\right) = 0.66075\ mol\ N$	1.000

The empirical formula is **BaN**.

Element	Grams in 100 g	Moles in 100 g	Relative # of moles
Ba	93.634 g	$93.634\ g\ Ba \times \left(\dfrac{1\ mol\ Ba}{137.327\ g\ Ba}\right) = 0.68183\ mol\ Ba$	1.500
N	100.000 g - 93.634 g = 6.336 g	$6.366\ g\ N \times \left(\dfrac{1\ mol\ N}{14.00674\ g\ N}\right) = 0.45450\ mol\ N$	1.000

The empirical formula is **Ba₃N₂**.

1-63. In this problem, we are given the masses (not percentages) of two of the elements in a sample of definite (not hypothetical) size. The third mass, that of oxygen, is found by difference:

mass oxygen = total mass - mass Na - mass S

= 65.32 g - 21.14 g - 14.74 g = 29.44 g O

1-63. (continued)

Element	Grams in sample	Moles in sample	Relative # of moles
Na	21.14 g	$21.14 \text{ g Na} \times \left(\dfrac{1 \text{ mol Na}}{22.9898 \text{ g Na}} \right) = 0.9195 \text{ mol Na}$	2.000
S	14.74 g	$14.74 \text{ g S} \times \left(\dfrac{1 \text{ mol S}}{32.066 \text{ g S}} \right) = 0.4597 \text{ mol S}$	1.000
O	29.44 g	$29.44 \text{ g O} \times \left(\dfrac{1 \text{ mol O}}{15.9994 \text{ g O}} \right) = 1.840 \text{ mol N}$	4.003

Within the precision of the measurements, the empirical formula is **Na_2SO_4**.

1-65. In a problem of this sort:

moles C in sample = moles C in CO_2 = moles of CO_2 after combustion

moles H in sample = moles H in H_2O = 2 × moles H_2O after combustion

$$n \, C = 3.701 \text{ g } CO_2 \times \left(\frac{1 \text{ mol } CO_2}{44.021 \text{ g } CO_2} \right) \times \left(\frac{1 \text{ mol C}}{1 \text{ mol } CO_2} \right) = 0.0841 \text{ mol C}$$

$$n \, H = 1.082 \text{ g } H_2O \times \left(\frac{1 \text{ mol } H_2O}{18.015 \text{ g } H_2O} \right) \times \left(\frac{2 \text{ mol H}}{1 \text{ mol } H_2O} \right) = 0.1201 \text{ mol H}$$

$n \, H : n \, C = 0.1201 : 0.08410 = 1.428 : 1.$

To determine the nearest whole number ratio equal to this value, we note that $0.427 \approx {}^3/_7$.
Therefore:

moles H : moles C = $(1 + {}^3/_7) : 1 = ({}^{10}/_7) : 1 = 10 : 7$.

To the precision of the data, the empirical formula is **C_7H_{10}**.

1-67. All of the carbon in the sample produces CO_2. We can therefore relate the number of grams of CO_2 produced with the number of moles C initially present. A similar relationship exists between grams of H_2O and moles of H, and grams of SO_2 and moles of S.

$$n \, C = n \, O = n \, CO_2 = 0.013252 \text{ g } CO_2 \times \left(\frac{1 \text{ mol } CO_2}{44.010 \text{ g CO}} \right) = 3.01 \times 10^{-4} \text{ mol C}$$

$$n \, H = 2 \times n \, H_2O = 2 \times 0.005427 \text{ g } H_2O \times \left(\frac{1 \text{ mol } H_2O}{18.0153 \text{ g } H_2O} \right) = H \times 10^{-4} \text{ mol H}$$

$$n \, S = n \, SO_2 = 0.009646 \text{ g } SO_2 \times \left(\frac{1 \text{ mol } SO_2}{64.065 \text{ g } SO_2} \right) = 1.51 \times 10^{-4} \text{ mol S}$$

The relative numbers of moles in the compound are:

moles C : moles H : moles S = 3.01 : 6.02 : 1.50. So the empirical formula is **C_2H_4S**.

1-69. The molecular formula of a compound is always an integral multiple of the empirical formula, so we can write:

molecular formula = (empirical formula)$_n$ where n = 1,2,3....

The same integer n relates the molar mass of a compound with the "empirical mass". The empirical mass of C_3H_5 is 41.07 g mol^{-1}.

$$\text{Molar mass} = n\,(\text{Empirical mass}),$$

$$n = \frac{\text{Molar mass}}{\text{Empirical mass}} = \frac{410.7 \text{ g mol}^{-1}}{41.07 \text{ g mol}^{-1}} = 10.00$$

Therefore: molecular formula = $(C_3H_5)_{10}$ = **$C_{30}H_{50}$.**

1-71. The mass of a sample is the product of the volume and the density:

$$m \text{ AsF}_3 = 5.12 \text{ cm}^3 \text{ AsF}_3 \times \left(\frac{2.163 \text{ g AsF}_3}{1 \text{ cm}^3 \text{ AsF}_3} \right) = 11.07 \text{ g} = \mathbf{11.1 \text{ g}}$$

$$\text{molar mass of AsF}_3 = (74.9216\,\frac{\text{g As}}{\text{mol As}}) + 3 \times (18.9984\,\frac{\text{g F}}{\text{mol F}}) = 131.9168\,\frac{\text{g AsF}_3}{\text{mol AsF}_3}$$

$$n \text{ AsF}_3 = 11.07 \text{ g AsF}_3 \times \left(\frac{1 \text{ mol AsF}_3}{131.9168 \text{ g AsF}_3} \right) = \mathbf{0.0840 \text{ mol AsF}_3}$$

1-73. Density is defined as the mass of a substance per unit volume. Therefore, the density in this case is 49.4 g / 537 cm^3 = 0.0920 g cm^{-3}.

1-75. This problem requires two conversions: from kg to g and from g to cm^3:

$$34.5 \text{ kg} \times \left(\frac{1000 \text{ g}}{1 \text{ kg}} \right) \times \left(\frac{1 \text{ cm}^3 \text{ Hg}}{13.6 \text{ g Hg}} \right) = 2540 \text{ cm}^3$$

1-77 In this problem, conversions use density (the ratio of mass to volume), molar mass (the ratio of mass and chemical amount) and Avogadro's number (the ratio of chemical amount and the amount of particles):

$$\left(\frac{1 \text{ cm}^3 \text{ Lu}}{9.84 \text{ g Lu}} \right) \times \left(\frac{174.967 \text{ g Lu}}{1 \text{ mol Lu}} \right) \times \left(\frac{1 \text{ mol Lu}}{6.0221 \times 10^{23} \text{ atom Lu}} \right) = 2.95 \times 10^{-23}\,\frac{\text{cm}^3}{\text{atom Lu}}$$

1-79 The official answer is correct, because the steel wool and the sawdust are themselves solids. The misunderstanding arises from confusion about heterogeneous mixtures. There are no holes in the steel wool, though the steel wool is spread out to enclose some gas. Similarly, the "pouring" of sawdust is not a property of liquids only. The important difference is that liquids, unlike sawdust, would also spread out uniformly over the bottom of a beaker.

1-81. If the rose-hip vitamin C is simply pure vitamin C, then there could be no difference. But the naturally-derived material may in fact be impure and consist of a mixture of vitamin C and small amounts of other substances. These impurities may have some separate health effects.

1-83. (a)

MATERIAL	CLASSIFICATION
Soft-wood chips	A complex heterogeneous mixture.
Water	A pure compound with the molecular formula H_2O.
Iron	An element (Fe)
Sodium Hydroxide	A pure compound with the formula NaOH.

(b) The iron vessel is a sealed system. All of the mass in it at the start remains in it at the end. The total mass at the end is the same as that at the beginning: $17.2 + 150.1 + 22.43 = $ **189.7 kg.**

1-85. The lightest of the objects are the grebes: the krulls are listed as heavier than the grebes and the grebes are listed as having a fraction of the mass of the stoats. The listed ratios are the basis for the calculation of the mass of a stoat and of a krull:

$$\text{mass of grebe} = 0.825 \times \text{mass of stoat}$$
$$1.000 \text{ g} = 0.825 \times \text{mass of stoat}$$
$$1.000 \text{ g} / 0.825 = \text{mass of stoat} = \textbf{1.21 g}$$
$$\text{mass of krull} = 4.80 \times \text{mass of grebe}$$
$$= 4.80 \times 1.000 \text{g} = \textbf{4.80 g}$$

$$\text{mass of collection} = (17 \times 4.80 \text{ g}) + (19 \times 1.00 \text{ g}) + (11 \times 1.21 \text{ g}) = \textbf{114 g}$$

1-87. 1. *Matter consists of indivisible atoms.* There are two ways scientists have seen that atoms are divisible: in spontaneous radioactive decay (for example with all isotopes of uranium or radium) and in "atom smashing" events, for example in a nuclear reactor.

2. *All atoms of a given chemical element are identical in mass and in all other properties.* Isotopes were not discovered until almost 100 years after Dalton's work. Isotopes have virtually identical chemical properties but have different masses.

3. *Different chemical elements have different kinds of atoms, and in particular, such atoms have different masses.* This statement (so far) needs no modification or extension.

1-87 (continued)

4. *Atoms are indestructible, and retain their identity in chemical reactions.* It is still true that there are no instances where atoms are destroyed in a chemical reaction. Atoms are destroyed in nuclear reactions (which we do not consider chemical reactions), discussed in Chapter 15 of the text.

5. *The formation of a compound from its elements occurs through combining atoms of unlike elements in small whole-number ratios.* Some solid compounds in fact do have fractional elemental composition. A few of these were known in Dalton's time, but experimental imprecision prevented their recognition.

1-89. This suggestion is entirely consistent with Gay-Lussac's observations. The student would write the equation for the decomposition of water as:

$$2\ H_4O_2 \quad\rightarrow\quad 2\ H_4 \quad+\quad O_4$$

In volumes, this is the same as saying:

2 volumes water $\rightarrow$ 2 volumes hydrogen + 1 volume oxygen.

Compare this with the "conventional" equation:

$$2\ H_2O \quad\rightarrow\quad 2\ H_2 \quad+\quad O_2,$$

which predicts the same relative volumes of the gases.

1-91. The mass fraction of Zr in both gems must be the same. In the first gem this is:

$$\text{mass fraction Zr} = \frac{5.474}{5.474+1.685+3.840} = 0.4977.$$

In the second gem, the mass fraction of Zr is the same, so the mass of Zr is

$$\text{mass Zr} = 43.67 \times 0.4977 = \textbf{21.73 g}$$

1-93. (a) The molecular mass of UF_6 is $238.03 + 6(18.998) = 352.02$, so the molar mass is 352.02 g mol^{-1}. The average mass of one molecule of UF_6 is:

$$\left(\frac{352.02 \text{ g } UF_6}{1 \text{ mol } UF_6}\right) \times \left(\frac{1 \text{ mol } UF_6}{6.0221\times10^{23} \text{ molecule } UF_6}\right) = 5.8454\times10^{-22}\ \frac{\text{g}}{\text{molecule } UF_6}$$

(b) The mass of uranium used to determine the molar mass of UF_6 is *average* mass for this element, taking into account the relative masses and abundances of the isotopes. There is no molecule that has this average mass. The actual masses of molecules of UF_6 will be distributed amongst several possible values, determined by the isotope of U that is present.

1-95. The ratios of the isotopes can be used to get the relative fractional amounts of the elements. First, we note that in a typical sample there are 498.7 parts of ^{16}O, 0.1832 parts ^{17}O, and 1.000 part ^{18}O, for a total of 499.8832 parts. Then the fractional abundances are:

$^{16}O = {}^{498.7}/_{499.8832} = 0.99763$ $\qquad$ $^{17}O = {}^{0.1832}/_{499.8832} = .0003665$

$^{18}O = {}^{1.000}/_{499.8832} = 0.0020005$

Then these fractional abundances can be applied to the 10 million atom sample:

No. ^{18}O atoms $= 10,000,000 \times 0.0020005$ $\quad = \quad$ **20,005 ^{18}O atoms**

No. ^{17}O atoms $= 10,000,000 \times 0.0003665$ $\quad = \quad$ **3,665 ^{17}O atoms**

No. ^{16}O atoms $= 10,000,000 \times 0.9976$ $\quad = \quad$ **9,976,000 ^{16}O atoms**

1-97. (a) This problem reminds us that the relative masses of the elements are not arbitrary, regardless of the mass scale or the counting numbers that we use.

32 margs of sulfur is actually 32×4.8648 g $= 155.67$ g of sulfur. This corresponds to

$$155.67\,\text{g }^{32}S \times \left(\frac{1\text{ mol }^{32}S}{31.972\text{ g }^{32}S}\right) = 4.8689\text{ mol }^{32}S.$$

The number of ^{32}S atoms in this quantity of ^{32}S is:

$$4.8689\text{ mol }^{32}S \times \left(\frac{6.0221\times10^{23}\text{ atom }^{32}S}{1\text{ mol }^{32}S}\right) = 2.9321\times10^{24}\frac{\text{atom }^{32}S}{1\text{ mol }^{32}S}$$

Therefore, $N_{or} =$ **2.9321 X 10²⁴.**

(b) On earth, the atomic mass of P is 30.9738 g/mol, so 6.0221×10^{23} atoms of P weigh 30.9738 g. Therefore, an average P atom weighs:

$$\left(\frac{30.9738\text{ g P}}{1\text{ mol P}}\right) \times \left(\frac{1\text{ mol P}}{6.0221\times10^{23}\text{ atoms P}}\right) = 5.1434\times10^{-23}\text{ g}.$$

Finally, we can convert from our units to the units on planet HTRAE because we know the values of their counting unit N_{or} and their mass unit, the marg:

$$\left(\frac{5.1434\times10^{-23}\text{ g}}{\text{atom P}}\right) \times \left(\frac{2.9321\times10^{24}\text{ atoms P}}{\text{elom }^{32}P}\right) \times \left(\frac{1\text{ marg}}{4.8648\text{ g}}\right) = 31.000\frac{\text{marg P}}{\text{elom P}}$$

A chemist on HTRAE would write the relative atomic mass of phosphorus as **31.000.**

1-99. The blue line corresponds to iridium and the red line to hydrogen. The data given indicate that the volume per gram of the elements is 0.04423 cm³ g⁻¹ for Ir and 12,100 cm³ g⁻¹ for hydrogen. Clearly, hydrogen occupies much more volume per gram than Ir.

1-101. The mass of reactant OsO_4 is 17.6 mg, or 0.0176 g OsO_4.

$$0.0176 \text{ g } OsO_4 \times \left(\frac{1 \text{ mol } OsO_4}{254.228 \text{ g } OsO_4}\right) \times \left(\frac{1 \text{ mol Os}}{1 \text{ mol } OsO_4}\right) \times \left(\frac{190.23 \text{ g Os}}{1 \text{ mol Os}}\right)$$

$$= 0.0132 \text{ g Os} = \textbf{13.2 mg Os}$$

1-103.

For $InCl_2$: $\dfrac{114.818 \text{ g In}}{185.723 \text{ g } InCl_2} = 0.61823 = \textbf{61.82\% In}$

For $In_3[In_2Cl_9]$ (= In_5Cl_9): $\dfrac{574.09 \text{ g In}}{893.164 \text{ g } In_5Cl_9} = 0.64276 = \textbf{64.28\% In}$

The compositions of the two compounds are quite similar.

1-105. The mass of potassium is the difference between the mass of the sample and the mass of the cesium:

mass K in sample = mass sample - mass Cs

$$= 0.03151 \text{ g} - 0.02346 \text{ g} = 0.00805 \text{ g}$$

Element	Grams in sample	Moles in sample	Relative # of moles
Cs	0.02346	$0.02346 \text{ g Cs} \times \left(\dfrac{1 \text{ mol Cs}}{132.91 \text{ g Cs}}\right) = 1.765 \times 10^{-4} \text{ mol Cs}$	1.000
K	0.00805	$0.00805 \text{ g K} \times \left(\dfrac{1 \text{ mol K}}{39.098 \text{ g K}}\right) = 2.06 \times 10^{-4} \text{ mol K}$	1.167

The K:Cs ratio of 1.167:1 is equivalent to 7:6. Thus, the formula is $\textbf{K}_7\textbf{C}_6$.

1-107. This real sample contains all of the 0.7200 g of Ba and an additional mass of 0.1678 g of O. Then:

Element	grams in sample	moles in sample	relative # of moles
Ba	0.7200 g	$0.7200 \text{ g Ba} \times \left(\dfrac{1 \text{ mol Ba}}{137.327 \text{ g Ba}}\right) = 0.0052439 \text{ mol Ba}$	1.0000
O	0.1678g	$0.1678 \text{ g O} \times \left(\dfrac{1 \text{ mol O}}{15.9994 \text{ g O}}\right) = 0.010488 \text{ mol O}$	2.0000

The compound is $\textbf{BaO}_2$ (barium peroxide).

CHAPTER 2 - CHEMICAL EQUATIONS AND REACTION YIELDS

2-1.   The following chemical equations can be balanced by "inspection." The first two are the answers alone. For (c) and (d) the method is developed as in Example 2–1.

(a) $N_2 + O_2 \rightarrow 2\,NO$

(b) $2\,N_2 + O_2 \rightarrow 2\,N_2O$

(c)   ___ K_2SO_3 + ___ $HCl \rightarrow$ ___ KCl + ___ H_2O + ___ SO_2

The key to this method is to avoid "traps" that occur when a particular element is present in more than one substance without a coefficient. We have to start somewhere, of course, and we begin by putting a "1" in front of any substance. Here the first substance is chosen, because it is the most complex substance *and it has elements that only occur in one other place.*

__1__ K_2SO_3 + ___ $HCl \rightarrow$ __ KCl + ___ H_2O + ___ SO_2

Next, assign a coefficent to another substance to balance one of the elements in K_2SO_3. We choose KCl, where we need 2 KCl to balance the 2 K's in K_2SO_3:

__1__ K_2SO_3 + ___ $HCl \rightarrow$ __2__ KCl + ___ H_2O + ___ SO_2

Next, repeat the process with the S in 1 K_2SO_3 and 1 SO_2:

__1__ K_2SO_3 + ___ $HCl \rightarrow$ __2__ KCl + ___ H_2O + __1__ SO_2

The Cl in KCl and HCl is next:

__1__ K_2SO_3 + __2__ $HCl \rightarrow$ __2__ KCl + ___ H_2O + __1__ SO_2

We finish with the H in HCl and H_2O:

__1__ K_2SO_3 + __2__ $HCl \rightarrow$ __2__ KCl + __1__ H_2O + __1__ SO_2

We clean up the writing now by omitting the unit ("1") coefficients:

$$K_2SO_3 + 2\,HCl \rightarrow 2\,KCl + H_2O + SO_2$$

Note that there was no place where we had to deal with the oxygens. These are balanced as a result of the others being done. The final check is that both sides have 2 K, 1 S, 3 O, 2 H, and 2 Cl.

(d)   ___ NH_3 + ___ O_2 + ___ $CH_4 \rightarrow$ ___ HCN + ___ H_2O

We start with the NH_3:

__1__ NH_3 + ___ O_2 + ___ $CH_4 \rightarrow$ ___ HCN + ___ H_2O

The HCN can be next, for there is 1 N in 1 NH_3 and 1 N in HCN. Note that we should *not* touch the H just yeat, since H occurs in four different substances!

__1__ NH_3 + ___ O_2 + ___ $CH_4 \rightarrow$ __1__ HCN + ___ H_2O

The C and be done from the HCN and the CH_4:

__1__ NH_3 + ___ O_2 + __1__ $CH_4 \rightarrow$ __1__ HCN + ___ H_2O

2-1. (continued)

Now we are ready to to the H. The equation as written has 7 H accounted for on the reactant side and 1 H accounted for on the product side (note that the H_2O doesn't have a coefficient yet, so it is not included in this accounting). We need 6 H on the product side, provided by 3 H_2O:

__1__ NH_3 + ___ O_2 + __1__ CH_4 → __1__ HCN + __3__ H_2O

We finish with the oxygen. There are no O's accounted for, yet, on the reactant side. But there are 3 O's accounted for on the product side. We need 3 O's from the O_2 on the reactant side. Since each O_2 gives 2 O's, we need 3 / 2 O_2:

__1__ NH_3 + **3/2** O_2 + __1__ CH_4 → __1__ HCN + __3__ H_2O

We are now ready to multiply through both sides to clear up fractions:

$$2\ NH_3 + 3\ O_2 + 2\ CH_4 \rightarrow 2\ HCN + 6\ H_2O$$

2-3. This problem is a great place to become familiar with the algebraic method described in the Chapter and illustrated in Example 2–2. The algebraic method can be used "from the start," as is shown here for (a) and (i), or from a point where the inspection method "fails," as in the text examples.

(a) ___ H_2 + ___ P_4 → ___ PH_3

We assign a "1" to the first substance, then assign algebraic coefficients to each substance:

__1__ H_2 + __y__ P_4 → __z__ PH_3

This gives us two simultaneous equations for each element:

For H: $2 \times 1 = 3\,z$

For P: $4\,y = z$

We solve the first equation for z: $z = 2/3$

We use this result to solve the second equation for y:

$$4\,y = 2\,/\,3$$
$$y = 2\,/\,12 = 1\,/\,6$$

For the whole equation:

__1__ H_2 + __1/6__ P_4 → __2/3__ PH_3

Multiply through by 6 to clear fractions:

$$6\ H_2 + P_4 \rightarrow 4\ PH_3$$

2-3. (continued)

(b) $$2\,K + O_2 \rightarrow K_2O_2$$

(c) $$PbO_2 + Pb + 2\,H_2SO_4 \rightarrow 2\,PbSO_4 + 2\,H_2O$$

(d) $$2\,BF_3 + 3\,H_2O \rightarrow B_2O_3 + 6\,HF$$

(e) $$2\,KClO_3 \rightarrow 2\,KCl + 3\,O_2$$

(f) $$2\,K_2O_2 + 2\,H_2O \rightarrow 4\,KOH + O_2$$

(g) $$3\,PCl_5 + 5\,AsF_3 \rightarrow 3\,PF_5 + 5\,AsCl_3$$

(h) $$2\,KOH + K_2Cr_2O_7 \rightarrow 2\,K_2CrO_4 + H_2O$$

(i) $$\underline{1}\,P_4 + \underline{w}\,NaOH + \underline{x}\,H_2O \rightarrow \underline{y}\,PH_3 + \underline{z}\,NaHPO_3$$

For P:	$4 \times 1 = y + z$
For Na:	$w = 2z$
For O:	$w + x = 3z$
For H:	$w + 2x = 3y + z$

We have four unknowns and four equations. We are ready to carry out a systematic solution. We can begin by substitutiin "$2z$" for "w":

For P:	$4 \times 1 = y + z$
For Na:	$2z = 2z$ (we are done with this equation now)
For O:	$2z + x = 3z$ or: $x = z$
For H:	$2z + 2x = 3y + z$

We next substitute x for z in all places:

For P:	$4 \times 1 = y + z$
For Na:	$2z = 2z$ (we are done with this equation now)
For O:	$z = z$ (we are done with this equation now)
For H:	$2z + 2x = 3y + z$ or $y = z$

Since we have now established that $y = z$ we can return to the first equation and solve for an actual variable: $\qquad 4 \times 1 = z + z$

$$4 = 2z$$

$$\mathbf{2 = z}$$

From this: $y = 2$; $x = 2$ and (from the second of the original examples): $w = 4$

This gives: $\mathbf{1}\,P_4 + \mathbf{4}\,NaOH + \mathbf{2}\,H_2O \rightarrow \mathbf{2}\,PH_3 + \mathbf{2}\,NaHPO_3$

Cleaning up: $P_4 + 4\,NaOH + 2\,H_2O \rightarrow 2\,PH_3 + 2\,NaHPO_3$

A check shows 4 P, 4 Na, $(4 + 2) = 6$ O, and $4 + 2 \times 2 = 8$ H

Chapter 2

2-5. Here the algebraic method is laid out for problem (a), which is an example of a combustion reaction, and for (d), which leads to a complex set of coefficients that the inspection method will have trouble with.

(a) $\underline{\quad} C_6H_6 + \underline{\quad} O_2 \rightarrow \underline{\quad} CO_2 + \underline{\quad} H_2O$

In a problem like we almost always start with either the carbon-containing reactant. To minmize problems with fractions later on, we start with a **2** for the the C_6H_6:

$$\underline{\textbf{2}}\ C_6H_6 + \underline{\textbf{\textit{w}}}\ O_2 \rightarrow \underline{\textbf{\textit{x}}}\ CO_2 + \underline{\textbf{\textit{y}}}\ H_2O$$

For C: $2 \times 6 = x$

For H: $2 \times 6 = 2\,y$

For O: $2\,w = x + z$

Note that the C and H can be solved direction: $x = 12$ and $y = 6$. Then solving for w is simple, too: $2\,w = 2x + z = 26 + 6 = 30$

$$w = 30 / 2 = 15$$

$$2\ C_6H_6 + 15\ O_2 \rightarrow 12\ CO_2 + 6\ H_2O$$

(b) $2\ F_2 + H_2O \rightarrow OF_2 + 2\ HF$

(c) $CaC_2 + 2\ H_2O \rightarrow Ca(OH)_2 + C_2H_2$

(d) $\underline{\quad} O_2 + \underline{\quad} NH_3 \rightarrow \underline{\quad} NO_2 + \underline{\quad} H_2O$

In this case we have two elements that occur once on each side—the N (in NH_3 and NO_2) and the H (in NH_3 and H_2O). Since the H is present in different amounts in its two substances (three times in NH_3 and twice in H_2O), we are better off starting with that. We know that there won't be a "1" for either substances. So let's start with "2" for NH_3:

$$\underline{\textbf{\textit{w}}}\ O_2 + \underline{\textbf{2}}\ NH_3 \rightarrow \underline{\textbf{\textit{x}}}\ NO_2 + \underline{\textbf{\textit{y}}}\ H_2O$$

For O: $2\,w = 2\,x + y$

For H: $2 \times 3 = 2\,y$

For N: $2 \times 1 = x$

Both H and N are simple in this case, with $y = 3$ and $x = 2$. We then get, for the O:

$$2\,w = 2 \times 2 + 3 = 7$$

$$w = 7/2$$

$$7/2\ O_2 + 2\ NH_3 \rightarrow 2\ NO_2 + 3\ H_2O$$

Multiply through by 2 to remove the fraction:

$$7\ O_2 + 4\ NH_3 \rightarrow 4\ NO_2 + 6\ H_2O$$

(e) $2\ Al + 6\ NaOH \rightarrow 3\ H_2 + 2\ Na_3AlO_3$

2-7.   (a) $Fe_2O_3 + 3\,H_2 \rightarrow 2\,Fe + 3\,H_2O$

        (b) $3\,Fe_2O_3 + H_2 \rightarrow 2\,Fe_3O_4 + H_2O$

2–9.   (a) This is a deceptively simple case, but is compicated by the presence of both N and O in three or four different compounds. We can start with H, since it is present in just the water and the nitric acid. We put down a "2" for the nitric acid:

$$\underline{w}\ NO_2 + \underline{x}\ H_2O \rightarrow \underline{2}\ HNO_3 + \underline{z}\ NO$$

Then we get:

N: $w = 2 + z$

O: $w + x = 2 \times 3 + z$

H: $2x = 2$

We can take care of x easily, for the H: $x = 1$. For the N and O we then get:

N: $w = 2 + z$

O: $2w + 1 = 6 + z$

                                     $2w = 5 + z$

Now we have two equations and two unknowns. We substitute in the equation for N into the the equation for O: $2\,(2 + z) = 5 + z$

                                 $4 + 2z = 5 + z$

                                      $z = 1$

And, finally: $w = 3$

  Giving: **$3\,NO_2 + 1\,H_2O \rightarrow 2\,HNO_3 + 1\,NO$**

(b) ___ KOH + ___ V + ___$KClO_3 \rightarrow$ ___ $K_3HV_2O_7$ + ___ KCl + ___ H_2O

For this equation, the one substance, $K_3HV_2O_7$, is so complex we want to anchor it writh away with a "1", then do the rest:

$$\underline{v}\ KOH + \underline{w}\ V + \underline{x}\ KClO_3 \rightarrow \underline{1}\ K_3HV_2O_7 + \underline{y}\ KCl + \underline{z}\ H_2O$$

K: $v + x = 3 \times 1 + y$

O: $v + 3x = 1 \times 7 + z$

H: $v = 1 \times 1 + 2z$

V: $w = 2 \times 1$

Cl: $x = y$

Only w is easy in this case: $w = 2$. And it doesn't help in any other case, because w isn't in any other compound!

2-9.   (continued)

For the others, we have:

K: $\qquad\qquad\qquad v + x = 3 + y$

O: $\qquad\qquad\qquad v + 3x = 7 + z$

H: $\qquad\qquad\qquad v = 1 + 2z$

Cl: $\qquad\qquad\qquad x = y$

We can use the simplest equation (for Cl) to get to work on theother three:

K: $\qquad\qquad\qquad v + y = 3 + y$

$\qquad\qquad\qquad\qquad v = 3$

H: $\qquad\qquad\qquad 3 = 1 + 2z$

$\qquad\qquad\qquad\qquad z = 1$

O: $\qquad\qquad\qquad 3 + 3x = 7 + 1$

$\qquad\qquad\qquad\qquad x = 5/3$

Cl: $\qquad\qquad\qquad y = x = 5/3$

This gives:

__3__ KOH + __2__ V + __5/3__ $KClO_2$ → __1__ $K_3HV_2O_7$ + __5/3__ KCl + __1__ H_2O

Mulitplying through by "3" results in:

9 KOH + 6 V + 5 $KClO_2$ → 3 $K_3HV_2O_7$ + 5 KCl + 3 H_2O

As a check, both sides have 11 K's, 16 O's, 6 H's, 4 V's, and 5 Cl's.

(c) Note that in this one the "CN" can and should be treated as a single unit. This is one that is then easy to do by inspection, starting with the Ag_2S:

1 Ag_2S + __ KCN + __ O_2 + __ H_2O → __ $KAg(CN)_2$ + __ S + __ KOH

Do the S and Ag:

1 Ag_2S + __ KCN + __ O_2 + __ H_2O → 2 $KAg(CN)_2$ + 1 S + __ KOH

Next do the CN:

1 Ag_2S + 4 KCN + __ O_2 + __ H_2O → 2 $KAg(CN)_2$ + 1 S + __ KOH

And the K:

1 Ag_2S + 4 KCN + __ O_2 + __ H_2O → 2 $KAg(CN)_2$ + 1 S + 2 KOH

The H:

1 Ag_2S + 4 KCN + __ O_2 + 1 H_2O → 2 $KAg(CN)_2$ + 1 S + 2 KOH

And the O:

1 Ag_2S + 4 KCN + 1/2 O_2 + 1 H_2O → 2 $KAg(CN)_2$ + 1 S + 2 KOH

2-9. (continued)

Mulitply through by "2":

$$2\ Ag_2S + 8\ KCN + O_2 + 2\ H_2O \rightarrow 4\ KAg(CN)_2 + 2\ S + 4\ KOH$$

(d) $4\ Cl_2 + NaI + 8\ NaOH \rightarrow NaIO_4 + 8\ NaCl + 4\ H_2O$

(e) $2\ Cr(CO_6)\ (s) + 4\ KOH\ (aq) \rightarrow$

$$KHCr_2(CO)_{10}\ (aq) + K_2CO_3(aq) + KHCOO\ (aq) + H_2O\ (l)$$

2-11. The final mass amounts are rounded off to the proper number of significant figures. In each case, the mass of the reactants is the same as the mass for the products.

(a) Reactants: $1\ mol\ CO_2 \times \dfrac{44.0095\ g\ CO_2}{1\ mol\ CO_2}$ = 44.0095 g

$1\ mol\ C \times \dfrac{12.0107\ g\ C}{1\ mol\ C}$ = $\underline{12.0107}$ g

TOTAL = **56.0202** g

Products: $2\ mol\ CO \times \dfrac{28.0101\ g\ CO}{1\ mol\ CO}$ = **56.0202** g

(b) Reactants: $2\ mol\ C_6H_{14} \times \dfrac{86.17536\ g\ C_6H_{14}}{1\ mol\ C_6H_{14}}$ = 172.35072 g

$19\ mol\ O_2 \times \dfrac{31.9988\ g\ O_2}{1\ mol\ O_2}$ = $\underline{607.9772}$ g

TOTAL = **780.3279** g

Products: $12\ mol\ CO_2 \times \dfrac{44.0095\ g\ CO_2}{1\ mol\ CO_2}$ = 528.1140 g

$14\ mol\ H_2O \times \dfrac{18.01528\ g\ H_2O}{1\ mol\ H_2O}$ = $\underline{252.21392}$ g

TOTAL = **780.3279** g

(c) Reactants: $1\ mol\ Mn_2O_7 \times \dfrac{221.8719\ g\ Mn_2O_7}{1\ mol\ Mn_2O_7}$ = 221.8719 g

$1\ mol\ H_2O \times \dfrac{18.01528\ g\ H_2O}{1\ mol\ H_2O}$ = $\underline{18.01528}$ g

TOTAL = **239.8872** g

Products: $2\ mol\ HMnO_4 \times \dfrac{119.94359\ g\ HMnO_4}{1\ mol\ HMnO_4}$ = **239.8872** g

2-13. The chemical equation indicates that for every 4 moles of H_2O_2 there is one mole of PbS,

so:
$$\frac{x \text{ mol } H_2O_2}{2.47 \text{ mol PbS}} = \frac{4 \text{ mol } H_2O_2}{1 \text{ mol PbS}}$$

$$x \text{ mol } H_2O_2 = \left(\frac{4 \text{ mol } H_2O_2}{1 \text{ mol PbS}}\right) \times 2.47 \text{ mol PbS}$$

$$= \mathbf{9.88 \text{ mol } H_2O_2}$$

2-15. As indicated in the textbook, one can work this kind of problem step-wise or in a single calculation (after calculating the molar masses of the substances). Here, (a) and (b) are done step-wise and (c) and (d) are done in a single calculation using the calculated molar masses.

(a) Molar mass of CH_4: $\quad 12.011 + 4(1.00794) = 16.042 \text{ g mol}^{-1}$.

Molar mass of O_2: $\quad 2(15.9994) = 31.9988 \text{ g mol}^{-1}$.

Chemical amount of O_2: $\quad 1.000 \text{ g } O_2 \times \left(\frac{1 \text{ mol } O_2}{31.9988 \text{ g } O_2}\right) = 0.031251 \text{ mol } O_2$

Chemical amount of CH_4: $0.031251 \text{ mol } O_2 \times \left(\frac{1 \text{ mol } CH_4}{2 \text{ mol } O_2}\right) = 0.015626 \text{ mol } CH_4$

Mass of CH_4: $\quad 0.015626 \text{ mol } CH_4 \times \left(\frac{16.042 \text{ g } CH_4}{1 \text{ mol } CH_4}\right) = \mathbf{0.2507 \text{ g } CH_4}.$

(b) Molar mass of $TiCl_2$: $\quad 47.88 + 2(35.4527) = 118.785 \text{ g mol}^{-1}$

Molar mass of $TiCl_4$: $\quad 47.88 + 4(35.4527) = 189.691 \text{ g mol}^{-1}$

Chemical amount of $TiCl_4$: $1.000 \text{ g } TiCl_4 \times \left(\frac{1 \text{ mol } TiCl_4}{189.691 \text{ g } TiCl_4}\right) = 5.2717 \times 10^{-3} \text{ mol}$

Chemical amount of $TiCl_2$: $\quad 5.2717 \times 10^{-3} \text{ mol} \times \left(\frac{1 \text{ mol } TiCl_2}{1 \text{ mol } TiCl_4}\right) = 5.2717 \times 10^{-3} \text{ mol}$

Mass of $TiCl_4$: $\quad 5.2717 \times 10^{-3} \text{ mol } TiCl_2 \times \left(\frac{118.785 \text{ g } TiCl_2}{1 \text{ mol } TiCl_2}\right) = \mathbf{0.6262 \text{ g } TiCl_2}.$

(c) Molar mass of Na_3VO_4: $\quad 3(22.990) + 50.942 + 4(15.9994) = 183.910 \text{ g mol}^{-1}$

Molar mass of H_2O: $\quad 2(1.0079) + 15.9994 = 18.0153 \text{ g mol}^{-1}$

Done in one line, the calculation for mass of Na_3VO_4 is:

$$1.000 \text{ g } H_2O \times \left(\frac{1 \text{ mol } H_2O}{18.0153 \text{ g } H_2O}\right) \times \left(\frac{2 \text{ mol } Na_3VO_4}{1 \text{ mol } H_2O}\right) \times \left(\frac{183.910 \text{ g } Na_3VO_4}{1 \text{ mol } Na_3VO_4}\right) = \mathbf{20.42 \text{ g } Na_3VO_4}$$

2-15. (continued)

(d)   Molar mass of K_2O_2: $2(39.0983) + 2(15.9994) = 110.1954$ g mol^{-1}.
        Molar mass of H_2O: $2(1.00794) + 15.9994 = 18.0153$ g mol^{-1}

Done in one line, the calculation for grams of K_2O_2 is:

$$1.000 \, \text{g} \, H_2O \times \left(\frac{1 \text{ mol } H_2O}{18.0153 \text{ g } H_2O} \right) \times \left(\frac{2 \text{ mol } K_2O_2}{2 \text{ mol } H_2O} \right) \times \left(\frac{110.1954 \text{ g } K_2O_2}{1 \text{ mol } K_2O_2} \right) = \textbf{6.117 g } \mathbf{K_2O_2}$$

2-17. When a problem requires the determination of the chemical or mass amounts of *two or more* substances based on one other key substance, it is easiest to calculate the chemical amount of the key substance, then use this amount in calculations of the other two.

Molar mass of KCl: $39.0983 + 35.4527$ $= 74.551$ g mol^{-1}

Molar mass of KNO_3: $39.0983 + 14.00674 + 3(15.9994)$ $= 101.103$ g mol^{-1}

Molar mass of Cl_2: $2(35.4527)$ $= 70.9054$ g mol^{-1}.

The chemical amount of KNO_3, the key substance in this case, is:

$$567 \, \text{g} \, KNO_3 \times \left(\frac{1 \text{ mol } KNO_3}{101.1032 \text{ g } KNO_3} \right) = 5.6081 \text{ mol } KNO_3$$

The mass of KCl that will give this amount of KNO_3 is:

$$5.6081 \text{ mol } KNO_3 \times \left(\frac{4 \text{ mol } KCl}{4 \text{ mol } KNO_3} \right) \times \left(\frac{74.551 \text{ g } KCl}{1 \text{ mol } KCl} \right) = \textbf{418 g KCl}$$

The mass of Cl_2 by-product is:

$$5.6081 \text{ mol } KNO_3 \times \left(\frac{2 \text{ mol } Cl_2}{4 \text{ mol } KNO_3} \right) \times \left(\frac{70.9054 \text{ g } Cl_2}{1 \text{ mol } Cl_2} \right) = \textbf{199 g } \mathbf{Cl_2}$$

2-19. The chemical equation indicates that for every one mole of C_2H_6 there are two moles of H_2. By Avogadro's law, this means the ratio of volume of C_2H_6 gas to liters of H_2 gas is also 1 : 2. Therefore:

$$\frac{x \, \text{L} \, C_2H_6}{14.2 \, \text{L} \, H_2} = \frac{1 \, \text{L} \, C_2H_6}{2 \, \text{L} \, H_2} \implies x \, \text{L} \, C_2H_6 = \left(\frac{1 \, \text{L} \, C_2H_6}{2 \, \text{L} \, H_2} \right) \times 14.2 \, \text{L} \, H_2 = \textbf{7.10 L } \mathbf{C_2H_6}$$

2-21. The formula $K_2Zn_3[Fe(CN)_6]_2$ indicates that there are *twelve* carbons per formula unit. Therefore, the formation of one mole of $K_2Zn_3[Fe(CN)_6]_2$ would require at least twelve moles of K_2CO_3. Then, for the mass of $K_2Zn_3[Fe(CN)_6]_2$

$$= 18.6 \, \text{g} \, K_2CO_3 \times \left(\frac{1 \text{ mol } K_2CO_3}{138.20 \text{ g } K_2CO_3} \right) \times \left(\frac{1 \text{ mol } K_2Zn_3[Fe(CN)_6]_2}{12 \text{ mol } K_2CO_3} \right) \times \left(\frac{698.3 \text{ g } K_2Zn_3[Fe(CN)_6]_2}{1 \text{ mol } K_2Zn_3[Fe(CN)_6]_2} \right)$$

$$= \textbf{7.83 g } \mathbf{K_2Zn_3[Fe(CN)_6]_2}$$

2-23. We can find the x by determining the number of moles of water that were driven off in producing the final amount of Na_3RhCl_6.

$$m \ H_2O = \text{original mass } Na_3RhCl_6 \bullet x \ H_2O - \text{mass of final } Na_3RhCl_6$$
$$= 2.1670 - 1.3872 = 0.7798 \ g$$

$$n \ H2O \text{ given off} = 0.7798 \ g \ H_2O \times \left(\frac{1 \ \text{mol} \ H_2O}{18.015 \ g \ H_2O} \right)$$

$$= 0.04329 \ \text{mol} \ H_2O$$

$$n \ Na_3RhCl_6 \text{ remaining} = 1.3872 \ g \ Na_3RhCl_6 \times \left(\frac{1 \ \text{mol} \ Na_3RhCl_6}{384.6 \ g \ Na_3RhCl_6} \right)$$

$$= 0.003607 \ \text{mol}$$

$$x = \frac{\text{moles} \ H_2O \ \text{given off}}{\text{moles} \ Na_3RhCl_6 \ \text{remaining}}$$

$$= \frac{0.04329}{0.003607} = \mathbf{12}$$

2-25. The words of the problem give the information needed for the balanced chemical equation relating Si_2H_6 and SiO_2: $6 \ Si_2H_6 + 21 \ O_2 \rightarrow 12 \ SiO_2 + 18 \ H_2O$
In this problem the measured amount of disilane starting material is given as a volume. With this, the density, and the molar mass of disilane, we calculate the mass of silica:

$$25.0 \ cm^3 \ Si_2H_6 \times \left(\frac{2.78 \times 10^{-3} \ g \ Si_2H_6}{1 \ cm^3 \ Si_2H_6} \right) \times \left(\frac{1 \ \text{mol} \ Si_2H_6}{62.220 \ g \ Si_2H_6} \right) \times \left(\frac{12 \ \text{mol} \ SiO_2}{6 \ \text{mol} \ Si_2H_6} \right) \times \left(\frac{60.084 \ g \ SiO_2}{1 \ \text{mol} \ SiO_2} \right)$$

$$= \mathbf{0.134 \ g \ SiO_2}.$$

2-27. The problem is one where we seem to start with information about the sodium nitrate, but in fact that is what we need to determine. We have a final and a beginning mass. This lets us get to the mass of the *oxygen* produced.

Reaction stoichiometry: $2 \ NaNO_3 \rightarrow O_2 + 2 \ NaNO_2$

Mass of oxygen produced: $3.4671 \ g - 2.9073 \ g = 0.5598 \ g$.

Mass of sodium nitrate reacted:

$$0.5598 \times \frac{1 \ \text{mol} \ O_2}{31.9988 \ g \ O_2} \times \frac{2 \ \text{mol} \ NaNO_3}{1 \ \text{mol} \ O_2} \times \frac{84.9947 \ g \ NaNO_3}{1 \ \text{mol} \ NaNO_3} = 2.974 \ g \ NaNO_3$$

For the purity of the sodium nitrate:

$$\frac{2.974 \ g \ NaNO_3 \ \text{reacted}}{3.4671 \ g \ \text{sample}} \times 100\% = \mathbf{85.77\%}$$

2-29. We start by calculating how many cheeseburgers might be made using all of each ingredient. The ingredient that gives the fewest cheeseburgers limits the process.

Patties: $\qquad$ 600 patties $\times \left(\dfrac{1 \text{ cheeseburger}}{2 \text{ patties}} \right) = 300$ cheeseburgers

Cheese: $\qquad$ 400 cheese slices $\times \left(\dfrac{1 \text{ cheeseburger}}{1 \text{ cheese slice}} \right) = 400$ cheeseburgers

Tomato: $\qquad$ 1000 tomato slices $\times \left(\dfrac{1 \text{ cheeseburger}}{4 \text{ tomato slices}} \right) = 250$ cheeseburgers

Buns: $\qquad$ 350 buns $\times \left(\dfrac{1 \text{ cheeseburger}}{1 \text{ hamburger bun}} \right) = 350$ cheeseburgers

The least product comes from the supply of **tomato slices**. That will make **250 cheeseburgers.** For these, 500 hamburger patties are required, so **100 hamburger patties** will be left over.

2-31.

Parts	# available - dozens	# available - units	conversion to tables	# of tables that can be made
legs	25.25	303	$\dfrac{1 \text{ table}}{4 \text{ legs}}$	75
leaves	20.00	240	$\dfrac{1 \text{ table}}{2 \text{ leaves}}$	120
frames	11.00	132	$\dfrac{1 \text{ table}}{1 \text{ frame}}$	132
tops	7.00	84	$\dfrac{1 \text{ table}}{1 \text{ top}}$	84

The least amount of product is given by the legs, so the legs are the "limiting reactant", and a maximum of 75 tables can be assembled.

2-33. The molar masses we need for this calculation are:

NH_3: 17.0306 g mol^{-1} $\qquad$ O_2: 31.9988 g mol^{-1} $\qquad$ H_2O: 18.0153 g mol^{-1}

The amount of NH_3 present in the reaction mixture is:

$$34.0 \text{ g } NH_3 \times \left(\frac{1 \text{ mol } NH_3}{17.03056 \text{ g } NH_3} \right) = 1.996 \text{ mol } NH_3$$

If it all reacted, the yield of H_2O would be:

$$1.996 \text{ mol } NH_3 \times \left(\frac{6 \text{ mol } H_2O}{4 \text{ mol } NH_3} \right) = 2.995 \text{ mol } H_2O$$

25

2-33. (continued)

The amount of O_2 present in the reaction mixture is:

$$50.0 \text{ g } O_2 \times \left(\frac{1 \text{ mol } O_2}{31.9988 \text{ g } O_2} \right) = 1.5626 \text{ mol } O_2$$

If it all reacted, the yield of H_2O would be:

$$1.5626 \text{ mol } O_2 \times \left(\frac{6 \text{ mol } H_2O}{3 \text{ mol } O_2} \right) = 3.1252 \text{ mol } H_2O$$

The NH_3 would give *less* of the H_2O product; hence, **NH_3 is the limiting reactant.** The mass of water that could form is: $2.995 \text{ mol } H_2O \times \left(\dfrac{18.0153 \text{ g } H_2O}{1 \text{ mol } H_2O} \right) = $ **53.9 g H_2O**

2-35. (a) The 35.12 g of $Ba(OH)_2 \bullet 8 \ H_2O$ (molar mass 315.464 g mol^{-1}) can provide 23.18 g of $BaCl_2$ (molar mass 208.232 g mol^{-1}). The 13.75 g of NH_4Cl (molar mass 53.4912 g mol^{-1}) can provide 26.76 g of $BaCl_2$. The lesser amount of product is from the **$Ba(OH)_2 \bullet 8 \ H_2O$,** which is the limiting reactant.

(b) This was answered in the previous problem—**23.18 g $BaCl_2$.**

(c) The 35.12 g of $Ba(OH)_2 \bullet 8 \ H_2O$ requires 11.91 g of NH_4Cl. The left over NH_4Cl is: 13.75 g – 11.91 g = **1.84 g.**

2-37. The theoretical yield is always controlled by the limiting reactant. A limiting reactant calculation is done first, and then the theoretical yield is calculated from the limiting reactant. In tabular format, the limiting reactant calculation is:

Reactant	Molar mass	Amount available - grams	Amount available - moles	Conversion factor to moles H_2O	Moles H_2O formed
H_2	2.0158 g mol^{-1}	2.000	0.9922 mol H_2	$\dfrac{1 \text{ mol } H_2O}{1 \text{ mol } H_2}$	0.9922 mol H_2O
O_2	31.9988 g mol^{-1}	1.000	0.03125 mol O_2	$\dfrac{2 \text{ mol } H_2O}{1 \text{ mol } O_2}$	0.06250 mol H_2O

The limiting reactant is O_2, so the theoretical yield is 0.06250 mol H_2O. This will equal (for a molar mass of 18.0153 g mol^{-1} for H_2O) **1.126 g H_2O**

2-39. The amounts of the reactants are given as *volumes*. Because the temperature and pressure are constant, the amount of product can be predicted based on the law of combining volumes.

Calculating from the C_2H_4: $40.0 \text{ L } C_2H_4 \times \left(\dfrac{4 \text{ L } CO_2}{2 \text{ L } C_2H_4}\right) = 80 \text{ L } CO_2$

Calculating from O_2: $20.0 \text{ L } C_2H_4 \times \left(\dfrac{4 \text{ L } CO_2}{6 \text{ L } O_2}\right) = 13.3 \text{ L } CO_2$

The O_2 is the limiting reactant. The theoretical yield of CO_2 is **13.3 L CO_2**.

2-41. The necessary molar masses are: Fe_2O_3: 159.69 g mol^{-1} Fe: 55.847 g mol^{-1}

Chemical amount of Fe_2O_3:

$$433.2 \text{ g } Fe_2O_3 \times \left(\frac{1 \text{ mol } Fe_2O_3}{159.69 \text{ g } Fe_2O_3}\right) = 2.713 \text{ mol } Fe_2O_3$$

Theoretical yield of Fe in moles :

$$2.713 \text{ mol } Fe_2O_3 \times \left(\frac{2 \text{ mol Fe}}{1 \text{ mol } Fe_2O_3}\right) = 5.426 \text{ mol Fe}$$

The theoretical yield of Fe in g: $5.426 \text{ mol Fe} \times \left(\dfrac{55.847 \text{ g Fe}}{1 \text{ mol Fe}}\right) = $ **303.0 g Fe**

The percentage yield: $\left(\dfrac{254.3 \text{ g Fe}}{303.0 \text{ g Fe}}\right) = 0.8393 = $ **83.93%**

2-43. (a) If 8.0% of the cookies are broken (and eaten) then 92.0% can be sold. You must make enough cookies so that 92% of them will equal 200 dozen (or 2400) cookies. We let x equal the number of cookies. Then:

$(0.92)x = 2400$ cookies $\Rightarrow$ $x = $ **2609 cookies**

(b) 2609 cookies made - 2400 cookies delivered = **209 cookies** eaten

2-45. $319 \text{ mL solution} \times \dfrac{0.375 \text{ mmol sucrose}}{1 \text{ mL solution}} = 120. \text{ mmol}$

2-47. We will solve for the initial volume of the undiluted acid in this problem.

$$V_i = \frac{c_f V_f}{c_i} = \frac{0.1 \text{ M} \times 10.0 \text{ L}}{12 \text{ M}} = 0.083 \text{ L} = \textbf{83 mL}$$

Chapter 2

2-49. We calculate the theoretical yield based on the molarity and volume of barium chloride, using the balanced chemical equation:

$$BaCl_2 \ (aq) + H_2SO_4 \ (aq) \rightarrow BaSO_4(s) + 2 \ HCl \ (aq)$$

$$0.1000 \ L \times \frac{0.2516 \ mol \ BaCl_2}{1 \ L} \times \frac{1 \ mol \ BaSO_4}{1 \ mol \ BaCl_2} \times \frac{233.3906 \ g \ BaSO_4}{1 \ mol \ BaSO_4} = \textbf{5.8721 g BsSO}_4$$

The percentage yield is: $\frac{actual \ yield}{theoretical \ yield} \times 100\% = \frac{4.9852 \ g}{5.8721 \ g} \times 100\% = \textbf{84.90\%}$

2-51. This is a stoichiometric calculation with the added feature that the amount of the key reactant is given as a volume

$$Moles \ OCl^- \ reacting \ = 51.6 \ L \times \frac{0.650 \ mol \ OCl^-}{1 \ L} = 33.5 \ mol \ OCl^-$$

$$mol \ N_2H_4 = mol \ OCl^- \ = 33.5 \ mol \ N_2H_4$$

$$33.5 \ mol \ N_2H_4 \times \frac{32.045 \ g \ N_2H_4}{1 \ mol \ N_2H_4} \ = 1070 \ g \ N_2H_4 = \textbf{1.07 kg N}_2\textbf{H}_4.$$

2-53. This is another stoichiometric calculation, with a volume of a reactant needed:

$$mol \ PbO_2 = 15.9 \ g \ \times \left(\frac{1 \ mol \ PbO_2}{239.20 \ g \ PbO_2}\right) = 0.0665 \ mol \ PbO_2$$

$$mol \ HNO_3 = (0.0665 \ mol \ PbO_2) \times \left(\frac{4 \ mol \ HNO_3}{2 \ mol \ PbO_2}\right) = 0.133 \ mol \ HNO_3$$

$$volume \ HNO_3 = 0.133 \ mol \ HNO_3 \times \left(\frac{1 \ L}{7.91 \ mol \ HNO_3}\right) = \textbf{0.0168 L HNO}_3$$

This is equal to **16.8 mL** of the HNO_3 solution

2–55. This is a limiting reactant problem. The reaction is:

$$2 \ AgNO_3 + CaCl_2 \rightarrow 2 \ AgCl + Ca(NO_3)_2$$

For the $AgNO_3$:

$$0.0215 \ L \times \frac{0.150 \ mol \ AgNO_3}{1 \ L} \times \frac{2 \ mol \ AgCl}{2 \ mol \ AgNO_3} \times \frac{143.321 \ g \ AgCl}{1 \ mol \ AgCl} = 0.462 \ g \ AgCl$$

For the $CaCl_2$:

$$0.0150 \ L \times \frac{0.100 \ mol \ CaCl_2}{1 \ L} \times \frac{2 \ mol \ AgCl}{1 \ mol \ CaCl_2} \times \frac{143.321 \ g \ AgCl}{1 \ mol \ AgCl} = 0.430 \ g \ AgCl$$

The $CaCl_2$ is limiting and only **0.430 g AgCl** can form.

2-57. Water is one portion of the compound. Its amount can be determined by considering three different values: (i) the number of moles of compound, (ii) the number of moles of water per mole of compound, and then (iii) the number of molecules of water per mole of water. We start with the molar mass of one mole of $MgSO_4 \bullet 7H_2O$, which is 247 g mol^{-1} (to three significant digits).

(i) moles of compound: $55.5 \text{ g } MgSO_4 \bullet 7H_2O \times \left(\dfrac{1 \text{ mol } MgSO_4 \bullet 7H_2O}{246.476 \text{ g } MgSO_4 \bullet 7H_2O} \right)$

$$= 0.225 \text{ mol } MgSO_4 \bullet 7H_2O$$

(ii) moles of water: $0.225 \text{ mol } MgSO_4 \bullet 7H_2O \times \left(\dfrac{7 \text{ mol } H_2O}{1 \text{ mol } MgSO_4 \bullet 7H_2O} \right)$

$$= 1.58 \text{ mol } H_2O = \textbf{1580 mmol}$$

(iii) molecules of water: $1.576 \text{ mol } H_2O \times 6.0221 \times 10^{23} \dfrac{\text{molecules } H_2O}{\text{mol } H_2O}$

$$= \textbf{9.49} \times \textbf{10}^{\textbf{23}} \textbf{ molecules of } \textbf{H}_2\textbf{O}.$$

2-59. We will need to know the mass of one mole of the drug. This is:

$$1 \text{ mol } AuC_4H_4O_4Na_2S = 196.96654 + (4 \times 12.011) + (4 \times 1.0074)$$
$$+ (4 \times 15.9994) + (2 \times 22.9898) + 32.066 = 391.09 \text{ g}$$

Within this amount of the drug, there is *one* mole of gold, which weighs 196.96654 g.

The mass of Au administered over the course of the therapy is:

$$\left(\dfrac{0.0500 \text{ g } AuC_4H_4O_4Na_2S}{\text{week}} \right) \times \left(\dfrac{196.96654 \text{ g Au}}{391.09 \text{ g } AuC_4H_4O_4Na_2S} \right) \times 20 \text{ weeks}$$

$$= 0.504 \text{ g Au} \times \left(\dfrac{1000 \text{ mg}}{1 \text{ g}} \right) = \textbf{504 mg Au.}$$

Because the subscripts in the molecular formula are integral multiples of the subscripts in the empirical formula, the *ratios* of the amounts of elements given by an empirical formula and by a molecular formula are *identical*. This is true for chemical amounts as well as for the masses of the elements. If there is one gram of Au per *x* grams of the compound when we calculate using the empirical formula, there will be one gram of Au per *x* grams of compound using the molecular formula. Because this problem relies on the ratios of the masses of Au and the drug, the answer would be the **same** with the empirical or the molecular formula.

Chapter 2

2-61. Molar mass of Al_2O_3: $\quad 2(26.98156) + 3(15.9994) = 101.9613$ g mol^{-1}.

Molar mass of Na_3AlF_6: $\quad 3(22.9898) + 26.98159 + 6(18.998) = 209.939$ g mol^{-1}

The mass of 287 metric tons is the same as 287 millinon grams, or 2.87×10^8 grams

Done in one line, the calculation for the number of grams of Na_3AlF_6 is:

$$2.87 \times 10^8 \text{ g } Al_2O_3 \times \left(\frac{1 \text{ mol } Al_2O_3}{101.96 \text{ g } Al_2O_3}\right) \times \left(\frac{2 \text{ mol } Na_3AlF_6}{1 \text{ mol } Al_2O_3}\right) \times \left(\frac{209.94 \text{ g } Na_3AlF_6}{1 \text{ mol } Na_3AlF_6}\right) =$$

$$\mathbf{1.18 \times 10^9 \text{ g} = 1.18 \times 10^3 \text{ metric tons}}$$

2-63. The number of molecules of water and TCDD must be determined separately.

$$1 \text{ kg } H_2O \times \left(\frac{1000 \text{ g } H_2O}{1 \text{ kg } H_2O}\right) \times \left(\frac{1 \text{ mol } H_2O}{18.0153 \text{ g } H_2O}\right) \times \left(\frac{6.0221 \times 10^{23} \text{ molecules } H_2O}{1 \text{ mol } H_2O}\right)$$

$$= 3 \times 10^{25} \text{ molecules } H_2O$$

$$0.6 \,\mu\text{g TCDD} \times \left(\frac{10^{-6} \text{ g TCDD}}{1 \,\mu\text{g TCDD}}\right) \times \left(\frac{1 \text{ mol TCDD}}{322.0 \text{ g TCDD}}\right) \times \left(\frac{6.0221 \times 10^{23} \text{ molecules TCDD}}{1 \text{ mol TCDD}}\right)$$

$$= 1 \times 10^{15} \text{ molecules TCDD}.$$

Percentage of TCDD molecules: $\left(\dfrac{1 \times 10^{15}}{3 \times 10^{25}}\right) = 3 \times 10^{-11} \times 100\% = \mathbf{3 \times 10^{-9}\%}$.

2-65. The balanced reaction for the chlor-alkali process is

$2 NaCl + 2 H_2O \rightarrow H_2 + 2 NaOH + Cl_2$.

For simplicity, since the reactant and the product are both present *in metric tons*, we do not convert to and from grams. The mass *ratio* of $Cl_2 : 2 NaOH$ is 70.906 : 79.994 regardless of the units of mass. Therefore:

$$660 \text{ metric tons } Cl_2 \times \left(\frac{79.994 \text{ tons NaOH}}{70.906 \text{ tons } Cl_2}\right) = \mathbf{745 \text{ metric tons NaOH}}.$$

2-67. We work this problem as a molar mass problem, with kilo- amounts of the grams *and* the moles: $\quad 88 \text{ kg } CO_2 \times \dfrac{1 \text{ kmol } CO_2}{44.0098 \text{ kg } CO_2} = 2.0 \text{ kmol } CO_2$.

The amount of oxygen contained is:

$$2.0 \text{ kmol } CO_2 \times \frac{2 \text{ kmol O}}{1 \text{ kmol } CO_2} \times \frac{1000 \text{ mol O}}{1 \text{ kmol O}} \times \frac{6.022 \times 10^{23} \text{ atoms O}}{1 \text{ mol O}} = \mathbf{2.4 \times 10^{27} \text{ atoms O}}$$

2-69. (a) $\mathbf{2 \text{ KClO}_3 + 4 \text{ HCl} \rightarrow 2 \text{ KCl} + 2 \text{ H}_2\text{O} + 1 \text{ Cl}_2 + 2 \text{ ClO}_2}$

$4 \text{ KClO}_3 + 4 \text{ HCl} \rightarrow 4 \text{ KCl} + 2 \text{ H}_2\text{O} + 1 \text{ Cl}_2 + 2 \text{ ClO}_2$

(b) $\mathbf{1 \text{ Ba(OH)}_2 + 1 \text{ H}_2\text{O}_2 + 2 \text{ ClO}_2 \rightarrow 1 \text{ Ba(ClO}_2)_2 + 2 \text{ H}_2\text{O} + 1 \text{ O}_2}$

$4 \text{ Ba(OH)}_2 + 2 \text{ H}_2\text{O}_2 + 8 \text{ ClO}_2 \rightarrow 4 \text{ Ba(ClO}_2)_2 + 6 \text{ H}_2\text{O} + 3 \text{ O}_2$

(c) This one is different in that it is possible to balance by either leaving out all chromium-containing species (by assigning a coefficient of "0" to $CrCl_2$ and K_2CrO_4 or by leaving out all iodine-containing species (by assigning a coefficient of "0" to KI and KIO_4):

$KI + 8\,KOH + 4\,Cl_2 \rightarrow KIO_4 + 8\,KCl + 4\,H_2O$

$CrCl_2 + 8\,KOH + 2\,Cl_2 \rightarrow K_2CrO_4 + 6\,KCl + 4\,H_2O$

2-71. The balanced chemical reaction, in words, is:

Aspartame + 2 water [yields] aspartic acid + methanol + phenylalanine.

In formulas:

$C_{14}H_{18}N_2O_5 + 2\,H_2O \rightarrow C_4H_7NO_4 + CH_3OH + C_wH_xN_yO_z$

The values for w, x, y, and z are obtained by using the fact that the number of C, H, N, and O atoms must balance.

As written, we have:

14 C / 22 H / 2 N / 7 O [yields] 5 C / 11 H / 1 N / 5 O + w C / x H / y N / z O

The values for w, x, y, and z are w = 9, x = 11, y = 1, z = 2.

The formula for phenylalanine is **$C_9H_{11}NO_2$**.

2-73. The key insight required for this problem is that the nitrogen in 3'-methylphthalanilic acid appears *entirely* in the *m*-toluidine. And the 3'-methylphtalanilic acid is 5.49% N, meaning that for every 100 g of the acid there are 5.49 g N. Working from 5.23 g of *m*-toluidine:

$$5.23\text{ g }m\text{-toluidine} \times \left(\frac{14.00674\text{ g N}}{1\text{ mol N}}\right) \times \left(\frac{1\text{ mol N}}{1\text{ mol m-toluidine}}\right) \times \left(\frac{1\text{ mol }m\text{-toluidine}}{107.16\text{ g }m\text{-toluidine}}\right) = 0.684\text{ g N}$$

This is the amount of N in the mass of the fruit set. The mass of the fruit set is:

$$0.684\text{ g N} \times \left(\frac{100\text{ g fruit set}}{5.49\text{ g N}}\right) = \mathbf{12.5\text{ g}}\text{ fruit set}$$

2-75.

Element	grams in sample	moles in sample	relative # of moles
C	48.044 g	$48.044\text{ g C} \times \left(\frac{1\text{ mol C}}{12.011\text{ g C}}\right) = 4.000\text{ mol C}$	1
H	12.09 g	$12.09\text{ g H} \times \left(\frac{1\text{ mol H}}{1.00794\text{ g H}}\right) = 12.00\text{ mol H}$	3

The empirical formula is **CH_3**.

2-77. This does work. It is known as putting all reactants on a "per reaction" basis. To see how this works, we can look at the reaction:

$$2 C_2H_6 + 7 O_2 \rightarrow 4 CO_2 + 6 H_2O$$

Let's lay out the calculation for the limiting reactant based on the presence of 10 g of C_2H_6 and 20 g of O_2:

mol CO_2 available from C_2H_6:

$$10.\ g\ CH \times \frac{1\ mol\ C_2H_6}{30.07\ g\ C_2H_6} \times \frac{4\ mol\ CO_2}{2\ mol\ C_2H_6} = 0.67\ mol\ CO_2$$

mol CO_2 available from O_2:

$$20.\ g\ O_2 \times \frac{1\ mol\ O_2}{31.999\ g\ O_2} \times \frac{4\ mol\ CO_2}{7\ mol\ C_2H_6} = 0.36\ mol\ CO_2$$

But, we can rewrite the calculations to determine the same answer, in another way:

mol CO_2 available from C_2H_6:

$$10.\ g\ CH \times \frac{1\ mol\ C_2H_6}{30.07\ g\ C_2H_6} \times \frac{4\ mol\ CO_2}{2\ mol\ C_2H_6}$$

$$= \left(\mathbf{10.\ g\ CH \times \frac{1\ mol\ C_2H_6}{30.07\ g\ C_2H_6} \times \frac{1}{2\ mol\ C_2H_6}} \right) \times 4\ mol\ CO_2$$

$$= \mathbf{0.166} \times 4\ mol\ CO_2 = 0.67\ mol\ CO_2$$

mol CO_2 available from O_2:

$$20.\ g\ O_2 \times \frac{1\ mol\ O_2}{31.999\ g\ O_2} \times \frac{4\ mol\ CO_2}{7\ mol\ C_2H_6}$$

$$= \left(\mathbf{20.\ g\ O_2 \times \frac{1\ mol\ O_2}{31.999\ g\ O_2} \times \frac{1}{7\ mol\ C_2H_6}} \right) \times 4\ mol\ CO_2$$

$$= \mathbf{0.0892} \times 4\ mol\ CO_2 = 0.36\ mol\ CO_2$$

The expression and the quantity in **bold** in both cases is the operation suggested by the student. Note that in both cases, it is simply multiplied by the *same* value-4 mol CO_2-to get the number of mol of CO_2 that can form. Therefore, this expression alone, as suggested by the student, is enough to determine the limiting reactant.

2-79. (a) $Nb + 4 NbCl_5 \rightarrow 5 NbCl_4$

(b) Potential yield from Nb:

$$0.50\ g\ Nb \times \left(\frac{1\ mol\ Nb}{92.606\ g\ Nb} \right) \times \left(\frac{5\ mol\ NbCl_4}{1\ mol\ Nb} \right) = 0.027\ mol\ NbCl_4$$

Potential yield from $NbCl_5$:

$$5.0\ g\ NbCl_5 \times \left(\frac{1\ mol\ NbCl_5}{269.87\ g\ NbCl_5} \right) \times \left(\frac{5\ mol\ NbCl_4}{4\ mol\ NbCl_5} \right) = 0.023\ mol\ NbCl_4$$

The limiting reagent is **$NbCl_5$.**

(c) This wouldn't matter, because Nb_2Cl_{10} has the same empirical formula as $NbCl_5$.

2-81. The chemical reaction for the combustion of octane is:

$$2\,C_8H_{18} + 25\,O_2 \rightarrow 16\,CO_2 + 18\,H_2O$$

We assume the car gets 30 miles per gallon, in which case 20 fewer miles saves 0.67 gallons of gasoline. This equals approximately 2.7 liters, or 2700 cm^3.

The density of octane should be around that given in the Chapter 1 for benzene, another hydrocarbon, say 0.85 g cm^{-3}, so this volume of gasoline contains approximately 2700 × 0.85 = 2300 g of octane. Octane has a molar mass of 114 g mol^{-1}, so the gasoline contains roughly 2300/114 = 20 moles of octane.

This much octane would yield $(^{16}/_2) \times 20 = 160$ moles of CO_2 per week, or 8300 moles CO_2 per year. The mass of 8300 moles of CO_2 in pounds is:

$$8300 \ \text{mol } CO_2 \times \left(\frac{44 \text{ g } CO_2}{1 \text{ mol } CO_2}\right) \times \left(\frac{1 \text{ lb}}{453 \text{ g}}\right) = 800 \ \text{lb } CO_2.$$

Within the (considerable) assumptions of this calculation, the article is correct.

2-83. (a) $CF_2Cl_2 + 2\,Na_2C_2O_4 \rightarrow 2\,NaF + 2\,NaCl + C + 4\,CO_2$

(b) We can work in kilomoles in this case. The mass of one kmol of freon is 120.913 kg / kmol. The mass of one kmol of sodium oxalate is 133.998 kg / kmol. Therefore:

$$147 \times 10^3 \text{ kg } CF_2Cl_2 \times \frac{1 \text{ kmol } CF_2Cl_2}{120.913 \text{ kg } CF_2Cl_2} \times \frac{2 \text{ kmol } Na_2C_2O_4}{1 \text{ kmol } CF_2Cl_2} \times \frac{133.998 \text{ kg } Na_2C_2O_4}{1 \text{ kmol } Na_2C_2O_4}$$

$$= \mathbf{3.26 \times 10^5 \ kg \ Na_2C_2O_4}$$

2-85 The balanced equation for this process is: $PF_3 + 3\,H_2O \rightarrow H_3PO_3 + 3\,HF$

$$\text{mol } PF_3 = 0.077 \text{ g } PF_3 \times \frac{1 \text{ mol } PF_3}{87.968 \text{ g } PF_3} = 0.000875 \text{ mol } PF_3$$

$$\text{mol } H_3PO_3 = \text{mol } PF_3 = 0.000875 \text{ mol } H_3PO_3$$

$$\text{mol } HF = 3 \times \text{mol } PF_3 = 0.002626 \text{ mol } HF$$

$$[H_3PO_3] = \frac{0.000875 \text{ mol } H_3PO_3}{0.872 \text{ L}} = \mathbf{0.0010 \ M \ H_3PO_3}$$

$$[HF] = \frac{0.002626 \text{ mol } HF}{0.872 \text{ L}} = \mathbf{0.0030 \ M \ HF}$$

2-87. The chemical amount of K_2O product is:

$$8.34 \text{ g } K_2O \times \left(\frac{1 \text{ mol } K_2O}{94.196 \text{ g } K_2O}\right) = 0.8854 \text{ mol } K_2O$$

The amount of K required is:

$$0.08854 \text{ mol } K_2O \times \left(\frac{10 \text{ mol K}}{6 \text{ mol } K_2O}\right) \times \left(\frac{39.0983 \text{ g K}}{1 \text{ mol K}}\right) = \mathbf{5.77 \ g \ K}$$

2-87. (continued)

The amount of KNO_3 required is:

$$0.08854 \text{ mol } K_2O \times \left(\frac{2 \text{ mol } KNO_3}{6 \text{ mol } K_2O} \right) \times \left(\frac{101.103 \text{ g } KNO_3}{1 \text{ mol } KNO_3} \right) = \textbf{2.98 g } \mathbf{KNO_3}$$

2-89. The relevant reaction is: $BaCl_2 + (NH_4)_2SO_4 \rightarrow BaSO_4 + 2 \text{ } NH_4Cl$

We calculate the mass of ammonium sulfate that is needed to react with this amount of $BaCl_2$ and compare that to the original mass of impure ammonium sulfate:

$$0.03271 \text{ L } BaCl_2 \times \frac{0.5148 \text{ mol } BaCl_2}{1 \text{ L } BaCl_2} \times \frac{1 \text{ mol } (NH_4)_2 SO_4}{1 \text{ mol } BaCl_2} \times \frac{132.041 \text{ g } (NH_4)_2 SO_4}{1 \text{ mol } (NH_4)_2 SO_4}$$

$$= 2.223 \text{ g } (NH_4)_2SO_4$$

For the percentage purity: $\dfrac{2.223 \text{ g } (NH_4)_2 SO_4}{2.513 \text{ g sample}} \times 100\% = \textbf{88.48\%}$

CHAPTER 3 - CHEMICAL PERIODICITY

AND THE FORMATION OF SIMPLE COMPOUNDS

3-1. We can predict a new formula based on a formula for a compound in the same group of the periodic table. Here, the matches are:

Element	Group	Related Compound	Predicted Formula
P	V	NH_3	PH_3
Cl	VII	HF	HCl
Si	IV	CH_4	SiH_4
S	VI	H_2O	H_2S

3-3. In this problem (compared to #3-1) there are no compounds to refer to for the listed elements. Instead, we just look at the group numbers. A binary compound will have a formula A_aX_x. One set of values of a and x that will be stable is given by the formula:

a = (8-Group # for X)	x = (Group # for A)

(a)   Rubidium - Group I   Arsenic - Group V
Formula: $Rb_{(8-5)}As_1 \equiv Rb_3As_1 \equiv$ **Rb_3As**

(b)   Strontium - Group II   Fluorine - Group VII
Formula: $Sr_{(8-7)}F_2 \equiv Sr_1F_2 \equiv$ **SrF_2**.

(c)   Calcium - Group II   Sulfur - Group VI
Formula: $Ca_{(8-6)}S_2 \equiv Ca_2S_2 \equiv$ **CaS**.

3-5. The tabulated melting points are shown at right. The periodic pattern has a consistent increase in the boiling point up to Group IV then a rapid drop to much lower values for Groups V, VI, VII, and VIII.

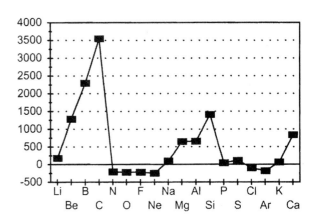

3-7. (a) The formation of a 1:2 chloride and 1:1 oxide indicates a **Group II** element.

(b) This formula problem can be solved using a hypothetical 100.00 g sample. This sample contains 74.5 grams of Cl, which (at 35.4527 g mol^{-1} for Cl) is equal to 2.10 mol Cl. This means there must be 1.05 mol of the unknown element present, weighing 25.5 g.

The molar mass of the element is : $\dfrac{25.5\,\text{g}}{1.05\,\text{mol}}$ = 24.3 g mol^{-1}. It is **magnesium (Mg)**.

3-9. (a) Six elements that are gases at room temperature are nitrogen, oxygen, fluorine, chlorine, helium, and neon.

(b) Yes, these elements are clustered at the top right of the periodic table.

3-11. The simple averages of the quantities are:

Melting point: **1250 °C**   Boiling point: **2386 °C**   Density: **3.02 g cm^{-3}**

3-13. Hydrogen forms compounds of the elements on the right side of the periodic table of the formula H$_x$A, where x = (8 - Group # of element A). Therefore:

Arsenic:	Group V	H$_{(8-4)}$As ≡ **H$_3$As** (also written as AsH$_3$)
Iodine:	Group VII	H$_{(8-7)}$I ≡ **HI**
Lead:	Group IV	H$_{(8-4)}$Pb ≡ **H$_4$Pb** (also written as PbH$_4$)
Tellurium:	Group VI	H$_{(8-6)}$Te ≡ **H$_2$Te** (also written as TeH$_2$).

3-15.

Element	Core electrons	Valence electrons	Lewis dot symbol
Sb	46	5	•S̈b•̣
Br	28	7	:C̈l:•
B	2	3	•B•̣
Ra	86	2	•Ra•

3-17. Xe   Group VIII   8 valence e⁻

$$:\!\overset{\displaystyle ..}{\underset{\displaystyle ..}{Xe}}\!:$$

Se²⁻   Group VI + 2 e-   8 valence e⁻

$$\left[:\!\overset{\displaystyle ..}{\underset{\displaystyle ..}{Se}}\!:\right]^{2-}$$

Se⁺   Group VI - 1 e-   5 valence e⁻

$$\left[\cdot\!\overset{\displaystyle ..}{\underset{\displaystyle \cdot}{Se}}\cdot\right]^{1+}$$

Be⁺   Group II - 1 e-   1 valence e⁻

$$\left[Be\cdot\right]^{1+}$$

3-19.

	Total e-	Core e-	Valence e-
Xe	54	46	8
Se²⁻	36	28	8
Se⁺	33	28	5
Be⁺	3	2	1

3-21. (a) CaS, calcium sulfide

Before: $\cdot Ca\cdot$ $:\!\overset{..}{\underset{\cdot}{S}}\!\cdot$

After: $\left[Ca\right]^{2+}\left[:\!\overset{..}{\underset{..}{S}}\!:\right]^{2-}$

(b) CsI, cesium iodide

Before: $Cs\cdot$ $:\!\overset{..}{\underset{\cdot}{I}}\!\cdot$

After: $\left[Cs\right]^{1+}\left[:\!\overset{..}{\underset{..}{I}}\!:\right]^{1-}$

(c) Fr₂S, francium sulfide

Before: 2 $Fr\cdot$ $:\!\overset{..}{\underset{\cdot}{S}}\!\cdot$

After: $2\left[Fr\right]^{1+}\left[:\!\overset{..}{\underset{..}{S}}\!:\right]^{2-}$

(d) GeCl₄, germanium chloride

Before: $Ge\cdot$ 4 $:\!\overset{..}{\underset{..}{Cl}}\!\cdot$

After: $\left[Ge\right]^{4+}$ 4$\left[:\!\overset{..}{\underset{..}{Cl}}\!:\right]^{-}$

(e) Li₂O, lithium oxide

Before: 2 $Li\cdot$ $:\!\overset{..}{\underset{\cdot}{O}}\!\cdot$

After: $2\left[Li\right]^{1+}\left[:\!\overset{..}{\underset{..}{O}}\!:\right]^{2-}$

37

3-23. (a) Aluminum sulfide (b) Cesium selenide (c) Magnesium oxide
(d) Calcium nitride (e) Cesium oxide (f) Potassium bromide

3-25. (a) Iodine and fluorine each have seven valence electrons. Hence, the Lewis structure must show $3 \times 7 = 21$ valence electrons. **Only 20 are shown.**

(b) Carbon has four valence electrons, oxygen six, and nitrogen five. The Lewis structure must therefore have $4 + 6 + 5 = 15$ electrons. **Sixteen are shown.**

(c) Carbon has four valence electrons, oxygen six, and hydrogen one. The Lewis structure must therefore have $4 + 6 + (2 \times 1) = 12$ electrons. **Fourteen are shown.**

3-27. The formula for formal charges is:

FORMAL CHARGE = Group No. - 2 ×(No. of lone pairs) - (No. of bonds)

Formula	ATOM	Group No.	No. of lone pairs	No. of bonds	Formal charge
(a) SO_4^{2-}	S	6	0	4	+2
	O	6	3	1	-1
(b) $S_2O_3^{2-}$	central S	6	0	5	+1
	outer S	6	2	2	0
	O	6	3	1	-1
(c) SbF_3	Sb	5	1	3	0
	F	7	3	1	0
(d) SCN^-	S	6	3	1	-1
	C	4	0	4	0
	N	5	1	3	0

3-29.

Formula	ATOM	Group No.	No. of lone pairs	No. of bonds	Formal charge
HNO	H	1	0	1	0
	N	5	1	3	0
	O	6	2	2	0
HON	H	1	0	1	0
	O	6	1	3	1+
	N	5	2	2	1-

The first formula has no formal charges on any of the atoms. It is therefore the favored structure.

3-31. This problem is solved by noting the formal charge, the number of bonds, and the number of lone pairs on the element Z in each case. We can then determine the Group Number for the element, and then we can name the group.

Group No. = formal charge + (2 × No. of lone pairs) + (No. of bonds)

The determination of the formal charge on Z requires that we determine the formal charge on each of the oxygens and (in (d)), hydrogen. Then we determine the formal charge on Z by subtracting the formal charges on the other atoms from the total charge per atom of Z:

Formal charge on Z = total charge per Z - formal charges on other atoms

(a)
Formal charges on oxygens = 0

Total charge on molecule = 0

Formal charge on Z = 0 - (2 × 0) = 0

Group No. of Z: = 0 + (2 × 0) + 4 = 4

The element comes from **Group IV** of the periodic table.

(b)
Formal charges on outer oxygens = -1

Formal charges on inner oxygen = 0

Total charge on molecule = 0

Formal charge on Z = 1/2 × [0 - (6 × -1) - (0)] = + 3

Group # of Z: = 3 + (2 × 0) + 4 = + 7

The element comes from **Group VII** of the periodic table (the halogens).

3-31. (continued)

(c)

$$\text{Formal charges on singly bonded oxygens} = -1$$
$$\text{Formal charges on doubly bonded oxygen} = 0$$
$$\text{Total charge on molecule} = -1$$
$$\text{Formal charge on Z} = -1 - (2 \times -1) - (0) = +1$$
$$\text{Group \# of Z:} = 1 + (2 \times 0) + 4 = +5$$

The element comes from **Group V** of the periodic table.

(d)

$$\text{Formal charges on outer oxygens} = -1$$
$$\text{Formal charges on inner oxygen} = 0$$
$$\text{Formal charges on hydrogen} = 0$$
$$\text{Total charge on molecule} = 0$$
$$\text{Formal charge on Z} = 0 - (3 \times -1) - (0) = +3$$
$$\text{Group \# of Z:} = 3 + (2 \times 0) + 4 = +7$$

The element comes from **Group VII** of the periodic table (the halogens).

3-33.
(a) H—Se—H

(b) H——Sb——H with H below

(c) H——O——I

(d) :C≡N—C≡N:

3-35.
:F–O–F:

3-37.

H—N—C—N—H structure with O above C

N-H: 1.01×10^{-10} m
C=O: 1.20×10^{-10} m
N-C: 1.47×10^{-10} m

3-39. (a)

There are multiple possible Lewis structures for ClO_3^-, which differ in that one has an octet around each of the atoms while the other has an octet for all the outer atoms and an expanded octet for the chlorine, but fewer formal charges. Both of these have two other resonance structures. As noted in the text for $POCl_3$, which Lewis structure is "correct" is ambiguous, and this is a limitation of the Lewis model.

(b)

(c)

3-41.

3-43.

3-45.

3-47. (a)   (b)   (c)

Chapter 3

3-49. (a) Selinium tetrafluoride   (b) Iodine tribromide
(c) Tetraphosphorus hexaoxide

3-51. Silver nitrate   **AgNO$_3$**
Calcium hypochloriate   Ca(OCl)$_2$
Potassium hydrogen sulfate   **KHSO$_4$**
Gallium sulfite   Ga$_2$(SO$_3$)$_3$
Potassium chlorate   **KClO$_3$**
Sodium hydrogen carbonate NaHCO$_3$

3-53. Sodium hydrogen tartrate

3-55. Tetraphosphorus decaoxide = P$_4$O$_{10}$
Phosphorus pentachloride = PCl$_5$
$$P_4O_{10} + 6\,PCl_5 \rightarrow 10\,POCl_3$$

3-57. (a) 4   (b) 3   (c) 6
(d) 4 (includes a lone pair on S)   (e) 5 (includes two lone pairs)

3-59. Lone pairs on the outer atoms are omitted; the outer atoms all have a complete octet.

(a)
tetrahedral

(b)
bent (angle < 120°)

(c)
octahedral

(d)
pyramidal

(e)
distoted "T"

3-61. (a)

SN = 6; Square planar
Derived from octahedron

(b) angle < 109.5°

SN = 4; Bent -
derived from tetrahedron

(c) angles < 109.5°

SN = 4; Pyramidal
derived from tetrahedron

(d) S$=$C$=$S

SN = 2; Linear

3-63. (a) Planar AB$_3$: BF$_3$ or SO$_3$ (b) Pyramidal AB$_3$: NF$_3$ or PCl$_3$

(c) Bent AB$_2^-$: NO$_2^-$ or NH$_2^-$ (d) Planar AB$_3^{2-}$: CO$_3^{2-}$

3-65. (a) The symmetric arrangement of four I atoms about carbon will make the molecule **non-polar.**

(b) The O atoms are asymmetric about the S. This molecule is **polar**.

(c) The symmetric arrangement of F atoms about S will make the molecule **non-polar.**

(d) This is a pyramidal molecule. It has a lone pair, which gives it a clear "top" and "bottom." The molecule will be **polar**.

(e) This distorted "T" shaped molecule has two adjacent lone pairs. It will be **polar**.

3-67. These molecules were considered from a Lewis-dot structure standpoint back in Problem No. 43. Acceptable Lewis-dot structures that like this:

Both have a central atom with S.N. = 3. Both structures are expected to be bent, and therefore both should be polar. The information about why the molecule is bent does not resolve the question of the structure.

3-69. (a) The Lewis dot structure is:

$$:\overset{(-1)}{\underset{\cdot\cdot}{N}}=\overset{(+1)}{N}=\overset{\cdot\cdot}{\underset{\cdot\cdot}{O}}: \longleftrightarrow :N\equiv\overset{(+1)}{N}-\overset{\cdot\cdot}{\underset{\cdot\cdot}{O}}:^{(-1)}$$

There are two reasonable resonance structures; both have no lone pairs on the central N; we expect the central N to have *S.N.* = 2 and the molecule to have a linear structure.

(b) The Lewis structures are no help in deciding the molecule's polarity. But because nitrogen is less electronegative than oxygen we can predict that the nitrogen end of the molecule is the positive end.

3-71. (a) Iron(III) oxide   (b) Titanium(IV) bromide   (c) Tungsten(VI) oxide
      (d) Lead(IV) chloride   (e) Manganese(III) fluoride   (f) Silver chlorate

3-73. (a) SnF_2 (b) Re_2O_7 (c) CoF_3
      (d) WCl_5 (e) $Cu(NO_3)_2$ (f) $Ni(BrO_4)_2 \bullet 6\ H_2O$

3-75. Rhenium (VII) sulfide = Re_2S_7 Rhenium (VI) sulfide = ReS_3

$$Re_2S_7 + H_2 \rightarrow H_2S + 2\ ReS_3$$

3-77. (a) The average of the boiling points of copper and gallium is 2485 $^{\circ}C$, which is a poor estimate for the boiling point of zinc (907 $^{\circ}C$).

(b) The information in this part of the problem about the melting points, not the boiling points, of the elements. But if we assume that the trend in boiling points matches the trend in melting points, then we predict that mercury would have a much lower boiling point than its neighbors. The same is true of zinc. We conclude that the boiling points of the zinc group are anomalously low.

(c) Based on the conclusion in (b), we expect the boiling point of cadmium to lie at a much lower temperature than the boiling points of silver and indium.

3-79. There is a fundamental error in that oxygen, a chalcogen, is *not* in the same chemical family as chlorine and fluorine, which are halogens. Hydrogen fluoride is strongly corrosive, but the compound of oxygen most similar to HF is H_2O (plain water)! Although oxygen *is* a corrosive gas, especially with metals and at elevated temperatures and pressures, the reasons the character advances are invalid.

3-81. (a) Na_3CO_3 is not expected to be stable. The stable compound related to this formula is Na_2CO_3, sodium carbonate.

(b) $MgCl$ is not expected to be stable. The stable compound related to this formula is $MgCl_2$, magnesium chloride.

(c) $(NH_4)_2SO_4$ is expected to be stable. Its name is ammonium sulfate.

(d) BaCN is not expected to be stable. The stable compound related to this formula is $Ba(CN)_2$, barium cyanide.

(e) $Sr(OH)_2$ is expected to be stable. Its name is strontium hydroxide.

3-83. (a) Ammonium phosphate, $(NH_4)_3PO_4$.

Potassium nitrate, KNO_3.

Ammonium sulfate, $(NH_4)_2SO_4$.

(b)

COMPOUND	$(NH_4)_3PO_4$	KNO_3	$(NH_4)_2SO_4$
Molar mass (g mol^{-1})	149.087	101.103	132.262
Mass of N in one mole of compound	42.202 g	14.0067 g	28.0135 g
Mass % N	28.307%	13.854%	21.18%
Mass of P in one mole of compound	30.974 g	0.00 g	0.00 g
Mass % P	20.776%	0.00%	0.00%
Mass of K in one mole of compound		39.0983 g	0.00 g
Mass % K		38.672%	0.00%

3-85. Both water and ammonia have one or more lone pairs. The hydrogen ion encountering one of them will "attach itself" to the lone pair, and the oxygen (in water) or nitrogen (in ammonia) will gain a formal charge, as indicated.

3-87.   (a)

$$\ddot{\text{O}}=\overset{\cdot\cdot}{\text{P}}-\overset{\cdot\cdot}{\underset{\cdot\cdot}{\text{Cl}}}:$$

    (b) No formal charges   Octet rule satisfied

3-89.   (a)   Central S   Central N

(b) Both structures have a separation of formal charge, but in the case with nitrogen in the center the negative charge is on an element (sulfur) to the right in the periodic table, where elements are known to prefer negative charges. It should be more likely.

3-91.

$$:\overset{\cdot\cdot}{\underset{\cdot\cdot}{\text{O}}}-\overset{\cdot\cdot}{\underset{\cdot\cdot}{\text{O}}}-\overset{\cdot\cdot}{\text{N}}=\overset{\cdot\cdot}{\text{O}}:$$

3-93.   (a)

SN of S = 6; square pyramid

    (b)

SN = 7 for Sb; pentagonal
pyramid derived from a bi-pyramid
In this case, the basic structure for SN = 7 is not known to
us. A reasonable structure is presented, based on
minimizing electron pair repulsion.

3-95.

SN = 4 for C; Bent -
derived from tetrahedron

The structure is derived from a tetrahedron, but the presence of two lone pairs on the carbon mean that the C-C-C angle is contracted to something *less* than 109.5°

3-97.

There are no resonance structures available in this case.

This isomer has the disadvantage of one resonance form possessing formal charge separation and the other an expanded octet. But it has the advantage of multiple resonance forms.

3-99. (a)

(b)

3-101. (a) There are two kinds of iodines and two kinds of oxygens.

2 iodines with six groups attached and zero lone pairs: $7 - 0 - 1/2 (12) = +1$

2 iodines with three groups attached and one lone pair : $7 - 2 - 1/2 (6) = +2$

6 oxygens in terminal positions: $6 - 6 - 1/2(2) = -1$

4 oxygens in bridging positions $6 - 4 - 1/2(4) = 0$

(b) All the O's in this case have four lone pair electrons, and zero formal charge.
The two I's with three groups attached have a formal charge of $7 - 2 - 1/2(8) = +1$ each
The two I's with six O's attached have a formal charge of $7 - 0 - 1/2(16) = -1$ each.

3-103. Recall that elemental sulfur has a formula of S_8.

(a) $10\ In + S_8 \rightarrow 2\ In_5S_4$

(b) Pentaindium tetrasulfide.

3-105. $Na_2[CuCl_4]$: transition metal is copper; ligands are 4 chlorides, Cl^-.

$[Co(H_2O)_6]Br_2$: transition metal is cobalt; ligands are six waters, H_2O.

$K_2[Pt(CN)_4]$: transition metal is platinum; ligands are four cyanides, CN^-.

$Cr(CO)_6$: transition metal is chromium; ligands are six carbon monoxides, CO.

3-107. K_2CuCl_4 and Na_3CoF_6

CHAPTER 4 - TYPES OF CHEMICAL REACTIONS

4-1. Strong electrolyte: A substance that, upon dissolution in water, gives a solution that conducts electricity well.

Weak electrolyte: A substance that, upon dissolution in water, gives a solution that conducts electricity poorly.

Nonelectrolyte: A substance that, upon dissolution in water, gives a solution that does not conduct electricity.

4-3. $Mg(ClO_4)_2 \ (s) \rightarrow Mg^{2+} \ (aq) + 2 \ ClO_4^- \ (aq)$

4-5. (a) This is a sugar, a low molar mass compound containing OH groups. We expect it to dissolve in water.
(b) Table 4-1 lists $KClO_4$ as slightly soluble. Between 0.01 and 1. g will dissolve in 100 g of water.
(c) This is a low molar mass compound containing OH groups. It will dissolve.

4-7. The original compounds, NaCl (sodium chloride) and KBr (potassium bromide) will be present, along with two new compounds formed by switching the anions and cations, NaBr (sodium bromide) and KCl (potassium chloride).

4-9. We can evaporate the water and then dissolve the LiCl in ethanol selectively. The undissolved material will be RbCl. We can filter the solution to separate the undissolved RbCl from the LiCl that is dissolved in ethanol. After evaporating the ethanol, we will also have pure LiCl.

4-11. The first two parts of the problem are worked in steps through (i) separation of all *aqueous* salts into ions and then (ii) elimination of common ions from the two sides of the equation. Common ions found in the intermediate equation (i) are marked in bold to make them clearer. The last two parts of the problem have just the answer.
(a) START: $\quad$ NaCl *(aq)* + $AgNO_3$ *(aq)* $\rightarrow$ AgCl *(s)* + $NaNO_3$ *(aq)*
(i) $\mathbf{Na^+}$ *(aq)* + Cl^- *(aq)* + Ag^+ *(aq)* + $\mathbf{NO_3^-}$ *(aq)*

$\rightarrow$ AgCl *(s)* + $\mathbf{Na^+}$ *(aq)* + $\mathbf{NO_3^-}$ *(aq)*

(ii) $\quad\quad\quad\quad$ Cl^- *(aq)* + Ag^+ *(aq)* $\rightarrow$ AgCl *(s)*

4-11. (continued)

(b) START: $K_2CO_3 \ (s) + 2 \ HCl \ (aq) \ \rightarrow 2 \ KCl \ (aq) + CO_2 \ (g) + H_2O \ (l)$

(i) $K_2CO_3 \ (s) + 2 \ H^+ \ (aq) + \mathbf{2 \ Cl^-} \ (aq)$

$$\rightarrow 2 \ K^+ \ (aq) + \mathbf{2 \ Cl^-} \ (aq) + CO_2 \ (g) + H_2O \ (l)$$

(ii) $K_2CO_3 \ (s) + 2 \ H^+ \ (aq) \ \rightarrow 2 \ K^+ \ (aq) + CO_2 \ (g) + H_2O \ (l)$

(c) $2 \ Cs \ (s) + 2 \ H_2O \ (l) \ \rightarrow 2 \ Cs^+ \ (aq) + 2 \ OH^- \ (aq) + H_2 \ (g)$

(d) $10 \ Cl^- \ (aq) + 2 \ MnO_4^- \ (aq) + 16 \ H^+ \ (aq)$

$$\rightarrow 5 \ Cl_2 \ (g) + 2 \ Mn^{2+} \ (aq) + 8 \ H_2O \ (l)$$

4-13. Balanced reaction: $NaBr \ (aq) + AgNO_3 \ (aq) \rightarrow AgBr \ (s) + NaNO_3 \ (aq)$

Net ionic reaction: $Br^- \ (aq) + Ag^+ \ (aq) \rightarrow AgBr \ (s)$

4-15. The solution prepared by dissolving solid barium sulfate in water contains only Ba^{2+} and SO_4^{2-} ions. The solution prepared from solutions of $BaCl_2$ and Na_2SO_4 will contain the unreacted spectator ions, Cl^- and Na^+, also.

4-17. (a) Balanced reaction: $Zn(NO_3)_2 \ (aq) + K_2S \ (aq) \ \rightarrow ZnS \ (s) + 2 \ KNO_3 \ (aq)$

Net ionic reaction: $Zn^{2+} \ (aq) + S^{2-} \ (aq) \ \rightarrow ZnS \ (s)$

(b) Balanced reaction:

$$2 \ AgClO_3 \ (aq) + CaCl_2 \ (aq) \ \rightarrow 2 \ AgCl \ (s) + Ca(ClO_3)_2 \ (aq)$$

Net ionic reaction: $2 \ Ag^+ \ (aq) + 2 \ Cl^- \ (aq) \ \rightarrow 2 \ AgCl \ (s)$

(c) Balanced reaction:

$$3 \ NaOH \ (aq) + Fe(NO_3)_3 \ (aq) \ \rightarrow Fe(OH)_3 \ (s) + 3 \ NaNO_3 \ (aq)$$

Net ionic reaction: $3 \ OH^- \ (aq) + Fe^{3+} \ (aq) \ \rightarrow Fe(OH)_3 \ (s)$

(d) Balanced reaction:

$$3 \ Ba(CH_3COO)_2(aq) + 2 \ Na_3PO_4(aq) \rightarrow Ba_3(PO_4)_2 \ (s) + 6 \ NaCH_3COO \ (aq)$$

Net ionic reaction: $3 \ Ba^{2+} \ (aq) + 2 \ PO_4^{3-} \ (aq) \rightarrow Ba_3(PO_4)_2 \ (s)$

4-19. (a) $Be^{2+} \ (aq) + 2 \ CH_3COO^- \ (aq) \rightarrow Be(CH_3COO)_2 \ (s)$

(b) $Ba^{2+} \ (aq) + SO_4^{2-} \ (aq) \rightarrow BaSO_4 \ (s)$

(c) $Ca^{2+} \ (aq) + 2 \ OH^- \ (aq) \rightarrow Ca(OH)_2 \ (s)$

(d) N. R.

4-21. Two precipitation reactions *could* occur given the starting solutions:

$$Na^+ \ (aq) + Cl^- \ (aq) \ \rightarrow NaCl \ (s)$$
$$K^+ \ (aq) + NO_3^- \ (aq) \ \rightarrow KNO_3 \ (s)$$

However, both of the possible solid products are soluble compounds. Though solutions of different salts were mixed, there were no insoluble substances generated.

4-23. The second solution should contain an anion that forms a soluble salt with magnesium and an insoluble salt with barium. A good candidate is sulfate, SO_4^{2-}. It must be delivered as a salt with a cation that will not precipitate nitrate. There are many soluble sulfates; sodium sulfate is a good choice.

4-25. (a) Hydrosulfuric acid   (b) Periodic acid

       (c) Carbonic acid   (d) Hydrobromic acid

4-27. (a) I_2O_5 – diiodine pentaoxide – acid anhydride

       (b) BaO – barium oxide – base anhydride

       (c) CrO_3 – chromium (VI) oxide – acid anhydride

       (d) N_2O – dinitrogen oxide – acid anhydride

4-29. Note that in all cases once we have figured out the formula of the product, the name of the product reminds us whether the anhydride is basic ("hydroxide" in the product) or is acidic ("acid" in the product).

       (a) MgO – base anhydride – $Mg(OH)_2$ – magnesium hydroxide

       (b) Cl_2O – acid anhydride – HOCl - hypochlorous acid

       (c) SO_3 – acid anhydride – H_2SO_4 – sulfuric acid

       (d) Cs_2O – base anhydride – CsOH – cesium hydroxide

4–31. This is a calculation from the mass of the $KHC_8H_4O_4$ to the moles of the NaOH. The moles of NaOH are then the basis of the concentration calculation, since we are given the volume of the NaOH.

For moles of NaOH:

$$0.5836 \ g \ KHC_8H_4O_4 \times \frac{1 \ mol \ KHC_8H_4O_4}{202.2236} \times \frac{1 \ mol \ NaOH}{1 \ mol \ KHC_8H_4O_4} = 0.002858 \ mol \ NaOH$$

For concentration: $\dfrac{0.002858 \ mol \ NaOH}{0.02372 \ L \ NaOH} = \mathbf{0.01205 \ mol \ L^{-1}}$

4-33. This is a calculation from the volume of NaOH to the moles of the HCl. The moles of HCl are then the basis of the concentration calculation, since we are given the volume of the HCl.

For moles of HCl:

$$0.02694 \text{ L NaOH} \times \frac{0.1025 \text{ mol NaOH}}{1 \text{ L NaOH}} \times \frac{1 \text{ mol HCl}}{1 \text{ mol NaOH}} = 0.002761 \text{ mol HCl}$$

For concentration: $\dfrac{0.002761 \text{ mol HCl}}{0.02500 \text{ L HCl}} = \mathbf{0.1105 \text{ mol L}^{-1}}$

4-35. For clarity we write the balanced equation and the net ionic equation for each (note that the weak acids are written in their undissociated form in the net ionic reaction).

(a) $2\,HNO_3\ (aq) + K_2CO_3\ (s)\ \rightarrow 2\,KNO_3\ (aq) + CO_2\ (g) + H_2O\ (l)$

$2\,HNO_3\ (aq) + K_2CO_3\ (s)\ \rightarrow 2\,K^+\ (aq) + 2\,NO_3^-\ (aq) + CO_2\ (g) + H_2O\ (l)$

(b) $HBr\ (aq) + Zn\ (s)\ \rightarrow H_2\ (g) + ZnBr_2\ (aq)$

$HBr\ (aq) + Zn\ (s)\ \rightarrow H_2\ (g) + Zn^{2+}\ (aq) + 2\,Br^-\ (aq)$

(c) $H_2SO_4\ (aq) + Zn(OH)_2\ (s)\ \rightarrow H_2O\ (l) + ZnSO_4\ (aq)$

$H_2SO_4\ (aq) + Zn(OH)_2\ (s)\ \rightarrow H_2O\ (l) + Zn^{2+}\ (aq) + SO_4^{2-}\ (aq)$

4-37. (a) $HClO_3\ (aq) + KOH\ (aq)\ \rightarrow H_2O\ (l) + K^+\ (aq) + ClO_3^-\ (aq)$

(b) $N_2O_5\ (g) + 2\,NaOH\ (aq)\ \rightarrow 2\,NO_3^-\ (aq) + 2\,Na^+\ (aq) + H_2O\ (l)$

(c) $(NH_4)_2SO_4\ (aq) + Ba(OH)_2\ (aq)\ \rightarrow 2\,NH_3\ (g) + 2\,H_2O\ (l) + BaSO_4\ (s)$

Note that this is both a precipitation and an acid-base reaction.

4-39. (a) $2\,HBr\ (aq) + Ca(OH)_2\ (s)\ \rightarrow 2\,H_2O\ (l) + Ca^{2+}\ (aq) + 2\,Br^-\ (aq)$

(b) $NH_3\ (aq) + H_2SO_4\ (aq)\ \rightarrow NH_4^+\ (aq) + HSO_4^-\ (aq)$

A further reaction can take place between ammonia and the hydrogen sulfate, *if the ammonia is present in excess*:

$NH_3\ (aq) + HSO_4^-\ (aq)\ \rightarrow NH_4^+\ (aq) + SO_4^{2-}\ (aq)$

(c) $LiOH\ (aq) + HNO_3\ (aq)\ \rightarrow Li^+\ (aq) + NO_3^-\ (aq) + H_2O\ (l)$

4-41. (a) $Ca(OH)_2\ (aq) + 2\,HF\ (aq) \rightarrow CaF_2\ (s) + 2\,H_2O\ (l)$

Acid: hydrofluoric acid   Salt: calcium fluoride

Base: calcium hydroxide

(b) $2\,RbOH\ (aq) + H_2SO_4\ (aq) \rightarrow Rb_2SO_4\ (s) + 2\,H_2O\ (l)$

Acid: sulfuric acid   Salt: rubidium sulfate

Base: rubidium hydroxide

(c) $Zn(OH)_2\ (s) + 2\,HNO_3\ (aq) \rightarrow Zn(NO_3)_2\ (aq) + 2\,H_2O\ (l)$

Acid: nitric acid   Salt: zinc nitrate

Base: zinc hydroxide

(d)   KOH *(aq)* + CH₃COOH *(aq)* → KCH₃COO *(aq)*+ H₂O *(l)*

Acid: acetic acid   Base: potassium hydroxide

Salt: potassium acetate

4-43. (a) PF₃ - phosphorus trifluoride   H₃PO₃ - phosphorus acid

HF - hydrofluoric acid

(b) PF₃ *(aq)* + 3 H₂O *(l)*→ H₃PO₃ *(aq)* + 3 HF *(aq)*

4-45. The salt is sodium sulfide, Na₂S. The other product is water.

4-47. The tomato sauce must contain an acid that reacts with aluminum metal. The net ionic reaction will be: 6H⁺ *(aq)* + 2 Al *(s)* → Al³⁺ *(aq)* + 3 H₂ *(g)* + 6 H₂O *(l)*

4-49. Some elements have their oxidation number fixed by the rules and some have their oxidation number determined by the other elements. The oxidation numbers must add to the total charge on the formula (rule 1). Here all compounds are neutral, to the oxidation numbers must add to zero.

(a) CH₂O. By rule 4: **Ox. No. H = +1** By rule 5: **Ox. No. O = –2** .

[Ox No. O] + 2 × [Ox. No. H] + [Ox. No. C] = 0

[Ox. No. C] = -([Ox No. O] + 2 × [Ox. No. H])

[Ox. No. C] = - (-2 + 2 × 1) = 0

Note that there is nothing wrong with the carbon (or any other element, for that matter) in a compound having an oxidation number of zero.

(b) H₂CO₃. By rule 4: **Ox. No. H = +1** By rule 5: **Ox. No. O = –2** .

3 × [Ox No. O] + 2 × [Ox. No. H] + [Ox. No. C]   = 0

[Ox. No. C] = –(3 × [Ox No. O] + 2 × [Ox. No. H])

[Ox. No. C] = – (3× -2 + 2× 1) = **+4**

(c) RbH. By rule 2: **Ox. No. Rb = +1**. By rule 4 **Ox. No. H = -1** (Rb is a metal).

(d) N₂O₅. By rule 5: **Ox. No. O = –2** .

2 × [Ox No. N] + 5 × [Ox. No. O] = 0

[Ox No. N] = –1/2 × (5 × [Ox. No. O]) = –1/2 × (5 × -2) = **+5**.

4-51. As₂O₃ *(aq)*+2 I₂ *(aq)*+2 H₂O *(l)* → As₂O₅ *(aq)*+4 HI*(aq)*

Oxidized: 2 As from +3 to +5...........+ 4 increase in oxidation number

Reduced: 4 I from 0 to -1.................–4 decrease in oxidation number.

We observe that the changes in oxidation number are equal.

4-53. (a) $\overset{+3}{2\,PF_2I}\,(l) + \overset{0}{2\,Hg}\,(l) \rightarrow \overset{+2}{P_2F_4}\,(g) + \overset{+1}{Hg_2I_2}\,(s)$

(b) $\overset{+5/-2}{2\,KClO_3}\,(aq) \rightarrow \overset{-1}{2\,KCl}\,(s) + \overset{0}{3\,O_2}\,(g)$

(c) $\overset{-3}{4\,NH_3}\,(g) + \overset{0}{5\,O_2}\,(g) \rightarrow \overset{+2/-2}{4\,NO}\,(g) + \overset{-2}{6\,H_2O}\,(g)$

(d) $\overset{0}{2\,As}\,(s) + \overset{+1}{6\,NaOH}\,(l) \rightarrow \overset{+3}{2\,Na_3AsO_3}\,(s) + \overset{0}{3\,H_2}\,(g)$

4-55. (a) Oxidized: 2 Hg from 0 to +1 **+2** change in oxidation numbers

Reduced: 2 P in 2 PF_2I's from +3 to +2 ....**−2** change in oxidation numbers

(b) Oxidized: 6 O in 2 $KClO_3$'s from -2 to 0 ....**+12** change in oxidation numbers

Reduced: 2 Cl in 2 $KClO_3$'s from +5 to -1..**−12** change in oxidation numbers

(c) Oxidized: 4 N in 4 NH_3's from -3 to +2......**+20** change in oxidation numbers

Reduced: 10 O in 5 O_2's from 0 to -2 ....**−20** change in oxidation numbers

(d) Oxidized: 2 As from 0 to +3 ....**+6** change in oxidation numbers

Reduced: 6 H in 6 NaOH's from +1 to 0.......**−6** change in oxidation numbers

4-57. The H and O atoms in this reaction are all in compounds where the rules predetermine their oxidation numbers to be +1 (H) and -2 (O). It is important to recognize the presence of selenate ion, SeO_4^{2-}, in the compound $Au_2(SeO_4)_3$.

REACTANTS: **Au** has an oxidation number of 0

Se in H_2SeO_4 has an oxidation number of +6.

PRODUCTS: **Se** in SeO_4^{2-} has an oxidation number of **+6**.

Se in H_2SeO_3 has an oxidation number of **+4**.

Au in $Au_2(SeO_4)_3$ has an oxidation number equal and opposite to 1/2 of the charge on 3 SeO_4^{2-} ions, or an oxidation number of **+3 on each Au**.

The species oxidized is the Au; *some* of the H_2SeO_4 is reduced.

4-59. The reaction: $2\,Al_2O_3\,(s) + 3\,C\,(s) \rightarrow 2\,Al\,(s) + 3\,CO_2\,(g)$

The species oxidized is C; the species reduced is Al_2O_3.

4-61. (a) $2\,HCl\,(g) + O_2\,(g) \rightarrow H_2O\,(g) + Cl_2\,(g)$ This is a displacement reaction.

(b) $H_2C{=}O\,(g) + H_2\,(g) \rightarrow CH_3OH\,(g)$ This is a hydrogenation reaction.

(c) $Mg\,(s) + HCl\,(aq) \rightarrow H_2\,(g) + MgCl_2\,(aq)$ This is a displacement reaction.

(d) $N_2\,(g) + 3\,H_2\,(g) \rightarrow 2\,NH_3\,(g)$ This is a combination reaction.

4-63. (a)

$$\overset{+1\,/\,+5\,/\,-2}{4\,HClO_3\,(aq)} \rightarrow \overset{+4\,/\,-2}{4\,ClO_2\,(g)} + \overset{0}{O_2\,(g)} + \overset{+1\,/\,-2}{2\,H_2O\,(l)}$$

$$\overset{+1\,/\,+5\,/\,-2}{2\,HClO_3\,(aq)} \rightarrow \overset{+1\,/\,-1}{2\,HCl\,(aq)} + \overset{0}{3\,O_2\,(g)}$$

$$\overset{+1\,/\,+5\,/\,-2}{4\,HClO_3\,(aq)} \rightarrow \overset{+1\,/\,+7\,/\,-2}{3\,HClO_4\,(g)} + \overset{+1\,/\,-1}{HCl\,(l)}$$

(b) Only the last has a single element going to two different oxidation numbers, one higher and one lower than the original oxidation number. It is the only disproportionation.

4-65. (a) $NH_4NO_2 \rightarrow N_2 + 2\,H_2O$

(b) N in ammonium has an oxidation number of -3; in nitrite it is +3. The "overall" oxidation number of N in ammonium nitrite is 0. This reaction is a redox reaction, since an atom is increasing its oxidation number and another is decreasing its oxidation number.

4-67.

$$Ba\,(s) + H_2\,(g) \rightarrow BaH_2\,(s)$$

$$BaH_2\,(s) + 2\,H_2O\,(l) \rightarrow Ba^{2+}\,(aq) + 2\,OH^-\,(aq) + 2\,H_2\,(g)$$

$$BaH_2\,(s) + 2\,HCl\,(aq) \rightarrow Ba^{2+}\,(aq) + 2\,Cl^-\,(aq) + 2\,H_2\,(g)$$

In the last case, the reaction of water with barium hydride causes the formation of hydroxide ion, which precipitates with the zinc, and aqueous barium ion, which precipitates with the sulfate:

$$BaH_2\,(s) + 2\,H_2O\,(l) + Zn^{2+}\,(aq) + SO_4^{2-}\,(aq)$$

$$\rightarrow BaSO_4\,(s) + Zn(OH)_2\,(s) + 2\,H_2\,(g)$$

4-69. (a) $\qquad H_2Te\,(aq) + H_2O\,(l) \rightarrow HTe^-\,(aq) + H^+\,(aq) \qquad$ acid-base

(b) $\qquad SrO\,(s) + CO_2\,(g) \rightarrow SrCO_3\,(s) \qquad$ Lewis acid-base

Both (c) and (d) are acid-base reactions

(c) $\qquad 2\,HI(aq) + CaCO_3\,(s) \rightarrow H_2O\,(l) + Ca^{2+}\,(aq) + 2\,I^-\,(aq) + CO_2\,(g)$

(d) $\qquad Na_2O\,(s) + 2\,NH_4Br\,(s) \rightarrow 2\,NH_3\,(g) + H_2O\,(l) + 2\,Na^+\,(aq) + 2\,Br^-\,(aq)$

4-71. We use the titration to determine the number of moles of Fe^{2+}.

$$0.02715\,L\,Cr_2O_7^{2-} \times \frac{0.01667\,mol\,Cr_2O_7^{2-}}{1\,L\,Cr_2O_7^{2-}} \times \frac{6\,mol\,Fe^{2+}}{1\,mol\,Cr_2O_7^{2-}} = 0.002716\,mol\,Fe^{2+}$$

For the mass, we assume that there is no appreciable difference between the molar mass of iron and that of the iron (II) ion. Then:

$$0.002716\,mol\,Fe \times \frac{55.845\,g\,Fe}{1\,mol\,Fe} = \mathbf{0.1516\,g\,Fe}$$

4-73. (a) Acetic acid is a weak acid and therefore a weak electrolyte. Its aqueous solutions will conduct electricity poorly.

(b) Acetone is a low molar mass compound and it contains no COOH groups. Its aqueous solutions will conduct electricity not at all.

(c) $HgCl_2$ is a chloride salt that (judging from Table 4-1) should be soluble. Its aqueous solutions will conduct electricity well.

(d) $NaClO_3$ is a simple ionic compound, and should be a strong electrolyte. Its aqueous solutions will conduct electricity well.

4-75. This problem consists of the generation of two different cations and two different anions; we must identify which combination might cause a

precipitate.

CATIONS	ANIONS	POSSIBLE SOLIDS
ammonium,	carbonate	$(NH_4)_2CO_3$
NH_4^+	CO_3^{2-}	$(NH_4)_2SO_4$
calcium,	sulfate,	$CaCO_3$
Ca^{2+}	SO_4^{2-}	$CaSO_4$

Tabulation in this manner helps one to systematically consider all possible combinations. For this set, the least soluble substance is calcium carbonate, $CaCO_3$. The appropriate ionic reaction is:

$$CO_3^{2-} \ (aq) + CaSO_4 \ (s) \rightarrow CaCO_3 \ (s) + SO_4^{2-} \ (aq)$$

The material left in solution will be ammonium sulfate, $(NH_4)_2SO_4$.

4-77. This fulfills the definition of a substance given in Chapter 1. There, the definition of a chemical substance was that it could not be "separated into portions that have different properties." In the case of water, its ionization is a fundamental property; dividing a sample of water into portions will always give a material with the same properties.

4-79. The heat comes from the reaction: $H^+ \ (aq) + OH^- \ (aq) \rightarrow 2 \ H_2O \ (l)$

The white precipitate comes from the reaction:

$$K^+ \ (aq) + ClO_4^- \ (aq) \rightarrow KClO_4 \ (s)$$

$KClO_4$ is one of the very few substances containing potassium *or* perchlorate that is not very soluble.

4-81. We are told that the hydrate of P_4O_{10} that is formed is HPO_3 What remains to know is that the anhydride available from HNO_3 is N_2O_5.

$$4\ HNO_3 + P_4O_{10} \rightarrow 4\ HPO_3 + 2\ N_2O_5$$

4-83. (a) The chemical equation for the reaction is:

$$2\ HBr\ (aq) + Ba(OH)_2\ (aq) \rightarrow BaBr_2\ (aq) + 2\ H_2O\ (l)$$

The limiting reactant is determined by a full stoichiometric calculation.

STARTING FROM HBr:

$$0.222\ L\ HBr \times \frac{0.0450\ mol\ HBr}{1\ L\ HBr} \times \frac{1\ mol\ BaBr_2}{2\ mol\ HBr} = 0.004995\ mol\ BaBr_2$$

STARTING FROM Ba(OH)$_2$:

$$0.125\ L\ Ba(OH)_2 \times \frac{0.0540\ mol\ Ba(OH)_2}{1\ L\ Ba(OH)_2} \times \frac{1\ mol\ BaBr_2}{1\ mol\ Ba(OH)_2} = 0.00675\ mol\ BaBr_2$$

We see that the HBr is limiting.

(b) Since there will be one mole of $Ba(OH)_2$ consumed for every mol of $BaBr_2$ formed, we know that the HBr will react with 0.004995 mol $Ba(OH)_2$, leaving

0.0675 mol $Ba(OH)_2$ – 0.004995 mol $Ba(OH)_2$ = 0.001755 mol $Ba(OH)_2$ unreacted. This will require:

$$0.001755\ mol\ Ba(OH)_2 \times \frac{2\ mol\ HBr}{1\ mol\ Ba(OH)_2} \times \frac{1\ L\ HBr}{0.0450\ mol\ HBr} = 0.0780\ L\ HBr$$

4-85 This is an unusual reaction because (i) the manganese ends up in two different oxidation numbers, both of which differ from the starting oxidation number and (ii) only some of the oxygen atoms change their oxidation number.

$$\overset{+1/+7/-2}{2\ KMnO_4\ (s)} \rightarrow \overset{+1/+6/-2}{K_2MnO_4\ (s)} + \overset{+4/-2}{MnO_2\ (s)} + \overset{0}{O_2\ (g)}$$

The analysis of the reaction is, nevertheless, straightforward:

Oxidized: 2 O from -2 to 0   +4 change in oxidation number

Reduced: 1 Mn from +7 to +6   −1 change in oxidation number

                   1 Mn from +7 to +4   −3 change in oxidation number

There are four electrons lost from O's and four gained by Mn's. The reaction is balanced with respect to the change in oxidation number.

4-87. Reactants:H: +1   O: -2   Xe: +6

Products:H: +1   O: -2 *and* 0   Xe: +8 *and* 0.

Some Xe in the hydrogen xenate ions is being reduced to Xe(0) and some is being oxidized to Xe(8+). Also, oxygens (on the hydrogen xenate, the hydroxide ion, or on both) are being oxidized from -2 to 0.

4-89. (a) $H_2C=CH_2 + Cl_2 \rightarrow CH_2ClCH_2Cl$ This can be compared to the addition of hydrogen to a carbon-carbon double bond. Note, however, that chlorination results in an oxidation of the carbon atoms, not a reduction.

(b) C_3H_8O *(g)* $\rightarrow C_3H_6O + H_2$ Here, in contrast to hydrogenation, hydrogen is formed. This results in a net oxidation of carbon.

(c) $H_2C=CH_2 + HCl \rightarrow CH_3CH_2Cl$ This is the addition of hydrogen chloride. It is similar to hydrogenation or chlorination, except there is no net change in the oxidation number of the carbons.

4-91. (a) The barium ion has a +2 oxidation number in all its compounds, and therefore the perxenate anion is XeO_6^{4-}. This will combine with americium, Am^{3+} in a neutral compound with the formula: **$Am_4(XeO_6)_3$**.

(b) Perxenic acid would be the compound formed by adding enough hydrogen ions to perxenate to neutralize its charge. The formula would be **H_4XeO_6**.

(c) Because perxenic acid has an even number of hydrogen ions, it is straightforward to determine that its anhydride would be: XeO_4.

4-93. The oxide in CaO acts as a strong Lewis base with many Lewis acids according to the general reaction:

$$O^{2-} + AO_x \rightarrow AO_{x+1}^{2-}$$

whereupon the new anion combines with the Ca^{2+} ion to form a new salt.

In this case: CaO *(s)* + SO_3 *(g)* $\rightarrow CaSO_4$ *(s)*

CaO *(s)* + SO_2 *(g)* $\rightarrow CaSO_3$ *(s)*

CHAPTER 5 - THE GASEOUS STATE

5-1. $NH_4HS\ (s) + HCl\ (g) \rightarrow NH_4Cl\ (s) + \mathbf{H_2S\ (g)}$

5-3. $NH_4HS\ (s) \rightarrow NH_3\ (g) + H_2S\ (g)$

5-5. Mix ammonium bromide with a base. For example:

$NH_4Br\ (s) + NaOH\ (aq) \rightarrow NH_3\ (g) + H_2O\ (l) + NaBr\ (aq)$

5-7. $2\ Na_2O_2\ (s) + 2\ CO_2\ (g) \rightarrow 2\ Na_2CO_3\ (s) + O_2\ (g)$. The gas is **oxygen**.

5-9. The situation described has the pressure of the gas equal to the pressure exerted by the water column. $P_{gas} = P_{water} = g\ d_{water}\ h_{water}$. The formula for the pressure exerted by the column of water ($P_{gas} = g\ d_{water}\ h_{water}$) requires densities in kg m^{-3} and heights in meters to yield pressure in standard units of pascals.

$$d_{water} = \left(\frac{1.00\,g}{1\,cm^3}\right) \times \left(\frac{10^6\,cm^3}{1\,m^3}\right) \times \left(\frac{1\,kg}{1000\,g}\right) = 1000\ \frac{kg}{m^3}$$

$$h_{water} = (44.5\ cm) \times \left(\frac{1 \times 10^{-2}\,m}{1\,cm}\right) = 0.445\ m$$

So:

$$P_{gas} = \left(9.80665\,\frac{m}{s^2}\right) \times \left(1000\,\frac{kg}{m^3}\right) \times (0.445\ m)$$

$$= 4364\ \frac{kg}{m-s^2} = 4364\ pascals$$

$$= (4364\ pascals) \times \left(\frac{1\ atm}{101,325\ pascals}\right) = \mathbf{0.0431\ atm}$$

5-11. Pressure is important to this problem only because the equal pressures exerted by the oil and the mercury mean the mass of the oil and the mass of the mercury must be the same. We combine this insight with the definition of density to obtain the height of the oil. The comment "identical tube" tells us that the cross-sectional areas of the tubes are the same.

$$\text{mass oil} = \text{mass of mercury}$$
$$d_{oil}\ V_{oil} = d_{Hg}\ V_{Hg}$$
$$d_{oil}\ h_{oil}\ area_{oil} = d_{Hg}\ h_{Hg}\ area_{Hg}$$
$$d_{oil}\ h_{oil} = d_{Hg}\ h_{Hg}$$
$$h_{oil} = \frac{d_{Hg} h_{Hg}}{d_{oil}} = \frac{\left(13.6\ g\ cm^{-3}\right) \times (305\ mm)}{\left(1.60\ g\ cm^{-3}\right)}$$

$$= \mathbf{2590\ mm = 2.59\ m}$$

5-13. We know the hundreds of kilometers of the earth's atmosphere exert a pressure of about 1.00 atm at sea level. The standard atmosphere (Table 5-2) is equivalent to the pressure exerted by 760 mm Hg. So the mercury ocean would only have to be **760 mm** (about 30 inches) deep.

5-15. $172.00 \text{ MPa} = 1.7200 \times 10^8 \text{ Pa}$

$$1.7200 \times 10^8 \text{ Pa} \times \left(\frac{1 \text{ atm}}{101,325 \text{ Pa}}\right) = \textbf{1697.5 atm}$$

$$1.7200 \times 10^8 \text{ Pa} \times \left(\frac{1 \text{ bar}}{100,000 \text{ Pa}}\right) = \textbf{1720.0 bar}$$

5-17. The trapped nitrogen exerts a pressure equal to that of the room pressure added to the pressure exerted by the extra height of Hg.

$$P_{gas} = P_{room} + P_{Hg}$$
$$= 1.000 \text{ atm} + (67.5 \text{ mm Hg}) \times \left(\frac{1 \text{ atm}}{760 \text{ mm Hg}}\right)$$
$$= 1.000 \text{ atm} + 0.0888 \text{ atm} = \textbf{1.09 atm}$$

5-19. The experiment has nitrogen, at a pressure of 4.00 atm ($= P_1$) and a volume of 3.00 L ($=V_1$), expanding to a volume of $3.00 + 6.00 = 9.00$ L ($=V_2$). Applying Boyle's Law:

$$P_2 V_2 = P_1 V_1$$
$$P_2 = \frac{P_1 V_1}{V_2} = \frac{(4.00 \text{ atm}) \times (3.00 \text{ L})}{9.00 \text{ L}} = \textbf{1.33 atm.}$$

We could also do this by noting that the volume change of 3.00 to 9.00 L represents an increase in volume by a factor of 3. Therefore, the pressure will decrease by a factor of 3, and $4/3 = 1.33$ atm.

5-21.

	V_1	P_1	V_2	P_2	Calculations
(a)	0.587 L	1.00 atm	**0.247 L**	2.37 atm	$V_2 = \dfrac{1.00 \text{ atm} \times 0.587 \text{ L}}{2.37 \text{ atm}} = 0.247 \text{ L}$
(b)	785 mL	1.75 atm	**1270 mL**	110.0 kPa	$P_2 = 110.0 \text{ kPa} \times \dfrac{1000 \text{ Pa}}{1 \text{ kPa}} \times \dfrac{1 \text{ atm}}{101325 \text{ Pa}}$ $= 1.0856 \text{ atm}$ $V_2 = \dfrac{1.75 \text{ atm} \times 785 \text{ mL}}{1.0856 \text{ atm}} = 1270 \text{ mL}$

5-21. (continued)

	V_1	P_1	V_2	P_2	Calculations
(c)	604 mL	995 torr	312 mL	**1930 torr**	$P_2 = \dfrac{604 \text{ mL} \times 995 \text{ torr}}{312 \text{ mL}} = 1930 \text{ torr}$
(d)	x mL	3.25 atm	$^x/_2$ mL	**6.50 atm**	$P_2 = \dfrac{3.25 \text{ atm} \times x \text{ mL}}{^x/_2 \text{ mL}} = 6.50 \text{atm}$

5-23.
$$^{\circ}C = \frac{5}{9}(^{\circ}F - 32.0) = \frac{5}{9}(98.6 - 32.0)$$
$$= 37.0 \;^{\circ}C$$
$$T \text{ (Kelvin)} = 273.15 + \;^{\circ}C = 273.15 + 37.0 = \textbf{310.2 K}$$

5-25. $\dfrac{V_2}{T_2} = \dfrac{V_1}{T_1}$; $\quad V_2 = \left(\dfrac{T_2}{T_1}\right) \times V_1 = 3 \times V_1 = 3 \times 3.50 = \textbf{11.5 L}$

5-27. Charles' Law is always easiest to use with the Kelvin as the temperature.
$$T_1 = 273.15 + \frac{5}{9}(^{\circ}F - 32) = 273.15 + \frac{5}{9}(100 - 32) = 310.9 \text{ K}$$
$$T_2 = 273.15 + \frac{5}{9}(0 - 32) = 255.4 \text{ K}$$

As above: $\quad V_2 = \left(\dfrac{T_2}{T_1}\right) \times V_1 = \left(\dfrac{255.4K}{310.9K}\right) \times 17.4 \text{ gills} = \textbf{14.3 gills}$

5-29. The same number of moles of CaC_2 reacts in both cases and the pressure stays constant. As a result, Charles' Law applies. Note $T_1 = 50^{\circ}C + 273 = 323$ K.
$$T_2 = 400^{\circ}C + 273 = 673 \text{ K}$$
$$V_2 = \left(\frac{T_2}{T_1}\right) \times V_1 = \left(\frac{673K}{323K}\right) \times 64.5L = \textbf{134 L}$$

5-31.

	T_1	V_1	T_2	V_2	Calculations
(a)	0.0°C = 273.15 K	164 L	100.0°C = 373.15 K	**224 L**	$V_2 = \left(\dfrac{373.15\text{ K}}{273.15\text{ K}}\right) \times 164 \text{ L}$ $= 224 \text{ L}$
(b)	400 K	26 quarts	200 K	**13 quarts**	$V_2 = \left(\dfrac{200\text{ K}}{400\text{ K}}\right) \times 26 \text{ quarts}$ $= 13 \text{ quarts}$
(c)	1000 K	1000 mL	153°C = 426.15K	**426 mL**	$V_2 = \left(\dfrac{426.15\text{ K}}{1000\text{ K}}\right) \times 1000 \text{ mL}$ $= 426 \text{ mL}$
(d)	200°C = 473.15K	y mL	**3030 K**	$6.4\,y$ mL	$T_2 = \left(\dfrac{6.4\,y\text{ mL}}{y\text{ mL}}\right) \times 473.15 \text{ K}$ $= 3030 \text{ K}$

5-33. It is useful to recognize that when one or more term in an ideal gas law calculation remains constant -- here it is the number of variable moles and the gas constant R—one analyzes the remaining variables as an equal pair of ratios in a "stripped" form of the idea gas law :

$$\frac{P_2V_2}{T_2} = \frac{P_1V_1}{T_1}$$

$$P_2 = \left(\frac{V_1}{T_1}\right) \times \left(\frac{T_2}{V_2}\right) \times P_1 = \left(\frac{V_1}{V_2}\right) \times \left(\frac{T_2}{T_1}\right) \times P_1$$

$$= \left(\frac{1}{3.25}\right) \times \left(\frac{2}{1}\right) \times 5.50 \text{ atm} = \mathbf{3.38 \text{ atm}}$$

5-35. **(a)** Presumably the temperature of the gas doesn't change when the bottle is filled. Then, this problem is a simple Boyle's Law problem.

$$P_2V_2 = P_1V_1 \implies P_2 = \left(\frac{V_1}{V_2}\right) \times P_1 = \left(\frac{20.6\text{ L}}{1.05\text{ L}}\right) \times 1.01 \text{ atm} = \mathbf{19.8 \text{ atm}}$$

5-35. (continued)

(b) Now we apply the ideal gas law. Under conditions of constant volume and number of moles: $\dfrac{P_2}{T_2} = \dfrac{P_1}{T_1}$

$$P_2 = \left(\frac{T_2}{T_1}\right)P_1 = \left(\frac{273.15 + 21.0}{273.15 - 20.0}\right) \times 19.8 = \textbf{23.0 atm}$$

5-37. The "pressure difference" across the walls refers to the difference in pressure inside and outside the bottle.

$$P_{inside} - P_{outside} = \text{pressure difference}$$
At bursting, for a $\quad P_{outside} = 0.98$ atm
$$P_{inside} - P_{outside} = 1.25 \text{ atm}$$
$$P_{inside} = 1.25 \text{ atm} + P_{outside} = 1.25 \text{ atm} + 0.98 \text{ atm}$$
$$= 2.23 \text{ atm}$$

This is P_2. We already know $P_1 = 0.98$ atm, $T_1 = 68°F = 293$ K.

Therefore: $\qquad \dfrac{T_2}{P_2} = \dfrac{T_1}{P_1}$

$$T_2 = \left(\frac{P_2}{P_1}\right) \times T_1 = \left(\frac{2.23 \text{ atm}}{0.98 \text{ atm}}\right) \times 293\,K = 667\,K$$

The bottle will burst at an oven temperature of 667 K. This is equal to **741°F.**

5-39. The only quantity given in the problem that cannot be plugged into the ideal gas law is the mass of O_2. This is equivalent to:

$$29.8 \text{ g } O_2 \times \left(\frac{1 \text{ mol } O_2}{31.9988 \text{ g } O_2}\right) = 0.931 \text{ mol } O_2$$

Then: $\qquad T = \dfrac{PV}{nR} = \dfrac{(2.00 \text{ atm})(10.0 \text{ L})}{(0.931 \text{ mol}) \times \left(0.08206\dfrac{L-atm}{mol-K}\right)} = \textbf{262 K}$

5-41. The handbook gives values for volume, the mass, and the molecular formula of the gas. But it omits any data on temperature and pressure. Given a pressure of 1 atm, and knowing that the molar mass of H_2Te is 129.62 g mol^{-1}, though, we can calculate the temperature of the gas:

$$T = \frac{PV}{nR} = \frac{(1.00\,atm) \times (1.00\,L)}{\left(\dfrac{6.234g}{129.62\,g\,mol^{-1}}\right) \times \left(0.08206\dfrac{L-atm}{mol-K}\right)} = \textbf{253 K}$$

This is equal to **-20°C.**

5-43. The data allow a calculation of the total number of molecules in the 15 cm^3 (= 0.015 L) sample, 21 °C (= 294.2 K). The ideal gas law will give us the number of moles of gas, and we convert separately to molecules.

$$n = \frac{PV}{RT} = \frac{(0.95 \text{ atm}) \times (0.015 \text{ L})}{\left(0.08206 \frac{\text{L atm}}{\text{mol K}}\right) \times (294.2 \text{ K})} = 5.91 \times 10^{-4} \text{ moles air}$$

The atoms of Kr is a fraction of this:

$$\text{atoms Kr} = (\text{molecules air}) \times (\text{fraction Kr})$$

$$= (5.91 \times 10^{-4} \times 6.0221 \times 10^{23} \text{ molecules air}) \times \frac{0.0001140}{100}$$

$$= \mathbf{4.1 \times 10^{14} \text{ Kr atoms}}.$$

5-45. This calculation requires the molar mass of SF_6, 146.05 g mol^{-1}. The ideal gas law yields density is in units of grams per liter, where the mass of the gas in grams is determined by mass = (moles, n) × (molar mass, $\mathcal{M}$)

$$\text{density} = \frac{n\mathcal{M}}{V}$$

The ideal gas law gives us $\dfrac{n}{V} = \dfrac{P}{RT}$, so:

$$\text{density} = \frac{P\mathcal{M}}{RT} = \frac{(0.942 \text{ atm})(146.05 \text{ g mol}^{-1})}{\left(0.08206 \frac{\text{L atm}}{\text{mol K}}\right) \times (288.15 \text{ K})} = \mathbf{5.82} \frac{\text{g}}{\text{L}}$$

5-47. The gas data permit a calculation of the molar mass,

$$\mathcal{M} = \frac{\text{mass of substance}, m}{\text{moles of substance}, n} = \frac{m}{n} = \frac{m}{\left(\dfrac{PV}{RT}\right)} = \frac{mRT}{PV}$$

$$= \frac{(1.55 \text{ g}) \times \left(0.08206 \frac{\text{L atm}}{\text{mol K}}\right) \times (273.15 \text{ K})}{(1.00 \text{ atm}) \times (0.174 \text{ L})} = 199.7 \text{ g mol}^{-1}.$$

The molar mass of "CF_2" is 50.01 g mol^{-1}. The molar mass for the compound is 199.7 g mol^{-1} = 4 × 50.01 g mol^{-1}. The molecular formula is $(CF_2)_4 = \mathbf{C_4F_8}$.

5-49. Note the 1:1 stoichiometry of HCl : NaCl. Therefore:

$$\text{mol HCl} = 2500 \text{ kg NaCl} \times \left(\frac{1000 \text{ g NaCl}}{1 \text{ kg NaCl}}\right) \times \left(\frac{1 \text{ mol NaCl}}{58.44 \text{ g NaCl}}\right) \times \left(\frac{1 \text{ mol HCl}}{1 \text{ mol NaCl}}\right)$$

$$= 42,800 \text{ mol HCl}$$

Then we may use the ideal gas law (note: 550 °C = 823.15 K):

$$V = \frac{nRT}{P} = \frac{(42,800 \text{ mol}) \times \left(0.08206 \frac{\text{L}-\text{atm}}{\text{mol}-\text{K}}\right) \times (823.15 \text{ K})}{0.97 \text{ atm}}$$

$$= 3.0 \times 10^6 \text{ L} \quad \text{(3 million liters--about the size of a small house)}.$$

5-51. (a) $2\,\text{Na}\,(s) + 2\,\text{HCl}\,(g) \rightarrow \text{H}_2\,(g) + 2\,\text{NaCl}\,(s)$

(b) Note that 20.0°C = 293.15 K and 655 torr is 0.862 atm.

$$\text{mol H}_2 = 6.24 \text{ g Na} \times \left(\frac{1 \text{ mol Na}}{22.990 \text{ g Na}}\right) \times \left(\frac{1 \text{ mol H}_2}{2 \text{ mol Na}}\right) = 0.1357 \text{ mol H}_2$$

$$V = \frac{nRT}{P} = \frac{(0.1357 \text{ mol}) \times \left(0.08206 \frac{\text{L}-\text{atm}}{\text{mol}-\text{K}}\right) \times (293.15 \text{K})}{0.862 \text{ atm}} = 3.79 \text{ L}$$

5-53. Note that 35.3°C = 308.45 K and 735.5 torr is 0.968 atm

$$\text{mol Cl}_2 = n = \frac{PV}{RT} = \frac{(0.968 \text{ atm}) \times (4.48 \text{ L})}{\left(0.08206 \frac{\text{L}-\text{atm}}{\text{mol}-\text{K}}\right) \times (308.45 \text{K})} = 0.1713 \text{ mol Cl}_2$$

$$\text{mol MnO}_2 = \text{mol Cl}_2 = 0.1713 \text{ mol}$$

$$\text{mass MnO}_2 = 0.1713 \text{ mol Cl}_2 \times \left(\frac{86.9368 \text{ g MnO}_2}{1 \text{ mol MnO}_2}\right) = 14.9 \text{ g MnO}_2$$

5-55. (a)

$$\text{mol S} = 2000 \text{ g S} \times \left(\frac{1 \text{ mol S}}{32.066 \text{ g S}}\right) = 62.4 \text{ mol S}$$

$$\text{mol H}_2\text{S} = 62.4 \text{ mol S} \times \left(\frac{2 \text{ mol H}_2\text{S}}{3 \text{ mol S}}\right) = 41.6 \text{ mol H}_2\text{S}$$

$$\text{Volume H}_2\text{S at STP} = 41.6 \text{ mol} \times 22.4 \frac{\text{L}}{\text{mol}} = 932 \text{ L H}_2\text{S}$$

(b) Note that we have 2 mol H_2S for 1 mol SO_2 in the chemical reaction. By Avogadro's law, we have 2 L H_2S for 1 L SO_2. Therefore

$$V_{SO_2} = \frac{V_{H_2S}}{2} = 466 \text{ L SO}_2$$

5-55. (continued) $m_{SO_2} = 41.6 \, \text{mol} \, H_2S \times \left(\dfrac{1 \, \text{mol} \, SO_2}{2 \, \text{mol} \, H_2S} \right) \times \left(\dfrac{64.065 \, \text{g} \, SO_2}{1 \, \text{mol} \, SO_2} \right)$

$$= 1330 \, \text{g} \, SO_2 = \textbf{1.33 kg SO}_2.$$

5-57. In a combustion problem like this, the moles of H_2O and CO_2 permit the determination of the empirical formula (see problem 2-37).

For this problem:

$$\text{mol C} = 1.930 \, \text{g} \, CO_2 \times \left(\frac{1 \, \text{mol} \, CO_2}{44.010 \, \text{g} \, CO_2} \right) \times \left(\frac{1 \, \text{mol} \, C}{1 \, \text{mol} \, CO_2} \right) = 0.0439 \, \text{mol C}$$

$$\text{mol H} = 1.054 \, \text{g} \, H_2O \times \left(\frac{1 \, \text{mol} \, H_2O}{18.015 \, \text{g} \, H_2O} \right) \times \left(\frac{2 \, \text{mol} \, H}{1 \, \text{mol} \, H_2O} \right) = 0.1170 \, \text{mol H}$$

(moles H : moles C) = $(0.1170 : 0.0438) = (2.67 : 1) = (^8/_3 : 1) = (8 : 3)$.

Therefore, the empirical formula is C_3H_8. The pressure, volume, and temperature data for the initial gas permit us to determine how many moles of C_xH_y we began with.

$$\text{mol} \, C_xH_y = \frac{PV}{RT} = \frac{(1.200 \, \text{atm}) \times (0.300 \, \text{L})}{\left(0.08206 \dfrac{L-atm}{mol-K} \right) \times (300 \, \text{K})} = 0.0146 \, \text{mol}.$$

This many moles of C_xH_y yielded 0.0438 mol of CO_2. So there are $^{0.0438}/_{0.0146} = 3.00$ mol of C per mole of C_xH_y. The molecular formula is **C_3H_8**.

5-59. $P_{O_2} + P_{N_2} = P_{TOTAL}$

$$P_{N_2} = P_{TOTAL} - P_{O_2} = 2.43 \, \text{atm} - 1.02 \, \text{atm} = \textbf{1.41 atm}$$

$$X_{N_2} = \frac{P_{N_2}}{P_{TOTAL}} = \frac{1.41}{2.43} = \textbf{0.580} \qquad X_{O_2} = \frac{P_{O_2}}{P_{TOTAL}} = \frac{1.02}{2.43} = \textbf{0.420}$$

5-61. $\text{mol} \, SO_2 = 263 \, \text{g} \, SO_2 \times \left(\dfrac{1 \, \text{mol} \, SO_2}{64.065 \, \text{g} \, SO_2} \right) = 4.105 \, \text{mol} \, SO_2$

$$\text{mol} \, O_2 = 837 \, \text{g} \, O_2 \times \left(\frac{1 \, \text{mol} \, O_2}{31.9988 \, \text{g} \, O_2} \right) = 26.157 \, \text{mol} \, O_2$$

$$\text{mol} \, SO_3 = 179 \, \text{g} \, SO_3 \times \left(\frac{1 \, \text{mol} \, SO_3}{80.0642 \, \text{g} \, SO_3} \right) = 2.236 \, \text{mol} \, SO_3$$

Total moles = $4.105 + 26.157 + 2.236 = 32.50 \, \text{mol}$

$$X_{SO_3} = \frac{2.236 \, \text{mol} \, SO_3}{32.50 \, \text{mol total}} = \textbf{0.0688}$$

$$V\% = 100 \times X_{SO_3} = \textbf{6.88\%}$$

$$P_{SO_3} = X_{SO_3} \, P_{TOTAL} = (0.0688) \times (0.950 \, \text{atm}) = \textbf{0.0654 atm}$$

5-63. Gases can freely interpenetrate, so in fact the nitrogen in the Martian atmosphere has the entire volume of the atmosphere available to it. Nevertheless, it is often convenient to speak of volume fractions for gases. For an ideal gas, the volume fraction will be equal to the mole fraction.

$$X_{N_2} = \text{volume fraction} = 2.7\% = \textbf{0.027}$$

$$P_{N_2} = X_{N_2} P_{TOTAL} = (0.027) \times (5.92 \times 10^{-3} \text{ atm}) = \textbf{1.6} \times \textbf{10}^{\textbf{-4}} \textbf{ atm}$$

5-65. (a) $n_{total} = 10.0 \text{ mol} + 12.5 \text{ mol} = 22.5 \text{ mol} \Rightarrow X_{CO} = \dfrac{10.0}{22.5} = \textbf{0.444}$

(b) 3 mol CO_2 is formed at the expense of *some* of the CO and O_2.

$$\text{mol CO}_{consumed} = 3 \text{ mol CO}_2 \times \left(\frac{1 \text{ mol CO}}{1 \text{ mol CO}_2} \right) = 3 \text{ mol CO}_{consumed}$$

$$\text{mol CO} = \text{mol CO}_{start} - \text{mol CO}_{consumed} = 10.0 - 3.0 = 7.0 \text{ mol CO}$$

$$\text{mol O}_{2 \text{ consumed}} = 3 \text{ mol CO}_2 \times \left(\frac{0.5 \text{ mol O}_2}{1 \text{ mol CO}_2} \right) = 1.5 \text{ mol O}_2 \text{ consumed}$$

$$\text{mol O}_2 = \text{mol O}_{2 \text{ start}} - \text{mol O}_{2 \text{ consumed}} = 12.5 - 1.5 = 11.0 \text{ mol O}_2$$

$$\text{mol}_{total} = 3.0 + 7.0 + 11.0 = 21 \text{ mol} \Rightarrow X_{CO} = \frac{7.0}{21} = \textbf{0.33}$$

5-67. As in Example 5-12, a common $3RT$ term is present in both parts of this problem. R is used in units appropriate for a root-mean-squared speed problem.

$$3RT = 3 \times \left(8.3145 \frac{kg - m^2}{s^2 - mol\text{-}K} \right) \times (300 \text{ K}) = 7{,}480 \frac{kg - m^2}{s^2 - mol}$$

(a) The molar mass of H_2 is 2.0159 g mol^{-1}, or 2.0159×10^{-3} kg mol^{-1}

$$u_{rms} = \sqrt{\frac{7{,}480 \dfrac{kg - m^2}{s^2 - mol}}{2.0159 \times 10^{-3} \dfrac{kg}{mol}}} = \textbf{1930} \frac{\textbf{m}}{\textbf{s}}$$

(b) Molar mass of SF_6 is 146.05 g mol, or 0.14605 kg mol^{-1}

$$u_{rms} (SF_6) = \sqrt{\frac{7{,}480 \dfrac{kg - m^2}{s^2 - mol}}{0.14605 \dfrac{kg}{mol}}} = \textbf{226} \frac{\textbf{m}}{\textbf{s}}$$

5-69. The molar mass of He is 4.0026×10^{-3} kg mol^{-1}

$$SUN: \quad u_{rms}(He) = \sqrt{\frac{3 \times \left(8.3145 \frac{kg-m^2}{s^2 \, mol \, K}\right) \times 6000K}{4.0026 \times 10^{-3} \frac{kg}{mol}}} = 6100 \frac{m}{sec}$$

$$\begin{array}{c} INTERSTELLAR \\ CLOUD: \end{array} \quad u_{rms} = \sqrt{\frac{3 \times \left(8.3145 \frac{kg-m^2}{s^2 \, mol \, K}\right) \times 100K}{4.0026 \times 10^{-3} \frac{kg}{mol}}} = 790 \frac{m}{sec}$$

The ratio of speeds is about 8 times, which is reasonable because u_{rms} is proportional to $\sqrt{T}$.

5-71. An increase in temperature "shifts" the speed distribution to higher velocities. Therefore, the percentage of ClO_2 molecules with speeds in excess of 400 m sec^{-1} will *exceed* 35.0%.

5-73. The relative rate of effusion is proportional to reciprocal of the square root of the molecular mass, so the relative rate can be used to determine the unknown molar mass:

$$\sqrt{\frac{M_{UNK}}{M_{CH_4}}} = \frac{\text{rate of effusion of } CH_4}{\text{rate of effusion of UNK}}$$

$$\sqrt{\frac{M_{UNK}}{16.0428 \text{ g mol}^{-1}}} = \frac{1.30 \times 10^{-8} \text{ mol s}^{-1}}{5.42 \times 10^{-9} \text{ mol s}^{-1}} = 2.398$$

$$\frac{M_{UNK}}{16.0428 \text{ g mol}^{-1}} = (2.398)^2 = 5.750$$

$$M_{UNK} = 5.750 \times 16.0428 \text{ g mol}^{-1} = \textbf{92.3 g mol}^{-1}$$

5-75. The number of moles of oxygen in the cylinder is:

$$6.80 \text{ kg O}_2 \times \left(\frac{1000 \text{ g O}_2}{1 \text{ kg O}_2}\right) \times \left(\frac{1 \text{ mol O}_2}{31.9988 \text{ g O}_2}\right) = 212.5 \text{ mol O}_2$$

Solving the van der Waal's equation for pressure gives:

$$P = \frac{nRT}{V-nb} - a\frac{n^2}{V^2}$$

$$P = \frac{(212.5 \text{ mol}) \times \left(0.08206 \frac{L-atm}{mol-K}\right) \times (293.15K)}{28.0L - (212.5 \text{ mol}) \times \left(0.03183 \frac{L}{mol}\right)} - 1.360 \frac{atm \, L^2}{mol^2} \frac{(212.5 mol)^2}{(28.0L)^2}$$

$$P = \frac{5111 \text{ L}-atm}{21.24 \text{ L}} - 78.3 \text{ atm} = 240.6 \text{ atm} - 78.3 \text{ atm} = 162.3 \text{ atm} = \textbf{162 atm}$$

5-75. (continued)

This compares to a pressure of 182 atm calculated from the ideal gas equation--a 12% difference . The pressure in psi is: 162.3 atm × 14.7 psi atm^{-1} = **2390 psi**

5-77. (a) $P = \dfrac{nRT}{V} = \dfrac{(0.7500 \text{ mol}) \times \left(0.08206 \dfrac{\text{L} - \text{atm}}{\text{mol} - \text{K}}\right)(240.0 \text{ K})}{1.000 \text{ L}} = \mathbf{14.77 \text{ atm}}$

(b) $P = \dfrac{nRT}{V - nb} - a\dfrac{n^2}{v^2}$

$= \dfrac{(0.7500 \text{ mol}) \times \left(0.08206 \dfrac{\text{L} - \text{atm}}{\text{mol} - \text{K}}\right)(240.0 \text{ K})}{1.000 \text{ L} - (0.7500 \text{ mol}) \times \left(0.04267 \text{ L mol}^{-1}\right)} - 3.592 \dfrac{\text{atm L}^2}{\text{mol}^2} \dfrac{(0.7500 \text{ mol})^2}{(1.000 \text{ L})^2}$

$= \dfrac{14.771 \text{ L} - \text{atm}}{0.9680 \text{ L}} - 2.0205 \text{ atm} = \mathbf{13.24 \text{ atm}}$

(c) Neither the ideal gas law nor the van der Waal's equation is exact. The van der Waal's equation gives a pressure that is closer to the experimental pressure; it is a better approximation.

5-79. Remember that the "m" in mbar refers to "millibar." So 866 mbar is 0.866 bar. We can convert this as we usually do using 1 bar = 0.986923 atm:

$$0.886 \text{ bar} \times \dfrac{0.986923 \text{ atm}}{1 \text{ atm}} = 0.874 \text{ atm}$$

5-81. The volume of the mixed gases is equal to the volume of the three flasks:

$$V_{\text{total}} = 5.00 \text{ L} + 4.00 \text{ L} + 3.00 \text{ L} = 12.00 \text{ L}$$

In this volume, there will be a pressure due the total number of moles of the gases.

$$n_{\text{total}} = n_{O_2} + n_{N_2} + n_{Ar}$$

$$\dfrac{P_{\text{total}} V_{\text{total}}}{RT} = \dfrac{P_{N_2} V_{N_2}}{RT} + \dfrac{P_{O_2} V_{O_2}}{RT} + \dfrac{P_{Ar} V_{Ar}}{RT}$$

The *RT* factor will cancel and we will have:

$$P_{\text{total}} V_{\text{total}} = P_{N_2} V_{N_2} + P_{O_2} V_{O_2} + P_{Ar} V_{Ar}$$

$$P_{\text{total}} = \dfrac{P_{N_2} V_{N_2} + P_{O_2} V_{O_2} + P_{Ar} V_{Ar}}{V_{\text{total}}}$$

$$= \dfrac{(2.51 \text{ atm} \times 5.00 \text{ L}) + (0.792 \text{ atm} \times 4.00 \text{ L}) + (1.23 \text{ atm} \times 3.00 \text{ L})}{12.00 \text{ L}}$$

$$= \mathbf{1.62 \text{ atm.}}$$

5-83. (a) The rule refers to *changes* in temperature and *changes* in pressure as if linear behavior applied:

$$\Delta P = C\,\Delta T, \quad \text{where } C = \frac{1 \text{ psi}}{10\ ^\circ F} = 0.0680\ \frac{\text{atm}}{^\circ F}$$

This assumes several things, all incorrect and all leading to some error in the exactness of the rule:

(1) That $X\,(= \frac{nR}{V})$ doesn't vary with T or P at constant V. With a rubber tire, that is unlikely.

(2) That the gas is ideal. This is unlikely with a tire pressure of roughly 3 atm.

(3) That the *gauge* pressure (measured by the tire gauge) depends only on the pressure inside the tire. It does not. It really depends on the difference of pressure inside and outside, and the pressure outside--atmospheric pressure--is always nearly a constant 14.7 psi.

(b) The rule suggests $\quad \Delta P = \left(\frac{1\text{psi}}{10\ ^\circ F}\right) \times 50\ ^\circ F = 5$ psi, so the Tire Council Rule suggests:

$$P = (28 - 5) \text{ psi} = \textbf{23 psi}.$$

Now let's do it more rigorously. The interior pressure is 28 psi + 14.7 = 42.7 psi.

The change in the pressure of an ideal gas can be obtained by a proportionality $T_{initial} = 273.15 + \frac{5}{9}\,(70 - 32) = 294$ K and $T_{final} = 273.15 + \frac{5}{9}\,(20 - 32) = 266$ K)

$$\frac{P_{final}}{T_{final}} = \frac{P_{initial}}{T_{initial}} \implies P_{final} = \frac{T_{final}}{T_{initial}} P_{initial} = \left(\frac{266 \text{ K}}{294 \text{ K}}\right)(42.7 \text{ psi}) = \textbf{38.6 psi}$$

$$P_{gauge} = P_{inside} - P_{outside} = 38.6 - 14.7 = \textbf{24 psi}$$

5-85. (a) We know the O_2 stoichiometry is indicated by the formula of the product, assuming Avogadro's hypothesis applies:

2 volumes CO_2 need 2 volumes O_2 4 volumes H_2O need 2 volumes O_2

2 volumes NO_2 need 2 volumes O_2 Therefore, **6 volumes** of O_2 are needed.

(b) The compound has a formula $C_xH_yN_z$, and the combustion reaction is:

$$6\,O_2 + C_xH_yN_z \rightarrow 2\,CO_2 + 4\,H_2O + 2\,NO_2$$

The C, H, and N on the right are all contained in the starting material:

2 CO_2 contain 2 C $\Rightarrow$ x=2   4 H_2O contain 8 H $\Rightarrow$ y = 8

2 NO_2 contain 2 N $\Rightarrow$ z = 2   The molecular formula is **$C_2H_8N_2$**.

5-87. The applicable equations are:

1: *Decomposition of sodium azide*

$$2\ NaN_3\ (s) \rightarrow 2\ Na\ (s) + 3\ N_2\ (g)$$

2: *Consumption of sodium*

$$10\ Na\ (s) + 2\ KNO_3\ (s) \rightarrow K_2O\ (s) + 5\ Na_2O\ (s) + N_2\ (g)$$

3: *Neutralization of potassium oxide*

$$K_2O\ (s) + SiO_2\ (s) \rightarrow K_2SiO_3\ (s)$$

4: *Neutralization of sodium oxide*

$$Na_2O\ (s) + SiO_2\ (s) \rightarrow Na_2SiO_3\ (s)$$

We can combine these into an overall reaction for the decomposition of sodium azide in the presence of potassium nitrate and silicon dioxide. This involves rewriting the reactions so the unwanted components (Na, K_2O, and Na_2O) cancel by subtraction:

5 × [$\qquad 2\ NaN_3\ (s) \rightarrow 2\ Na\ (s) + 3\ N_2\ (g) \qquad$]

+ 1 × [$10\ Na\ (s) + 2\ KNO_3\ (s) \rightarrow K_2O\ (s) + 5\ Na_2O\ (s) + N_2\ (g)$]

+ 1 × [$\qquad K_2O\ (s) + SiO_2\ (s) \rightarrow K_2SiO_3\ (s) \qquad$]

+ 5 × [$\qquad Na_2O\ (s) + SiO_2\ (s) \rightarrow Na_2SiO_3\ (s) \qquad$]

This is the same as writing:

[$\qquad 10\ NaN_3\ (s) \rightarrow 10\ Na\ (s) + 15\ N_2\ (g) \qquad$]

+ [$10\ Na\ (s) + 2\ KNO_3\ (s) \rightarrow K_2O\ (s) + 5\ Na_2O\ (s) + N_2\ (g)$]

+ [$\qquad K_2O\ (s) + SiO_2\ (s) \rightarrow K_2SiO_3\ (s) \qquad$]

+ [$5\ Na_2O\ (s) + 5\ SiO_2\ (s) \rightarrow 5\ Na_2SiO_3\ (s) \qquad$]

The net reaction is:

$$10\ NaN_3\ (s) + 2\ KNO_3\ (s) + 6\ SiO_2 \rightarrow K_2SiO_3\ (s) + 5\ Na_2SiO_3\ (s) + 16\ N_2$$

Checking, we have 10 Na, 2 K, 6 Si, 32 N, and 18 O on the left and on the right.

Now we are ready for a pair of straightforward stoichiometric calculations:

$$125.0\ g\ NaN_3 \times \frac{1\ mol\ NaN_3}{65.009988\ g\ NaN_3} \times \frac{2\ mol\ KNO_3}{10\ mol\ NaN_3} \times \frac{101.10324\ g\ KNO_3}{1\ mol\ KNO_3} = \mathbf{38.88\ g\ KNO_3}$$

$$125.0\ g\ NaN_3 \times \frac{1\ mol\ NaN_3}{65.009988\ g\ NaN_3} \times \frac{6\ mol\ SiO_2}{10\ mol\ NaN_3} \times \frac{60.0843\ g\ SiO_2}{1\ mol\ SiO_2} = \mathbf{69.32\ g\ SiO_2}$$

5-89. We know the decomposition of $CaCO_3$ proceeds: $CaCO_3 \rightarrow CO_2 + CaO$

Therefore mol CO_2 = mol CaO = mol $Ca(OH)_2$

$$mol\ Ca(OH)_2 = 8.47\ kg\ Ca(OH)_2 \times \left(\frac{1000\ g}{1\ kg}\right) \times \left(\frac{1\ mol\ Ca(OH)_2}{74.093\ g\ Ca(OH)_2}\right)$$

$$= 114.3\ mol\ Ca(OH)_2 = mol\ CO_2 = 114.3\ mol\ CO_2$$

$$V_{CO_2} = \frac{nRT}{P} = \frac{(114.3\ mol) \times \left(0.08206\frac{L\text{-}atm}{mol\text{-}K}\right) \times (950K)}{(0.976\ atm)} = \textbf{9130 L CO}_\textbf{2}$$

5-91. We use the final volume, pressure, and temperature of H_2 to determine moles of Na.

Note that: $2.6\ ft^3 \times \left(\frac{0.3048\ m}{ft}\right)^3 \times \left(\frac{10\ dm}{m}\right)^3 \times \left(\frac{1\ L}{1\ dm^3}\right) = 73.624\ L$

Then: $mol\ H_2 = \frac{PV}{RT} = \frac{(1.00\ atm)(73.62\ L)}{\left(.08206\frac{L\text{-}atm}{mol\text{-}K}\right) \times (273.15\ K)} = 3.285\ mol$

$mass\ Na = 3.285\ mol\ H_2 \times \left(\frac{2\ molNa}{1\ mol\ H_2}\right) \times \left(\frac{22.9898g\ Na}{1\ mol\ Na}\right) = 151.0\ g\ Na$

This is equal to $151.0\ g \times \frac{1\ lb}{453.6\ g} = 0.333\ lb\ Na$

$Mass\ \%\ Na = \frac{0.333\ lb\ Na}{1.0\ lb\ alloy} \times 100 = \textbf{33.3\%}$

5-93. The balanced chemical reaction is: $CS_2\ (g) + 3\ O_2\ (g) \rightarrow CO_2\ (g) + 2\ SO_2\ (g)$.

We can use the stoichiometry to set up a changes table (these tables will be more common in Chapter 8). The first line shows the pressures *before* the reaction; the second line shows the *change* in the pressures in the course of the reaction, and the final line shows the pressures *after* the reaction.

The relevant stoichiometric coefficients determine the relative magnitude of the changes. For example, a decrease of x atm in the partial pressure of CS_2 causes a decrease of *3x* atm in the partial pressure of O_2.

(all in atmospheres)	CS_2	O_2	CO_2	SO_2
Pressure Before	x	y	0	0
Pressure Change	$-x$	$-3x$	$+x$	$+2x$
Pressure After	0	$y-3x$	x	$2x$

72

5-93. (continued)

We know the total final pressure is 2.40 atm. We use Dalton's Law to determine y and x:

After the reaction: $P_{total} = P_{CS_2} + P_{O_2} + P_{CO_2} + P_{SO_2}$

2.40 atm $= 0 + (y - 3x) + x + 2x = y$

Before the reaction: $P_{total} = P_{CS_2} + P_{O_2} + P_{CO_2} + P_{SO_2}$

3.00 atm $= x + y + 0 + 0 = x + y$

$x = 3.00 - y = 3.00 - 2.40 = 0.60$ atm

We now have the initial pressure of the CS_2 -- 0.60 atm. Then:

$$\text{mol } CS_2 = \frac{PV}{RT} = \frac{(0.60\,\text{atm}) \times (10.0\,\text{L})}{\left(0.08206\,\dfrac{\text{L} - \text{atm}}{\text{mol} - \text{K}}\right) \times (373.15\,\text{K})} = 0.1959\,\text{mol}$$

The mass is: $(0.1959\,\text{mol}\,CS_2) \times \left(\dfrac{76.14\,\text{g}\,CS_2}{1\,\text{mol}\,CS_2}\right) = \textbf{15g } \boldsymbol{CS_2}$

5-95. The pressure depends on the mass and the speed of the particles. Molecules of UF_6 may be heavier, but on average they move much more slowly. The two effects (mass and average velocity) cancel each other exactly in the ideal gas.

5-97. Separation by effusion requires that a molecule containing ^{235}U *never* have the same mass as a molecule containing ^{238}U. But with a ready supply of ^{20}F and ^{19}F, we could have some $^{235}U(^{20}F)_3(^{19}F)_3$ (molecular mass about 352) and $^{238}U(^{19}F)_6$ (molecular mass also about 352). These molecules would have very, very slightly different masses, on the order of 0.001%. But even then effusion would not separate these.

CHAPTER 6 - CONDENSED PHASES AND PHASE TRANSITIONS

6-1. (a) For Cl_2 the bond is about 2.0×10^{-10} m long, and for molecular KCl we estimate about 2.5×10^{-10} m

(b) It is clear that the Cl-Cl bond is about half as strong as the K^+–Cl^- bond, so the book quoted in this problem is not correct. The error arises when different types of bonding are compared: an ionic bond has considerable strength because of the charge separation.

6-3. In *all* of the substances the dispersion force is present.

(a) KF – The K-F bond is strongly polar with **dominant ion-ion** interactions. KF will also have dipole-dipole interactions among KF molecules.

(b) HI – Unlike HF or even HCl, there is little hydrogen bonding with iodine. Still, the molecule is strongly polar and **dipole-dipole interactions dominate**. Dispersion forces are also relatively strong in this case.

(c) CH_3OH – An O-H bond is present, so **hydrogen bonding will dominate** intermolecular interactions, along with some dipole-dipole interactions.

(d) Rn – The only interatomic force available to a noble gas is the **dispersion force.**

(e) N_2 is symmetric and, therefore, non-polar. The sole intermolecular force available is the **dispersion force**.

6-5. The three forces available are:

Na^+--- Br^- ion-ion   *strongest*

Na^+---HBr   ion-dipole   *second strongest*

Na^+---Kr   ion-induced dipole   *least strong*

6-7.

As discussed in the solution to problem 6-3c above, methanol will have hydrogen bonds as the dominant intermolecular interaction. It is reasonable to expect the tetramer to be held together by hydrogen bonds.

6-9. The diffusion constant increases as it becomes easier for molecules to move around among the "cages" formed by their neighbors in a condensed phase -- a liquid or a solid. Compression should make these "cages" smaller and tighter, so compression should significantly decrease the diffusion constant of a liquid or a solid.

6-9. (continued)

The major determinants of the diffusion constant are the average speed and the rate of collision. The speed distribution will not be significantly affected by an increase in density. But the rate of collisions amongst gas molecules should increase dramatically when a gas is compressed. This will cause the diffusion constant of a gas to decrease when the density is increased.

6-11. The mean free path of molecules in a liquid must be much shorter than that for molecules in a gas. Molecules in a liquid are essentially in contact with each other, and even a small motion will cause a collision. Molecules in a gas are much farther apart, with a great deal of empty space that must be traversed before a collision.

6-13. By the ideal gas law, rearranged to yield volume:

$$V_{H_2} = \frac{n_{H_2} RT}{P} = \frac{(1.00 \ mol H_2) \times \left(0.08206 \frac{L-atm}{mol-K} \right) \times (16.0 \ K)}{(0.213 \ atm)} = 6.16 \ \textbf{liters}$$

This is **27%** of the volume of one mole of hydrogen gas at STP (22.4 liters).

6-15. The pressure of the hydrogen gas is only part of the total gas pressure; the pressure of the water vapor also contributes:

$$P_{water} + P_{C_2H_2} = P_{total}$$
$$P_{C_2H_2} = P_{total} - P_{water} = 756.2 \ torr - 55.3 \ torr = 700.9 \ torr$$

For the ideal gas law calculation we want the pressure to be in atmospheres:

$$700.9 \ torr \times \frac{1 \ atm}{760 \ torr} = 0.92222 \ atm$$

The ideal gas law and the molar mass of acetylene (26.038 g mol^{-1}) give the answer:

$$\frac{grams}{liter} = \left(\frac{grams}{mol} \right) \times \left(\frac{mol}{liter} \right) = \left(\frac{grams}{mol} \right) \times \left(\frac{P}{RT} \right)$$

$$= \left(\frac{26.038 grams}{mol} \right) \times \left(\frac{0.9222 \ atm}{0.08206 \frac{L-atm}{mol-K} \times 313.15 \ K} \right) = 0.9345 \frac{\textbf{grams}}{\textbf{liter}}$$

6-17. (a) The Chapter 3 statement also means that the molecular character of compounds increases with increasing oxidation number. Molecular substances have lower melting points, in general, than ionic substances. So we expect the melting points for to decrease with increasing oxidation number:

$$Mn_2O_7 \ (Mn \ 7+) > MnO_2 \ (Mn \ 2+) > Mn_3O_4 \ (Mn \ +^4/_3).$$

(b) The compound $KMnO_4$ contains the distinct K^+ and MnO_4^-. This means the compound is highly ionic and should have a high melting point, even though the *anion* MnO_4^- does have a lot of covalent character.

6-19. The vapor pressure of water at 90°C is approximatel**y 0.70** atm from the figure (more precisely, it is **0.69 atm**). This indicates that the air pressure at this altitude is 0.69 atm and that only 69% of the earth's atmosphere is above the explorer. Fully 31% of the earth's atmosphere is below the explorer.

6-21. The total pressure on the substance determines when it will boil. Heating it under reduced pressure will enable it to boil, and be distilled, at a lower temperature.

6-21. The trends in boiling and melting points indicate that the interatomic forces in iridium are much stronger than those of sodium. The surface tension of iridium will be higher than that of sodium.

6-23. Compression will favor the denser phase, but the denser (liquid) phase is already present. The sample will remain liquid; no phase change will occur.

6-25. (a) liquid   (b) gas
        (c) solid   (d) gas

6-27. (a) The temperature of the triple point must be above this; any point where liquid and gas are in equilibrium must always lie at a temperature equal to or higher than any point where the liquid and solid are in equilibrium.
(b) During this experiment, the temperature goes from well below to well above -84°C (= 189 K), where we know solid acetylene is in equilibrium with its vapor at a pressure of 1 atm. Therefore, the solid acetylene is sure to **sublime**.

6-29. The material in the tube will enter a "supercritical" phase beyond which there is no distinction between liquid and gas. This will be indicated by the disappearance of the meniscus between the liquid and gaseous phases. The pressure inside the tube will be very high.

6-31. We are told the mass of the components of the solution. We must convert these to moles to evaluate the mole fractions correctly.

$$\text{mol } H_2O = 164.6 \text{ g } H_2O \times \left(\frac{1 \text{mol } H_2O}{18.0153 \text{ g } H_2O}\right) = 9.137 \text{ mol } H_2O$$

$$\text{mol } H_2O_2 = 5.10 \text{ g} \times \left(\frac{1 \text{ mol } H_2O}{34.0147 \text{ g } H_2O}\right) = 0.150 \text{ mol } H_2O_2$$

$$\text{weight \% } H_2O_2 = 100 \times \left(\frac{5.10 \text{ g } H_2O}{164.6 \text{ g} + 5.10 \text{ g}}\right) = 3.01$$

$$\text{mole fraction } H_2O_2 = \left(\frac{0.150 \text{ mol } H_2O_2}{9.137 \text{ mol} + 0.150 \text{ mol}}\right) = 0.0161$$

$$\text{molality } H_2O_2 = \left(\frac{0.150 \text{ mol } H_2O_2}{0.1646 \text{ kg}}\right) = 0.911 \text{ mol kg}^{-1}$$

6-33. Density is the bridge between the mass of the solution and the volume of solution.

$$\left(\frac{1.0422 \text{ mol HCl}}{100.0 \text{ g solution}}\right) \times \left(\frac{1.1886 \text{ g solution}}{1 \text{ cm}^3 \text{ solution}}\right) \times \left(\frac{1000 \text{ cm}^3 \text{ solution}}{1 \text{ L solution}}\right) = 12.38 \frac{\text{mol}}{\text{L}} = 12.38 \text{ M}$$

6-35. One liter of solution is equal to 1000 cm^3 of solution. It weighs:

$$1000 \text{ cm}^3 \times 1.33 \text{ g cm}^{-3} = 1330 \text{ g}$$

A liter of solution contains 12.2 mol H_3PO_4. This is equal to a mass of H_3PO_4:

$$\text{g } H_3PO_4 = (12.2 \text{ mol } H_3PO_4) \times \left(\frac{97.995 \text{ g } H_3PO_4}{1.0 \text{ mol } H_3PO_4}\right) = 1195.5 \text{ g } H_3PO_4$$

Therefore, the mass of water in 1 liter of solution is:

$$\text{mass } (H_3PO_4) + \text{mass } (H_2O) = \text{total mass}$$
$$\text{mass } H_2O = \text{total mass} - \text{mass } (H_3PO_4)$$
$$= 1330 \text{ g} - 1195.5 = 134.5 \text{ g } H_2O$$

We now obtain molality: $\left(\frac{12.2 \text{ mol } H_3PO_4}{134.5 \text{ g } H_2O}\right) \times \left(\frac{1000 \text{ g } H_2O}{1 \text{ kg } H_2O}\right) = 90.7 \frac{\text{mol}}{\text{kg}}$

6-37. The mole fraction of the solute is determined using the known mass of the solvent and solute and computing their chemical amounts.

$$\text{mol } C_{13}H_{10}O = 15.0 \text{ g } C_{13}H_{10}O \times \left(\frac{1 \quad \text{mol } C_{13}H_{10}O}{182.224 \text{ g } C_{13}H_{10}O} \right) = 0.0823 \text{ mol}$$

$$\text{mol } C_3H_6O = 50.0 \text{ g } C_3H_6O \times \left(\frac{1 \text{ mol } C_3H_6O}{58.081 \text{ g } C_3H_6O} \right) = 0.8608 \text{ mol } C_3H_6O$$

$$X_{C_3H_6O} = \frac{\text{mol}_{C_3H_6O}}{\text{mol}_{C_3H_6O} + \text{mol}_{C_{13}H_{10}O}} = \frac{0.861}{0.861 + 0.0823} = 0.913$$

$$P_{C_3H_6O} = X_{C_3H_6O} P^o_{C_3H_6O} = (0.913) \times (0.3720 \text{ atm}) = \mathbf{0.340 \text{ atm}}$$

6-39. The vapor pressure change is 0.411 - 0.437 = -0.026 atm.

Therefore: $\qquad -X_2 P_i = \Delta P_i$

$$X_2 = -\frac{\Delta P_i}{P_i^o} = -\frac{(-0.026 \text{ atm})}{0.437} = \mathbf{0.059}$$

6-41. Molality of the sugar: $m = \dfrac{\Delta t_b}{K_b} = \dfrac{\left(0.30\,^{\circ}C\right)}{\left(0.512 \dfrac{^{\circ}C\text{-kg}}{\text{mol}}\right)} = 0.586 \text{ mol kg}^{-1}$

Moles of sugar in 200.0 g (= 0.2000 kg) H_2O:

$$\left(\frac{0.586 \text{ mol sugar}}{1 \text{ kg } H_2O} \right) \times (0.2000 \text{ kg } H_2O) = 0.1172 \text{ mol sugar}$$

Molar mass of sugar: $\mathcal{M} = \dfrac{39.8 \text{ g}}{0.1172 \text{ mol sugar}} = \dfrac{340 \text{ g sugar}}{1 \text{ mol sugar}} = \mathbf{340 \text{ g mol}^{-1}}$.

6-43. The substance that gives the largest number of moles of solute will give the lowest freezing point. This is the substance with the smallest molar mass—N_2H_4.

6-45. Assume a 100 g sample, made of 34 g sucrose and 66 g (0.066 kg) water.

$$\text{Moles of sucrose} = \frac{34 \text{ g}}{342.3 \text{ g mol}^{-1}} = 0.099 \text{ mol}$$

$$\text{Molality of sucrose} = \frac{0.099 \text{ mol}}{0.066 \text{ kg}} = 1.5 \frac{\text{mol}}{\text{kg}}$$

$$\Delta t_f = -K_f m = -1.86 \left(\frac{\text{kg-}^{\circ}C}{\text{mol}} \right) \times 1.5 \left(\frac{\text{mol}}{\text{kg}} \right) = -2.79\,^{\circ}C$$

$$t_f = \mathbf{-2.8\,^{\circ}C}$$

6-45. (continued)

The increasing concentration of sucrose will depress the freezing point further, until solid sucrose also begins to crystallize.

6-47. *If* it is true that acetone does not dissociate, then the freezing point depression would be equal to $K_f\,m$. Let us test that assumption against the experimental results. (Note: 50.00 g = 0.05000 kg):

$$\Delta t_f = -K_f m = -1.86\left(\frac{\text{kg-}^{\circ}\text{C}}{\text{mol}}\right)\times\left(\frac{1.55\ \text{g acetone}}{0.0500\ \text{kg H}_2\text{O}}\right)\times\left(\frac{1\ \text{mol acetone}}{58.080\ \text{g acetone}}\right)$$

$$= -0.993\ ^{\circ}\text{C}$$

This is very close to the observed freezing point. Acetone **does not** dissociate.

6-49. (a) If a substance does dissociate, then the molality of all ions will be some multiple x of the listed molality. In this case, where i is the number of dissociated ions and m is the molality of the dissolved formula unit:

$$\Delta t_f = -K_f m i \quad \Rightarrow \quad i = \frac{-\Delta t_f}{K_f m} = \frac{-\left(-4.218\ ^{\circ}\text{C}\right)}{\left(1.86\ \dfrac{\text{kg-}^{\circ}\text{C}}{\text{mol}}\right)\left(0.8402\ \dfrac{\text{mol}}{\text{kg}}\right)} = 2.7$$

Na_2SO_4 dissociates into, effectively, **2.7 particles.**

(b) Since the theoretical number of particles ν is 3, we can determine that the fractional degree of ionization α is $(i-1)/(\nu-1) = 1.7/2 = 0.85$.

6-51. The pressure is expressed here as a height of water (16.3 cm = 0.163 m). Recall that this hydrostatic pressure is (see problem 5-9):

$$P = dgh = \left(1000\frac{\text{kg}}{\text{m}^3}\right)\times\left(9.80665\frac{\text{m}}{\text{s}^2}\right)\times(0.163\ \text{m})$$

$$= 1598\ \frac{\text{kg}}{\text{m}^2\text{-s}^2} = 1598\ \text{Pa}$$

$$= 1598\ \text{Pa}\times\left(\frac{1\ \text{atm}}{101,325\ \text{Pa}}\right) = 1.577\times10^{-2}\ \text{atm}$$

The concentration of the polymer is then

$$c = \frac{\pi}{RT} = \frac{1.577\times10^{-2}\ \text{atm}}{\left(0.08206\dfrac{\text{L-atm}}{\text{mol-K}}\right)\times(288.15\text{K})} = 6.67\times10^{-4}\frac{\text{mol}}{\text{L}}$$

We are told that 1 L has 5.73 g of polymer. Therefore:

$$M = \frac{5.73\ \text{g}}{6.67\times10^{-4}\ \text{mol}} = 8,590\ \frac{\text{g}}{\text{mol}}$$

Chapter 6

6–53. This problem is similar to 6–45 above, except now we use the osmolality as a measure of the total concentration (in molality units) of all dissolved substances:

$$\Delta t_f = -K_f\, m = -1.86\left(\frac{\text{kg-}^\circ\text{C}}{\text{mol}}\right) \times \left(\frac{1.40\ \text{mol}}{1\ \text{kg}}\right) = -2.60\ ^\circ\text{C}$$

This means the expected melting point is **–2.60 °C**
For the boiling point:

$$\Delta t_b = -K_b\, m = -0.512\left(\frac{\text{kg-}^\circ\text{C}}{\text{mol}}\right) \times \left(\frac{1.40\ \text{mol}}{1\ \text{kg}}\right) = +0.72\,^\circ\text{C}$$

This means the expected boiling point is **+100.72 °C**
For the osmotic pressure we need the osmolarity, the measure of the total concentration in *molarity* units) of all dissolved substances. We can estimate this by assuming that the solution is dilute enough so one kg is one liter. In that case, the osmolarity is approximately the same value as the osmolality. Then:

$$\Pi = \text{osmolality} \times RT = \frac{1.40\ \text{mol}}{1\ \text{L}} \times \frac{0.08206\ \text{L atm}}{1\ \text{mol K}} \times 298\,\text{K} = \textbf{34 atm}$$

6-55. (a) We will need to know the number of moles of water in 1 L of water, since Henry's law deals with mole fractions. We assume that 1 L of water has a density of 1.00 g cm^{-3} or 1,000 g L^{-1}. Then the number of moles of water is:

$$\frac{1,000}{18.015} = 55.5\ \text{mol H}_2\text{O}.$$

Henry's Law states: $\quad P_{CO_2} = X_{CO_2} k_H$

$$X_{CO_2} = \frac{P_{CO_2}}{k_H} = \frac{5.0\,\text{atm}}{1.65\times 10^3\ \text{atm}} = 3.03 \times 10^{-3}$$

The mole fraction of CO$_2$ is so much smaller than the mole fraction of water that we can approximate: $\quad X_{CO_2} = \dfrac{\text{mol CO}_2}{\text{mol H}_2\text{O}}$

$$\text{mol CO}_2 = (X_{CO_2})(\text{mol H}_2\text{O})$$
$$= (3.03 \times 10^{-3})(55.5) = \textbf{0.17 mol CO}_2$$

(b) On a microscopic level, the removal of the cap allows the reaction
CO$_2$ *(aq)* → CO$_2$ *(g)* to shift from left to the right.

6-57. The methane evolved is equal to the number of moles of methane in 1.00 kg (= 1.00 L) water, which we know (see problem No. 57) contains 55.5 mol H$_2$O.
Because 1.00 mol of an ideal gas occupies 22.4 L at STP, the number of moles of methane gas is: $\dfrac{3.01\text{L}}{22.4\ \text{L mol}^{-1}} = 0.134\ \text{mol CH}_4$

6-57. (continued)

The Henry's Law constant is obtained by rearranging the basic equation:

$$P_{CH_4} = X_{CH_4}\, k_H$$

$$k_H = \frac{P_{CH_4}}{X_{CH_4}} = \frac{1.00\,atm}{(0.134\,mol/55.5\,mol)} = \frac{55.5}{0.134}\,atm = \textbf{414 atm}$$

6-59. The mole fractions in the liquid phase are needed. Note that with equal chemical amounts $X_{toluene} = X_{benzene} = 0.50$. The partial pressures for each are estimated by Raoult's Law:

$$P_{benzene} = X_{benzene}\, P^o_{benzene} = (0.50)(0.0987\,atm) = 0.0494\,atm$$

A similar calculation yields $P_{toluene} = 0.01445\,atm$.

The total pressure is $0.0494 + 0.01445 = 0.0638\,atm$.

6-59. (continued)

The mole fraction of benzene in the gas phase is obtained from the partial pressure:

$$X_{benzene} = \frac{P_{benzene}}{P_{total}} = \frac{0.0494}{0.0638} = \textbf{0.774}$$

6-61. (a) Keeping track of all the variables requires a table:

compound	mass	molar mass	moles	mole fraction
CCl_4	41.5 g	$153.8\ \frac{g}{mol}$	0.2698	$\frac{0.2698}{0.2698+0.2284} = \textbf{0.542}$
$C_2H_4Cl_2$	22.6g	$98.96\ \frac{g}{mol}$	0.2284	0.458

(b) To get the total vapor pressure, we need the mole fraction of both substances. That for $C_2H_4Cl_2$ is given in the table, too.

$$P_{TOTAL} = P_{CCl_4} + P_{C_2H_4Cl_2}$$
$$= \left(X_{CCl_4} \times P^o_{CCl_4}\right) + \left(X_{C_2H_4Cl_2} \times P^o_{C_2H_4Cl_2}\right)$$
$$= (0.542 \times 0.293\,atm) + (0.458 \times 0.209\,atm)$$
$$= 0.1588\,atm + 0.0957\,atm = \textbf{0.255 atm}$$

(c) $\quad X_{CCl_4} = P_{CCl_4}/P_{TOTAL} = 0.1598/0.255 = \textbf{0.623}$

6-63. A solution is homogenous with all molecules (or atoms) completely separated. With a colloid, some or many of these-especially atoms-form clusters. A colloid can be generated by reacting $AuCl_4^-$ (*aq*) with Sn^{2+} *(aq)*, both as solutions. A product, Au, results with very small Au clusters. These are, though, too small to be separated, for example, by filtering.

6-65. The cloudy character of the observation indicates that a heterogeneous mixture is present. The correct answer is that the water vapor (steam) arising from the tea kettle has condensed into a colloidal suspension of liquid water in air.

6-67. We need to use the information about the gas formed to find out how much sodium chlorate was actually present. This is a gas stoichiometry problem, but first we must correct for the pressure of the water.

The total pressure is 742 atm. This is equivalent to 0.9763 atm.
$$P_{water} + P_{O_2} = P_{total}$$
$$P_{O_2} = P_{total} - P_{water} = 0.9763 \text{ atm} - 0.03126 \text{ atm} = 0.9450 \text{ atm}$$

For the moles of oxygen:
$$n_{O_2} = \frac{PV}{RT} = \frac{0.9450 \text{ atm} \times 0.03512 \text{ L}}{0.08206 \frac{\text{L-atm}}{\text{mol-K}} \times 297.15 \text{ K}} = 0.0013612 \text{ mol}$$

For the moles of sodium chlorate:
$$n_{NaClO_3} = 0.0013612 \text{ mol O}_2 \times \frac{2 \text{ mol NaClO}_3}{3 \text{ mol O}_2} = 9.075 \times 10^{-4} \text{ mol}$$

The mass of sodium chlorate is
$$m_{NaClO_3} = 9.075 \times 10^{-4} \text{ mol} \times \frac{106.441 \text{ g}}{1 \text{ mol}} = 0.09659 \text{ g}$$

This means that the initial sample was $100\% \times \frac{0.09659 \text{ g}}{0.265 \text{ g}}$ $100\% \times \frac{0.09659 \text{ g}}{0.265 \text{ g}} = \textbf{36.4\%}$

6-69.
$$\text{mol H}_2\text{O} = \frac{PV}{RT} = \frac{(0.03126 \text{atm}) \times (110{,}000 \text{L})}{\left(0.08206 \frac{\text{L}-\text{atm}}{\text{mol}-\text{K}}\right) \times (298.15 \text{K})} = 140.5 \text{mol}$$

$$\text{g H}_2\text{O} = 140.5 \text{ mol H}_2\text{O} \times \left(\frac{18.0153 \text{ g H}_2\text{O}}{1 \text{mol H}_2\text{O}}\right) = 2530 \text{ g H}_2\text{O}$$

This is roughly 2.5 liters of liquid water, or almost 3 quarts.

6-71. The spacecraft (and planets, and stars) *are* slowly losing mass to space. It is not boiling away, which refers to the case of liquid only. The spacecraft is a solid, to the loss of substances is by the process of sublimation. But that the equilibrium state of the spacecraft (and planets, and stars) is completely vaporized says *nothing* about the *rate* of sublimation, which is vanishingly small.

6-73

			T_c (K)		P_c (atm)		$(V/n)_c$
			Calc.	Obs'd.	Calc.	Obs'd.	L mol^{-1}
O_2	1.360	0.03183	**154.3**	154.76	**49.72**	50.06	**0.09549**
CO_2	3.592	0.04267	**303.0**	304.20	**73.07**	72.86	**0.1280**
H_2O	5.464	0.03049	**647.1**	647.067	**217.7**	217.57	**0.09147**

The observed values for the critical temperature are taken from the experimental literature. The good agreement between calculated and observed values shows the quality of the van der Waal's equation.

6-75. (a) There are many temperatures when a substance changes state. The critical point is, in fact, a point when changes in state *cease*.

(b) This is a tautology—saying that the temperature at the critical point is the critical temperature, etc. It does not show an understanding of what happens at that point.

(c) A substance can be a liquid or a vapor at any point along the liquid / vapor coexistence curve of a phase diagram. There is liquid and vapor water present, for example, when a sample of water and water vapor are in equilibrium at normal room temperature and pressure.

(d) The critical point defines the edge of a region of the phase diagram. Changing the pressure moves us off that point—and perhaps to region where multiple phases exist. Consider the phase diagram for water. If we decrease the pressure we move to a region where temperature does define the phases present.

6-77. (a) We use the information about the molality of the CS_2 solution to determine the number of moles of S that dissolved, and then the mass of S in the original mixture.

$$1.171 \text{ kg } CS_2 \times \frac{0.0304 \text{ mol S}}{1 \text{ kg } CS_2} = 0.0355984 \text{ mol S}$$

$$0.0355984 \text{ mol S} \times \frac{32.066 \text{ g S}}{1 \text{ mol S}} = \textbf{1.14 g S}$$

The mass of the iron is therefore 195.4 − 1.14 = 194.3 g Fe

6-77. (continued)

(b) Assuming that the S and Fe do not cling to each other, we could remove the Fe with a magnet, or separate the two by tweezers. Also, in this case, floatation with a soapy mixture might be useful: although S is denser that water, small particles will float because of the surface tension of the water.

6-79. One liter of this solution weighs 1,007 g. Of this, 1%, or **10.07 g**, is $AgNO_3$.

The mass of Ag in this is $10.07 \text{ g AgNO}_3 \times \left(\dfrac{107.87 \text{ g Ag}}{169.875 \text{ g AgNO}_3} \right) = \textbf{6.39 g}$

6-81. Percent "relative humidity" is relative the vapor pressure of water in air saturated with water vapor over *pure* H_2O. The lower vapor pressure above the salt solution is a result of vapor pressure depression.

6-83. The data for each acid can be used to determine a value of K_f for the solvent. The formula for K_f is $K_f = \left(\dfrac{-\Delta t_f}{0.050 \text{ mol kg}^{-1}} \right)$.

SOLUTE	$t_f (°C)$	$\Delta t_f (°C)$	K_f (apparent)
HCl	-22.795	-0.330	6.60 °C -kg-mol^{-1}
H_2SO_4	-22.788	-0.323	6.46 °C -kg-mol^{-1}
$HClO_4$	-22.791	-0.326	6.52 °C -kg-mol^{-1}
HNO_3	-22.632	-0.167	3.34 °C -kg-mol^{-1}

The K_f value for nitric acid is much smaller than for the other acids. In fact, it's about one-half. This can be explained by suggesting that HNO_3 doesn't dissociate *at all* while the other acids all dissociate almost completely to one H^+ and one X^- ion. This affects the van't Hoff "*i*" term, and for HCl, H_2SO_4, and $HClO_4$ solutions this would give $i \times m = 2 \times 0.050 \text{ mol kg}^{-1} = 0.100 \text{ mol kg}^{-1}$. The values for K_f (observed) in these solutions are then 3.30, 3.23, and 3.26 °C -kg-mol^{-1}. The average value for K_f is then 3.28 °C -kg-mol^{-1}.

6-85. The high pressure inside the sealed bottle kept a fair amount of CO_2 dissolved in the water, enough to significantly lower the freezing point. When the pressure is released by opening the can, enough of the dissolved CO_2 escapes to lessen the freezing point lowering. This raised the freezing point *above* the actual temperature of the solution. The mixture freezes.

6-87. Consider the possible dissociation products per gallium:

$$GaCl_2 \rightarrow Ga^{2+} + 2Cl^- \qquad \text{3 ions}$$

$$Ga(GaCl_4) \rightarrow Ga^+ + GaCl_4^- \qquad \text{2 ions}$$

Because the different substances form different numbers of particles when they dissolve, any experiment that depends on a colligative property would work. The chapter has discussed four -- vapor pressure depression, boiling point elevation, freezing point depression, and osmotic pressure. Let's tabulate the different effects of dissolving 0.100 mol of "$GaCl_2$" in 1.00 L of water. We will assume that the molarity is equal to the molality in such a dilute solution. The mole fraction of the "$GaCl_2$" unit then is 1.80×10^{-3}.

PROPERTY	CONDITIONS	FORMULA	EFFECT OF "$GaCl_2$"	EFFECT OF "$Ga(GaCl_4)$"
Vapor pressure depression	$T = 30.0^{\circ}C$ $P^{o}_{H_2O} = 0.04187$ atm	$\Delta P_{H_2O} = -X_{\text{all ions}} P^{o}_{H_2O}$	ΔP (atm) = -2.3×10^{-4}	ΔP (atm) = -1.5×10^{-4}
Boiling point elevation	$K_b (H_2O) =$ 0.512 K kg mol^{-1}	$\Delta T_b = K_b m$	$\Delta T_b =$ 0.15 K	$\Delta T_b =$ 0.10 K
Freezing point depression	$K_f (H_2O) =$ 1.86 K kg mol^{-1}	$\Delta T_f = -K_f m$	$\Delta T_f =$ -0.56 K	$\Delta T_f =$ -0.37 K
Osmotic pressure	$T = 300$ K	$\pi = c RT$	$\pi = 7.38$ atm	$\pi = 4.92$ atm

6-89. Although the number of ions formed from each substance is the same, so that ν is 2 in both cases, the ions that form from $MgSO_4$ have two charges. Therefore, we expect there to be much less full ionization than for the singly charged ions formed from $NaCl$.

6-91. A 0.50 molal solution has 0.50 moles of sucrose (171 g) per 1,000 g of water. Also, a mixture of these amounts of water and sucrose has a volume of 1.104 L. Therefore, 0.50 molal solution has a molarity, c, of 0.453 mol L^{-1}. To get the "volume molality" we calculate that 1000 g of water is the same as 1002 mL, at 20°C, where the density of water is 0.998 g ml^{-1}. Therefore the volume molality is 0.499 mol L^{-1}.

6-91. (continued)

Now we calculate Π both ways:

$\Pi = cRT = (0.453 \text{ mol L}^{-1})(0.08206 \text{ L atm mol}^{-1} \text{ K}^{-1})(293 \text{ K}) = \textbf{10.9 atm}$

$\Pi = m'RT = (0.499 \text{ mol L}^{-1})(0.08206 \text{ L atm mol}^{-1} \text{ K}^{-1})(293 \text{ K}) = \textbf{12.0 atm}$

The value from the volume molality is more accurate.

6-93. We assume $M_{NaCl} = c_{erythrocyte}$, where $c_{erythrocyte}=$ concentration of all dissolved species. Re-arranging the freezing point depression and osmotic pressure equations

$$\pi = cRT \approx M\,RT \qquad \text{and} \qquad -K_f m = \Delta t_f \Rightarrow m = -\frac{\Delta t_f}{K_f}$$

$$\pi = -\frac{\Delta t_f}{K_f}RT = -\frac{(-0.406)}{(1.86)}(0.08206)(298) = \textbf{5.34 atm}$$

6-95. In water: $X_{N_2} = \dfrac{P_{N_2}}{K_{H,water}} = \dfrac{5.0 \text{ atm}}{8.57 \times 10^4 \text{ atm}} = \textbf{5.83} \times \textbf{10}^{\textbf{-5}}$

In CCl_4: $X_{N_2} = \dfrac{P_{N_2}}{K_{H,water}} = \dfrac{5.0 \text{ atm}}{3.52 \times 10^9 \text{ atm}} = \textbf{1.42} \times \textbf{10}^{\textbf{-9}}$

6-97. (a) If we expand the Raoult's Law terms in the total vapor pressure, we obtain (with $X_{MB} + X_{EA} = 1$) data for the mole fractions in the solution phase:

$X_{MB}P_{MB}^o + X_{EA}P_{EA}^o = P_{total}$

$(1 - X_{EA})P_{MB}^o + X_{EA}P_{EA}^o = P_{total}$

$$X_{EA}^{solution} = \frac{P_{total} - P_{MB}^o}{P_{EA}^o - P_{MB}^o} = \frac{0.2400 - 0.1443}{0.3713 - 0.1443} = \textbf{0.421}$$

(b) For the mole fraction in the *vapor* phase:

$$X_{EA}^{vap} = \frac{P_{EA}}{P_{total}} = \frac{X_{EA}^{sol'n} P_{EA}^o}{P_{total}} = \frac{(0.421)(0.3713 \text{ atm})}{(0.240 \text{ atm})} = \textbf{0.651}$$

6-99. This becomes a problem in the optimum protein concentration that will give an osmotic pressure of between 50 and 150 kPa. This is equivalent (since there are 101.325 kPa per atm) to between 0.50 to 1.50 atm. The concentration that would give this osmotic pressure (from $c = \Pi / RT$) is between 0.020 and 0.060 mol L^{-1}.

If we have 100 mL of water, we want between 0.0020 and 0.0060 mole of protein. If the molar mass is indeed 100,000 g mol^{-1}, this means we need between 200 and 600 grams of protein. Alas, this is much more than the amount of water present and it is unlikely that we can use this osmometer in this case.

CHAPTER 7 - CHEMICAL EQUILIBRIUM

7-1. (a) $K = \dfrac{\left(P_{H_2O}\right)}{\left(P_{H_2}\right)\left(P_{O_2}\right)^{1/2}}$ (b) $K = \dfrac{\left(P_{XeF_4}\right)}{\left(P_{F_2}\right)^2\left(P_{Xe}\right)}$

(c) $K = \dfrac{\left(P_{CO_2}\right)^{20}\left(P_{H_2O}\right)^{10}}{\left(P_{C_5H_5}\right)^4\left(P_{O_2}\right)^{25}}$

7-3. Chemical equation: $P_4\ (g) + 2\ O_2\ (g) + 6\ Cl_2\ (g) \rightarrow 4\ POCl_3\ (g)$

$$K = \dfrac{\left(P_{POCl_3}\right)^4}{\left(P_{P_4}\right)\left(P_{O_2}\right)^2\left(P_{Cl_2}\right)^6}$$

7-5. We must indicate the equilibrium expression before inserting the numbers. In this problem, we will do this *with the P_{ref} terms included* to emphasize the convention followed in the textbook and this manual. This also renders the equilibrium constant as a dimensionless number. In all subsequent problems, division by P_{ref} = 1 atm is assumed.

$$K = \dfrac{\left(\dfrac{P_{Al_3Cl_9}}{P_{ref}}\right)^2}{\left(\dfrac{P_{Al_2Cl_6}}{P_{ref}}\right)^3} = \dfrac{\left(\dfrac{1.02\times10^{-2}\,\text{atm}}{1\,\text{atm}}\right)^2}{\left(\dfrac{1.00\,\text{atm}}{1\,\text{atm}}\right)^3} = \mathbf{1.04\times10^{-4}}$$

7-7. The equilibrium expression is $K = \dfrac{P_{boat}}{P_{chair}}$. It does not matter what the units are for pressure , because all we need to know is the *ratio of the two pressures* -- and by Dalton's Law for gases the pressure ratio is equal to the ratio of the mole fractions or mole percentages. Since 6.42% of the molecules are in the chair form, there must be 93.58% of the molecules in the boat form.

Therefore: $\dfrac{P_{boat}}{P_{chair}} = \dfrac{X_{boat}P_{total}}{X_{chair}P_{total}} = \dfrac{X_{boat}}{X_{chair}} = \dfrac{93.58}{6.42} = \mathbf{14.6}$

7-9. To keep track of the partial pressures, it is important to lay out the relationships among the amounts of substances before, during, and after the reaction. An orderly way to do this is by a table of changes listing initial pressures, the change in pressures on the way to equilibrium, and the equilibrium pressures. The changes in pressure during the reaction are proportional to the stoichiometric coefficients for the substances.

87

7-9. (continued)

A negative change in pressure appears for the reactants, for they are being consumed. A positive change in pressure appears for the products, for they are being generated.

(all P in atm)	H_2	I_2	HI
Initial Pressure	5.73	5.73	0.00
Change in Pressure	$-x$	$-x$	$+2x$
Equilibrium Pressure	5.73-x	5.73-x	$2x$

These pressures all share a common variable x, which we can obtain in this case because we know the pressure of HI at the end of the reaction:

$$P_{HI} = 9.00 = 2x, \text{ so } x = 4.50 \text{ atm}$$

and: $$P_{I_2} = P_{H_2} = 5.73 - x = 5.73 - 4.50 = 1.23 \text{ atm}$$

These values are the equilibrium pressures, and they are used to determine the magnitude of the equilibrium constant:

$$K = \frac{\left(P_{HI}\right)^2}{\left(P_{H_2}\right)\left(P_{I_2}\right)} = \frac{(9.00)^2}{(1.23) \times (1.23)} = \mathbf{53.5}$$

7-11. (a) The question tells the amount of SO_2Cl_2 added as a liquid. This will completely vaporize as the system is warmed to the reaction temperature.

We must use the ideal gas law to determine the pressure of the SO_2Cl_2 at the start of the reaction. Rather than calculate the number of moles of SO_2Cl_2 in a separate step, we can use the relation $n = \dfrac{m}{\mathcal{M}}$ in the ideal gas law, where:

m = mass of SO_2Cl_2 and $\mathcal{M}$ = molar mass of SO_2Cl_2.

$$P = \frac{mRT}{\mathcal{M}V} = \frac{(3.174 \text{ g}) \times \left(0.08206 \dfrac{L-atm}{mol-K}\right) \times (373.15 \text{ K})}{\left(134.97 \dfrac{g}{mol}\right) \times (1.000 \text{ L})} = 0.7200 \text{ atm}$$

(all P in atm)	SO_2Cl_2	SO_2	Cl_2
Initial Pressure	0.7200	0.00	0.00
Change in Pressure	$-x$	$+x$	$+x$
Equilibrium Pressure	0.7200 - x	x	x

7-11. (continued)

We get x by noting that the equilibrium pressures are the partial pressures at the end of the reaction. Then:

$$P_{total} = P_{SO_2Cl_2} + P_{SO_2} + P_{Cl_2}$$

$$1.30 \text{ atm} = (0.7200 - x) + x + x = 0.7200 + x$$

$$1.30 - 0.72 = 0.58 \text{ atm} = x$$

$$P_{SO_2Cl_2} = 0.7200 - x = 0.7200 - 0.58 = \mathbf{0.14 \text{ atm}}$$

$$P_{Cl_2} = P_{SO_2} = \mathbf{0.58 \text{ atm}}$$

(b) The values calculated in part (a) are put into the expression for K:

$$K = \frac{\left(P_{SO_2}\right)\left(P_{Cl_2}\right)}{\left(P_{SO_2Cl_2}\right)} = \frac{(0.58) \times (0.58)}{(0.14)} = \mathbf{2.4}.$$

7-13. The equilibrium expression for the first equation is:

$$K_1 = \frac{\left(P_{H_2O}\right)^6 \left(P_{NO_2}\right)^4}{\left(P_{NH_3}\right)^4 \left(P_{O_2}\right)^7}$$

While that for the second equation is: $K_2 = \dfrac{\left(P_{H_2O}\right)^3 \left(P_{NO_2}\right)^2}{\left(P_{NH_3}\right)^2 \left(P_{O_2}\right)^{7/2}}.$

The first equilibrium constant is the square of the second: $\mathbf{K_1 = K_2{}^2}$.

7-15. The third reaction has XeO_4 *(g)* as a reactant; this also occurs as a reactant in the second reaction. The third reaction has 2 HF *(g)* as a reactant, while the first reaction has it as product. Therefore, we can try to obtain the third reaction by taking the second reaction and adding it to the *reverse* of the first reaction:

$$XeO_4 \text{ (g)} + XeF_6 \text{ (g)} \rightleftharpoons XcOF_4 \text{ (g)} + XcO_3F_2 \text{ (g)} \qquad K = K_2$$

$$XeOF_4 \text{ (g)} + 2HF \text{ (g)} \rightleftharpoons XeF_6 \text{ (g)} + H_2O \text{ (g)} \qquad K = \frac{1}{K_1}$$

$$XeO_4 \text{ (g)} + 2 \text{ HF (g)} \rightleftharpoons XeO_3F_2 \text{ (g)} + H_2O \text{ (g)} \qquad K = \frac{K_2}{K_1}$$

7-17. The equilibrium expression is: $K = \dfrac{(P_{NOBr})^2}{(P_{Br_2})(P_{NO})^2}$. The values for the initial partial pressures, when input into the same expression, give us the reaction quotient Q. If $Q > K$, then the reaction will go to the *left*. If the $Q < K$, then the reaction will proceed to the *right*.

89

Chapter 7

7-17. (continued)

(a) $Q = \dfrac{(0.855)^2}{(0.225)(0.300)^2} = 36.1 < K$; reaction proceeds *left* to *right*.

(b) $Q = \dfrac{(0.855)^2}{(0.300)(0.225)^2} = 48.1 < K$; reaction proceeds *left* to *right*.

(c) $Q = \dfrac{(1.70)^2}{(0.450)(0.600)^2} = 107 < K$; reaction proceeds *left* to *right*.

7-19. (a) The reaction quotient is determined by putting the actual pressures into the equilibrium expression. In this case:

$$Q = \frac{(P_{Al_3Cl_9})^2}{(P_{Al_2Cl_6})^3} = \frac{(1.02 \times 10^{-2})^2}{(0.473)^3} = 9.83 \times 10^{-4}$$

(b) From Problem 5, $K = 1.04 \times 10^{-4}$. Since $Q > K$, the reaction will go *right* to *left* and there will be net **consumption** of Al_3Cl_9.

7-21. The observations indicate that the reaction proceeds with consumption of Br_2. This means that the initial reaction mixture had a reaction quotient Q that was *less* than K: $Q < K$.

The value of K cannot be obtained from this data, but we can get a limit for it.

$$K > Q = \frac{(P_{HBr})^2}{(P_{H_2}) \times (P_{Br_2})} = \frac{(0.90)^2}{(0.40) \times (0.40)} = 5.0625.$$

K > 5.1

7-23. The partial pressure of phosgene is one component of the equilibrium expression:

$$K = \frac{(P_{COCl_2})}{(P_{CO}) \times (P_{Cl_2})};$$

Rearranging: $P_{COCl_2} = K \times (P_{CO}) \times (P_{Cl_2})$

$$= 0.20 \times 0.050 \times 0.0045 = 4.5 \times 10^{-5} \text{ atm}.$$

7-25. The equilibrium expression is: $K = \dfrac{(P_{SO_2})(P_{Cl_2})}{(P_{SO_2Cl_2})}$

(a) When the reaction begins, $P_{SO_2} = P_{Cl_2} = 0$, so $Q = 0 < K$ and the reaction will proceed to the *right*. Since two moles of gas are produced for every one mole of SO_2Cl_2 consumed, the total number of particles will increase as the reaction proceeds, and so the **pressure will increase**, too.

7-25. (continued)

(b)

(all P in atm)	SO_2Cl_2	SO_2	Cl_2
Initial Pressure	2.500	0.00	0.00
Change in Pressure	$-x$	$+x$	$+x$
Equilibrium Pressure	$2.500 - x$	x	x

At equilibrium: $K = \dfrac{x^2}{2.500 - x} = 2.40$

rearranging: $0 = x^2 + 2.40\,x - 6.00$

The values for x are determined by using the quadratic formula, with $a = 1$, $b = 2.40$, and $c = -6.000$:

$$x = \frac{-b \pm \sqrt{b^2 - 4ac}}{2a} = \frac{-2.40 \pm \sqrt{(2.40)^2 - (4 \times 1 \times -6.00)}}{2 \times 1} = 1.53 \text{ or } -3.93$$

Note that $x = -3.93$ would mean a *negative* pressure for the products, so this root is rejected as not physically reasonable. The only suitable root is $x = 1.53$ and:

$P_{SO_2} = P_{Cl_2} = \textbf{1.53 atm}$; $P_{SO_2Cl_2} = 1.50 - 1.045 = \textbf{0.97 atm}$

7-27. (a) The pressure of the benzyl alcohol before the reaction proceeds can be calculated from the ideal gas equation:

$$P_{C_7H_8O} = \frac{mRT}{\mathcal{M}V} = \frac{(1.20\text{ g}) \times \left(0.08296\,\frac{\text{L - atm}}{\text{mol - K}}\right) \times (523\,\text{K})}{\left(108.14\,\frac{\text{g}}{\text{mol}}\right) \times (2.00\text{ L})} = 0.238\text{ atm}$$

(all P in atm)	C_7H_8O	C_7H_6O	H_2
Initial Pressure	0.238	0.00	0.00
Change in Pressure	$-x$	$+x$	$+x$
Equilibrium Pressure	$0.238 - x$	x	x

At equilibrium: $K = \dfrac{x^2}{0.238 - x} = 0.558$

rearranging: $0 = x^2 + 0.558\,x - 0.1328$

91

7-27. (continued)

The values for x are determined by using the quadratic equation,

with $a = 1$, $b = 0.558$, and $c = -0.1328$:

$$x = \frac{-b \pm \sqrt{b^2 - 4ac}}{2a} = \frac{-0.558 \pm \sqrt{(0.558)^2 - (4 \times 1 \times -0.1328)}}{2 \times 1} = 0.180 \text{ or } -0.737$$

The root $x = -0.737$ is rejected as unreasonable. The only suitable root is $x = 0.180$, so $P_{C_7H_6O} = x = \mathbf{0.180 \text{ atm}}$

(b) The decrease in the pressure of the benzyl alcohol is x. This represents a fraction dissociated of $\dfrac{0.180}{0.238} = \mathbf{0.76}$. The final pressure of benzyl alcohol is

$P_{C_7H_8O} = 0.238 - 0.180 = 0.058$ atm.

7-29. For this problem, we are only told the total pressure of the $PCl_5 + PCl_3 + Cl_2$ mixture at equilibrium and the value of the equilibrium constant. But these facts are enough, given a knowledge of the reaction stoichiometry, to get an answer.

(all P in atm)	PCl_5	PCl_3	Cl_2
Initial Pressure	y	0.00	0.00
Change in Pressure	$-x$	$+x$	$+x$
Equilibrium Pressure	$y - x$	x	x

The total pressure is the sum of the partial pressures:

$$P_{total} = P_{PCl_5} + P_{PCl_3} + P_{Cl_2} = (y - x) + x + x = y + x$$

$$0.776 = y + x; \qquad y = 0776 - x$$

$$P_{PCl_5} = y - x = (0.776 - x) - x = 0.776 - 2x; \quad P_{Cl_2} = P_{PCl_3} = x$$

We can get x from the expression for K and these partial pressures:

$$K = \frac{P_{PCl_3} \times P_{Cl_2}}{P_{PCl_5}} = \frac{x^2}{0.776 - 2x} = 2.15 \implies 0 = x^2 + 4.30\,x - 1.668$$

The values for x are determined by using the quadratic equation:

$$x = \frac{-4.30 \pm \sqrt{(4.30)^2 - (4 \times 1 \times -1.668)}}{2 \times 1} = 0.358; \; -4.66$$

The root $x = -4.66$ is rejected as unreasonable. The only suitable root is $x = 0.358$, so: $\quad P_{PCl_5} = 0.776 - 2\,x = 0.776 - 2\,(0.358) = \mathbf{0.060 \text{ atm}}$

$$P_{Cl_2} = P_{PCl_3} = x = \mathbf{0.358 \text{ atm}.}$$

7-31. This problem has two reactants, *both* present at the start of the reaction.

The table tracking their pressures from start to equilibrium is:

(all P in atm)	Br_2	I_2	IBr
Initial Pressure:	0.0500	0.0400	0.00
Change in Pressure:	-x	-x	+2x
Equilibrium Pressure:	0.0500 - x	0.0400 - x	2x

These final pressures go into the equilibrium expression:

$$K = 322 = \frac{\left(P_{IBr}\right)^2}{\left(P_{Br_2}\right) \times \left(P_{I_2}\right)} = \frac{(2x)^2}{(0.0500 - x) \times (0.0400 - x)}$$

Rearranging: $0 = 318\,x^2 - 28.98\,x + 0.644$

For the quadratic formula: $a = 318$; $b = -28.98$; $c = +0.644$.

$$x = \frac{-(-28.98) \pm \sqrt{(-28.98)^2 - (4 \times 318 \times 0.644)}}{2 \times 318} = 0.0527;\ 0.0384$$

Here, both roots are positive, but the first one, 0.0527, is greater than 0.0400 or 0.0500, the initial partial pressures of I_2 and Br_2, respectively so it would give *negative* pressures of both gases, which is physically unreasonable.

So: P_{I_2} = 0.0400 - x = 0.0400 - 0.0384 = **0.0016 atm**

P_{Br_2} = 0.0500 - x = 0.0500 - 0.0384 = **0.0116 atm**

P_{BrI} = 2 x = 2 (0.0384) = **0.0768 atm**.

7-33. We are given *K* for this reaction, which is not at the temperature of problem 5:

$$K = \frac{\left(P_{Al_3Cl_9}\right)^2}{\left(P_{Al_2Cl_6}\right)^3} = 0.050$$

The table of changes for this problem will be:

(all P in atm)	Al_2Cl_6	Al_3Cl_9
Initial Pressure:	0.050	0
Change in Pressure:	−3x	+2x
Equilibrium Pressure:	0.050–3x	+2x

Chapter 7

When inserted into the equilibrium expression we obtain:

$$K = \frac{(2x)^2}{(0.050 - 3x)^3}$$

Expansion of this will yield a cubic equation, which is much harder to solve than a quadratic equation. Yet, with K relatively large, we cannot approximate that x will be small relative to 0.050! Therefore, as in Example 7-7, we will try likely values of x to see which gets us close to K. We know that x cannot exceed $0.050 / 3 = 0.0167$, since that would make the pressure of Al_2Cl_6 negative. We will try a value slightly smaller than that, 0.015.

For example, if $x = 0.015$ $\quad Q = \frac{(2 \times 0.015)^2}{(0.050 - 3 \times 0.015)^3} = \frac{0.0016}{(0.005)^3} = 12{,}800$

This is too high, so the initial value of x is too high. Let's go to a much smaller value of x, $x = 0.0050$: $\quad Q = \frac{(2 \times 0.0050)^2}{(0.035)^3} = \frac{1.0 \times 10^{-4}}{2.7 \times 10^{-5}} = 3.7$

This is too small, so x will be bigger than 0.0050. We next use $x = 0.010$:

$$Q = \frac{(2 \times 0.010)^2}{(0.02)^3} = \frac{4.0 \times 10^{-4}}{8.0 \times 10^{-6}} = 50$$

In this case, we have hit upon exactly the right answer! $x = 0.010$, so

$P_{Al_3Cl_9} = 2x = \textbf{0.020 atm}$ $\qquad P_{Al_2Cl_6} = 0.050 - 3x = \textbf{0.020 atm}$

7-35. Problem 7-17 refers to three different sets of initial partial pressures. All the gases are at the same temperature, $25°C$ (= 298.15 K), so the method used convert to concentration units is the same for each. Here is one example in detail-- the conditions for NO in part (a). We begin with a $P_{NO} = 0.300$ atm. The ideal gas law can be rewritten to obtain the concentration, $\frac{n}{V}$:

$$[NO] = \frac{n}{V} = \frac{P_{NO}}{RT} = \frac{(0.300 \text{ atm})}{\left(0.08206 \frac{\text{L-atm}}{\text{mol-K}}\right)(298.15 \text{ K})} = 1.23 \times 10^{-2} \text{ mol L}^{-1}$$

The others are worked in turn. Mathematically, the conversion *in this case* is:

$[X] = 0.0409\, P_X$. where the number refers to the $^1/_{RT}$ term.

7-35. (continued)

	[NO]	[Br₂]	[NOBr]
(a)	1.23×10^{-2} M	9.20×10^{-3} M	3.50×10^{-2} M
(b)	9.20×10^{-3} M	1.23×10^{-2} M	3.50×10^{-2} M
(c)	2.45×10^{-2} M	1.84×10^{-2} M	6.95×10^{-2} M

7-37. The value of the equilibrium constant given in problem No. 23 is related to units of *atmospheres*. Here we need units of mol L^{-1}. In the chemical equation, two moles of gas are in the reactants and one mole is in the product, so the sum of the stoichiometric coefficients is: $2 - 1 = +1$. For an equilibrium in units of mol L^{-1}:

$$\frac{[COCl_2]}{[CO] \times [Cl_2]} = K \times \left(\frac{RT}{P_{ref}}\right)^{+1} = 0.20 \times \left(\frac{\left(0.08206 \frac{L \text{-} atm}{mol \text{-} K}\right) \times (873.15\ K)}{1\ atm}\right) = 14.33 \frac{L}{mol}$$

$$[COCl_2] = [CO] \times [Cl_2] \times 14.33 \frac{L}{mol}$$
$$= \left(3.2 \times 10^{-4} \frac{mol}{L}\right) \times \left(1.3 \times 10^{-2} \frac{mol}{L}\right) \times \left(14.33 \frac{L}{mol}\right) = \mathbf{6.0 \times 10^{-5} \frac{mol}{L}}$$

7-39. Two equilibrations occur in this experiment: the first as the reactant H_2 and I_2 equilibrate, and the second after the reduction of the partial pressure of the HI. The first equilibrium state is computed in a way similar to Problem No. 9. *K* varies very slightly because of the difference in temperature.

For the first equilibration:

(all P in atm)	H₂	I₂	HI
Initial Pressure	5.73	5.73	0.00
Change in Pressure	-x	-x	+2x
Equilibrium Pressure	5.73-x	5.73-x	2x

$$K = \frac{\left(P_{HI}\right)^2}{\left(P_{H_2}\right)\left(P_{I_2}\right)} = \frac{(2x)^2}{(5.73 - x)^2} = 53.7$$

7-39. (continued)

This problem does not in fact require the quadratic equation, for both the top and the bottom of the fraction are squares, so we can rewrite:

$$\frac{(2x)^2}{(5.73-x)^2} = \left[\frac{2x}{5.73-x}\right]^2 = 53.7 \Rightarrow \frac{2x}{5.73-x} = \sqrt{53.7} = 7.328$$

Rearranging: $x = 4.501$

$P_{HI} = 2x = 9.00$; $P_{I_2} = P_{H_2} = 5.73 - x = 5.73 - 4.501 = 1.23$ atm

For the second equilibration, the preceding partial pressures are altered by the removal of 1.50 atm of HI, so we must do a second calculation, starting with the pressures for H_2 and I_2 derived above and $P_{HI} = 7.50$ atm

(all P in atm)	H_2	I_2	HI
Initial Pressure	1.23	1.23	1.50
Change in Pressure	-x	-x	+2x
Equilibrium Pressure	1.23-x	1.23-x	7.5 +2x

$$K = \frac{(P_{HI})^2}{(P_{H_2})(P_{I_2})} = \frac{(7.50+2x)^2}{(1.23-x)^2} = \left[\frac{7.002+2x}{1.229-x}\right]^2 = 53.7$$

$$\frac{7.50+2x}{1.23-x} = \sqrt{53.7} = 7.328$$

$$x = 0.161$$

So, at this point: $P_{HI} = 7.50 + 2x = $ **7.82 atm**

For completeness: $P_{I_2} = P_{H_2} = 1.23 - x = 1.07$ atm

This problem illustrates LeChatelier's principle: in response to the stress of removing product, the equilibrium responds by converting more of the reactants to product.

7-41. (a) The additional NO on the left of the equation will cause the equilibrium to shift to the **right**- consuming NO and making more NO_2 and N_2O.

(b) When the volume of the system is increased, the equilibrium concentrations will react to make *more* particles (think in terms of a release of crowding). In this system, there are more particles on the left than on the right, so the system will respond to an increase in volume by moving from **right to left.**

7-41. (continued)

(c) The reaction is exothermic, so heat appears as a "product." Adding heat by increasing the temperature is like adding a chemical product, so the reaction will proceed from **right to left.**

(d) The Ar does not cause a reaction, but it does increase the total volume that reacting species have to move around in. For the reacting particles, this is the same effect as in part (b), and the equilibrium will shift to a state of *more* particles, which in this case is from **right to left.**

(e) A non-reactant such as Ar can never affect an equilibrium system, unless it induces a change in volume, which is not the case here. There will be **no effect.**

7-43. There will be different strategies for the different steps. For the first reaction, one should work at *higher pressures* (to take advantage of the decrease in the number of gaseous particles) and at *lower temperatures* (to remove the heat evolved). For the second reaction, one should work at *lower pressures* (because there are more moles of gas among the products) and at *higher temperatures* (to give the reaction the heat it requires).

7-45. (a) A reaction that is driven by cooling - removing heat - is one that gives off heat as it proceeds. The reaction is **exothermic.**

(b) A reaction that is driven by decreasing the volume (and thereby increasing the total pressure) must have more a net **decrease** in the number of gas molecules.

7-47. (a) $K = \dfrac{\left(P_{H_2S}\right)^8}{\left(P_{H_2}\right)^8}$ (b) $K = \dfrac{\left(P_{H_2}\right)\left(P_{COCl_2}\right)}{\left(P_{Cl_2}\right)}$ (c) $K = P_{CO_2}$ (d) $K = \dfrac{1}{\left(P_{C_2H_2}\right)^3}$

7-49. (a) $K = \dfrac{\left[Zn^{2+}\right]}{\left[Ag^+\right]^2}$ (b) $K = \dfrac{\left[VO_3(OH)^{2-}\right]\left[OH^-\right]}{\left[VO_4^{3-}\right]}$ (c) $K = \dfrac{\left[HCO_3^-\right]^6}{\left(P_{CO_2}\right)^6\left[As(OH)_6^{3-}\right]^2}$

Chapter 7

7-51.

[N$_2$O$_4$]	[NO$_2$]^2	$K = \dfrac{[NO_2]^2}{[N_2O_4]}$
0.190×10^{-3}	0.784×10^{-5}	0.0412
0.686×10^{-3}	2.70×10^{-5}	0.0393
1.54×10^{-3}	5.27×10^{-5}	0.0342
2.55×10^{-3}	10.8×10^{-5}	0.0423
3.75×10^{-3}	13.7×10^{-5}	0.0363
7.86×10^{-3}	29.9×10^{-5}	0.0380
11.9×10^{-3}	44.1×10^{-5}	0.0370

(b) The average value for the K's calculated at the different points is **0.0384**.

7-53. The equilibrium expression for this reaction is: $\qquad K = \dfrac{(P_{H_2})(P_{CO})}{(P_{H_2O})}$

Evaluating Q by putting the listed values of the pressures into this equation, we obtain:

(a) $Q = 2.05 < K \Rightarrow$ reaction will go to **left to right**.

(b) $Q = 3.27 > K \Rightarrow$ reaction will go **right to left**.

7-55. (a) The pressure in the reaction flask is due to build up of the NH$_3$ (g) and the H$_2$Se (g), which will have the same pressure.

$$P_{total} = P_{NH_3} + P_{H_2Se} = 2x$$

$$x = \frac{P_{total}}{2} = \frac{0.0184}{2} = 0.0092 \, atm$$

For this reaction, we have: $K = (P_{NH_3})(P_{H_2Se}) = (0.0092)^2 = 8.46 \times 10^{-5}$.

(b) Rearranging the equilibrium expression:

$$P_{H_2Se} = \frac{K}{P_{NH_3}} = \frac{8.46 \times 10^{-5}}{0.0252} = 0.00336 \, atm$$

98

7-57. The table of changes relating the starting and final pressures is:

all P in atm	CO_2	H_2	CO
Initial pressure	x	x	0
Change in pressure	$-y$	$-y$	$+y$
Equilibrium pressure	$x - y =$ $x - 0.100$	$x - y =$ $x - 0.100$	$y = 0.100$

We solve for x in the equilibrium expression:

$$K = \frac{(P_{CO})}{(P_{CO_2})(P_{H_2})} = \frac{(0.100)}{(x - 0.100)(x - 0.100)} = \frac{0.100}{(x - 0.100)^2}$$

There is no need to work this problem as a quadratic formula (see also problem No. 37). Taking a square root of both sides and rearranging:

$$(x - 0.100)^2 = (P_{H_2})^2 = \frac{0.100}{K}$$

$$P_{H_2} = \sqrt{\frac{0.100}{K}} = \sqrt{\frac{0.100}{3.22 \times 10^{-4}}} = \sqrt{310.6} = 17.722$$

$$P_{CO_2} = P_{H_2} = \textbf{17.7 atm}$$

7-59. A critical piece of information here is that the volumes of the water and the organic solvent are *equal*. Their values are not given, but that is OK. We still know that a decrease in the concentration of iodine in the water by x mol L^{-1} will be matched by an *increase* in the concentration of iodine in the organic layer by the same amount x. In this case, the change in the concentration of the iodine in the water layer is $(1.00 \times 10^{-2}$ M$) - (1.30 \times 10^{-4}) = .00987$ M.

Therefore:
$$K = \frac{[I_2]_{CCl_4}}{[I_2]_{H_2O}} = \frac{0.00987}{0.00013} = \textbf{76}$$

7-61. (a) First reaction: $K_1 = [C_6H_5COOH]_{aq} = \left(2.00 \frac{g}{L}\right) \times \left(\frac{1 \text{ mol}}{122.1 \text{ g}}\right) = \textbf{0.0164}$

Second reaction: $K_2 = [C_6H_5COOH]_{ether} = \left(660 \frac{g}{L}\right) \times \left(\frac{1 \text{ mol}}{122.1 \text{ g}}\right) = \textbf{5.41}$

Chapter 7

7-61. (continued)

(b) The partition equilibrium is obtained by combining the two solubility equilibria :

$$C_6H_5COOH \; (s) \rightleftharpoons C_6H_5COOH \; (ether) \qquad\qquad K_2$$

$$C_6H_5COOH \; (aq) \rightleftharpoons C_6H_5COOH \; (s) \qquad\qquad \frac{1}{K_1}$$

$$C_6H_5COOH \; (aq) \rightleftharpoons C_6H_5COOH \; (ether) \qquad\qquad K_3 = \frac{K_2}{K_1} = \frac{5.41}{0.0164} = \textbf{330}$$

7-63. Relative amounts are directly proportional to the mole fractions, so we can say:

$$P_{BCl3} = \frac{90}{805} P_{total} \qquad\qquad P_{BF3} = \frac{470}{805} P_{total}$$

$$P_{BF2Cl} = \frac{200}{805} P_{total} \qquad\qquad P_{BFCl2} = \frac{45}{805} P_{total}$$

For the first reaction, these values allow the $\frac{P_{total}}{805}$ terms to cancel:

$$K_1 = \frac{\left(P_{BFCl2}\right)^3}{\left(P_{BCl3}\right)^2\left(P_{BF3}\right)} = \frac{\left(45\dfrac{P_{total}}{805}\right)^3}{\left(90\dfrac{P_{total}}{805}\right)^2\left(470\dfrac{P_{total}}{805}\right)} = \frac{(45)^3}{(90)^2(470)} = \textbf{0.024}$$

A similar calculation for the second reaction yields $K_2 = \dfrac{(200)^3}{(90)(470)^2} = \textbf{0.40}$

(b) The value of the equilibrium constant for the third reaction is obtained from the partial pressures; again, the $\frac{P_{total}}{805}$ will cancel:

$$K_3 = \frac{\left(P_{BFCl2}\right)\left(P_{BF2Cl}\right)}{\left(P_{BCl3}\right)\left(P_{BF3}\right)} = \frac{\left(45\dfrac{P_{total}}{805}\right)\left(200\dfrac{P_{total}}{805}\right)}{\left(90\dfrac{P_{total}}{805}\right)\left(470\dfrac{P_{total}}{805}\right)} = \frac{(45)(200)}{(90)(470)} = \textbf{0.21}$$

This value can also be obtained by a combination of K_1 and K_2.

7-65. (a) The relative concentrations [cis] and [trans] are a fraction of the total,

$$[cis] + [trans] = [total]:$$

$$[trans] = 0.736 \,[total] = 0.736\frac{(P_{total})}{RT} \qquad [cis] = 0.264 \,[total] = 0.264\frac{(P_{total})}{RT}$$

The fact that the concentration P_{total} isn't known doesn't matter in the actual equilibrium

calculation: $\quad K = \dfrac{P_{cis}}{P_{trans}} = \dfrac{0.736\dfrac{(P_{total})}{RT}}{0.264\dfrac{(P_{total})}{RT}} = \dfrac{0.736}{0.264} = \textbf{2.79}$

100

7-65. (continued)

(b) The initial pressure for the *cis* compound is, by the ideal gas law (and noting that $426.6^{\circ}C = 698.75\,K$):

$$P_{cis}(initial) = \frac{nRT}{V} = \frac{(0.525) \times \left(0.08206\,\frac{L\text{-}atm}{mol\text{-}K}\right) \times (698.75\,K)}{15.00\,L} = 2.007\,atm$$

All P in atm	P_{cis}	P_{trans}
Initial pressure	2.007	0
Change in pressure	-x	+x
Equilibrium pressure	2.007 - x	x

Then: $K = \dfrac{P_{trans}}{P_{cis}} = \dfrac{x}{2.007 - x} = 2.79 \qquad \Rightarrow \qquad x = 1.48$

$P_{cis} = 2.007 - 0.529 = 1.48\,atm \qquad\qquad P_{trans} = 0.53\,atm.$

7-67. For both (a) and (b) we have the same table of changes:

all P in atm	SO_2	O_2	SO_3
Initial pressure	2.0	3.0	0
Change in pressure	- 2y	- y	+ 2y
Equilibrium pressure	2.0 - 2y	3.0 - y	2y

These are then placed in the same equilibrium expression, which is solved for *y*:

$$K = \frac{\left(P_{SO_3}\right)^2}{\left(P_{SO_2}\right)^2\left(P_{O_2}\right)} = \frac{(2y)^2}{(2.0 - 2y)^2(3.0 - y)}$$

This will yield a cubic equation, and with such large values of *K we cannot reliably assume that y will be small!*. We will now (as in problem 7-33) try values of *y* to get a value for the reaction quotient *Q* as close as possible to *K*. Because we cannot use up all of the SO_2, we know that *y* must be less than 1. Let's start with 0.50, a half-way number.

$y = 0.50 \qquad\qquad Q = \dfrac{(2 \times 0.50)^2}{(2.0 - 2 \times 0.50)^2(3.0 - 0.50)} = 0.40$

Chapter 7

7-67. (continued)

This is too large, so we move down to 0.40:

$$y = 0.40 \qquad Q = \frac{(2 \times 0.40)^2}{(2.0 - 2 \times 0.40)^2 (3.0 - 0.40)} = 0.17$$

This is too large, so we work our way to the correct answer by picking a number half-way between 0.50 and 0.60:

$$y = 0.45 \qquad Q = \frac{(2 \times 0.45)^2}{(2.0 - 2 \times 0.45)^2 (3.0 - 0.45)} = 0.26$$

This is just a little too big, so we move down a bit :

$$y = 0.44 \qquad Q = \frac{(2 \times 0.44)^2}{(2.0 - 2 \times 0.44)^2 (3.0 - 0.44)} = 0.24$$

We must be close, in between these last two:

$$y = 0.0445 \qquad Q = \frac{(2 \times 0.445)^2}{(2.0 - 2 \times 0.445)^2 (3.0 - 0.445)} = 0.252$$

The best value we can get is $y = 0.0444$.

We then use this to obtain a value for the pressures of SO_2 and O_2, the final pressure of SO_2 is $2y$, or **0.89 atm**.

(b) In this case, the K is 0.70. The procedure is the same. But for the first trial, we note that $y = 0.50$ gave us $Q = 0.40$, which is now too low. So we test $y = 0.60$:

$$y = 0.60 \qquad Q = \frac{(2 \times 0.60)^2}{(2.0 - 2 \times 0.60)^2 (3.0 - 0.60)} = 0.94$$

Successive trials in this case converge to a value of $y = 0.566$ and a pressure of SO_3 **1.13 atm.**

(c) The amount of SO_2 that reacts is proportional to y. The percent that reacts is:

$$\% \text{ reacted} = \frac{P_{SO_2} \, (reacted)}{P_{SO_2} \, (initial)} \times 100\% = \frac{2y}{P_{SO_2} \, (initial)} \times 100\%$$

At $T = 1100$ K: $\dfrac{2(0.444)}{P_{SO_2} \, (initial)} \times 100\% = 44.4\%$

At $T = 523$ K, $\dfrac{2(0.566)}{P_{SO_2} \, (initial)} \times 100\% = 56.6\%$

7-69. (a) The tabular format works perfectly well even when the system starts with all possible chemical substances present. First, we note that looking at Q will tell us the direction of the approach to equilibrium:

$$Q = \frac{(P_{IBr})^2}{(P_{Br_2})(P_{I_2})} = \frac{(3.00)^2}{(2.00)(4.00)} = 1.125 < K = 131.$$

The reaction will proceed left to *right*, with loss of Br_2 and I_2 and gain of IBr.

(all P in atm)	Br_2	I_2	IBr
Initial pressure:	4.00	2.00	3.00
Change in pressure:	-x	-x	+2x
Equilibrium pressure:	4.00 - x	2.00 - x	3.00+2x

$$K = \frac{(P_{IBr})^2}{(P_{Br_2})(P_{I_2})} = \frac{(3.00 + 2x)^2}{(4.00 - x)(2.00 - x)} = \frac{9.00 + 12.00x + 4x^2}{8.00 - 6x + x^2}$$

For the purposes of clean algebraic manipulation, it is wise to rearrange to the quadratic formula before putting in the numerical value for K:

$$K(8.00 - 6x + x^2) = 9.00 + 12.00x + 4x^2$$

$$0 = (4 - K)x^2 + (6K + 12)x + (9 - 8K)$$

For the quadratic formula:

$$a = 4 - K = 4 - 131 = -127$$
$$b = 6K + 12 = 6(131) + 12 = 798$$
$$c = 9 - 8K = 9 - 8(131) = -1039$$

Processing through the quadratic formula (not shown here) gives two roots:

$$x = 1.84 \qquad x = 4.44$$

The second root is physically unreasonable, since it would give negative equilibrium pressures for Br_2 and I_2. But the first root is reasonable, so then we will have:

$$P_{Br_2} = 4.00 - 1.84 = \textbf{2.16 atm}$$
$$P_{I_2} = 2.00 - 1.84 = \textbf{0.16 atm}$$
$$P_{IBr} = 3.00 + 2(1.84) = \textbf{6.68 atm.}$$

Chapter 7

7-71.

$$Q = \frac{\left(P_{C_6H_{12}}\right)}{\left(P_{C_6H_6}\right)\left(P_{H_2}\right)^3}$$

$$Q = \frac{\left(\frac{n_{C_6H_{12}} RT}{V}\right)}{\left(\frac{n_{C_6H_6} RT}{V}\right)\left(\frac{n_{H_2} RT}{V}\right)^3} = \left(\frac{\left(n_{C_6H_{12}}\right)}{\left(n_{C_6H_6}\right)\left(n_{H_2}\right)^3}\right) \times \left(\frac{V}{RT}\right)^3$$

(a) If we begin with chemical amounts such that $Q = K$, then the addition of H_2 will cause Q to *decrease* to a value less than K. The reaction will go to the **right** in returning to equilibrium.

(b) A decrease in volume will *decrease* Q to a value less than K. The reaction will go to the **left** in returning to equilibrium.

7-73. For reaction #1, the value for [Cl_2] that is important is the actual concentration, corrected for the reaction to give HOCl and Cl⁻:

$$Cl_2\ (g) \rightleftharpoons Cl_2\ (aq) \qquad K_1 = \frac{[Cl_2]}{P_{Cl_2}} = \frac{0.061}{1.00} = \mathbf{0.061}$$

For reaction #2 we must recognize that a fourth aqueous species is formed: hydrogen ion. The stoichiometry for hydrogen ion is the same as for Cl⁻ and HOCl, so [H^+] = 0.030 M. The equilibrium constant then becomes:

$$K_2 = \frac{\left[H^+\right]\left[HOCl\right]\left[Cl^-\right]}{[Cl_2]} = \frac{(0.030)^3}{(0.061)} = \mathbf{4.43 \times 10^{-4}}$$

7-75. (a) The decomposition reaction is:

$$NH_4HS\ (s) \rightleftharpoons NH_3\ (g) + H_2S\ (g) \qquad K = \left(P_{NH_3}\right)\left(P_{H_2S}\right)$$

For the first equilibration, the only source of either NH_3 or H_2S is the decomposition reaction. Therefore, the final pressure is equal to twice the pressure of each of them. That is, before the addition of extra $NH_3\ (g)$,
$P_{NH_3} = P_{H_2S} = 0.659/2 = 0.3295$ atm and $K = (0.3295)^2 = \mathbf{0.1086}$.

(b) The system is then changed by increasing P_{NH_3} by some amount x. But rather than carefully tracking pressure changes with x's and y's, we just need to realize that the equilibrium partial pressure of H_2S will be at a value determined by the equilibrium partial pressure of NH_3 -- *which we are given.*

104

7-75. (continued)

We are told that at the end $P_{NH_3} = 0.750$ atm

Substituting: $K = (P_{NH_3})(P_{H_2S})$

$$P_{H_2S} = \frac{K}{P_{NH_3}} = \frac{0.1086}{0.750} = \mathbf{0.145\ atm}$$

(c) The additional NH_4S has no effect because it is a solid and cannot influence the equilibrium (so long as some of it is present).

7-77. We have two equations to work with, and two unknown pressures.

$$P_{total} = P_{CO_2} + P_{CO} \qquad \Rightarrow \qquad P_{CO_2} = P_{total} - P_{CO}$$

$$K = \frac{P_{CO_2}}{P_{CO}} = \frac{P_{total} - P_{CO}}{P_{CO}} \qquad \Rightarrow \qquad P_{CO} = \frac{P_{total}}{K+1} = \frac{2.50}{256.4} = \mathbf{0.00975\ atm}$$

$$P_{CO_2} = 2.50 - 0.00975 = \mathbf{2.49\ atm}$$

7-79. We have two equations to work with, and two unknowns. For simplicity:

$$[C_{17}H_{35}COOH] = [\text{monomer}] \qquad\qquad [(C_{17}H_{35}COOH)_2] = [\text{dimer}]$$

The initial concentration of the monomer is:

$$[\text{monomer}]_o = \left(\frac{15.0\ g}{284.48\ g\ mol^{-1}}\right) \times \left(\frac{1}{1.250\ L}\right) = 0.0422\ M$$

All concentrations in M	[monomer]	[dimer]
Initial concentration:	0.0422	0.00
Change in concentration:	-2x	+x
Equilibrium concentration	0.0422-2x	x

$$K = \frac{[\text{dimer}]}{[\text{monomer}]^2} = \frac{x}{(0.0422 - 2x)^2} = \frac{x}{0.00178 - 0.1688x + 4x^2}$$

$$0 = 4\,K\,x^2 + (-1-0.1688K)x + 0.00178K$$

$a = 4K = 160 \qquad\qquad b = -1-0.1688K = -7.752 \qquad\qquad c = 0.00178K = 0.0712$

Use of the quadratic formula yields the roots $x = 0.0123$ and 0.03613. Only the first is physically reasonable. Then:

[monomer]= $0.0422 - 2x = 0.0176$ M $\qquad\qquad$ [dimer] $= x = 0.0123$ M

7-81. This problem does *not* entail an equilibrium *between* PCB-2 and PCB-11. We are merely concerned with the how much they will, relative to one another, partition between octanol and water. Let's assume they both start off with a concentration x in the octanol layer. Then the table of changes becomes:

	$[PCB\text{-}2]_o$	$[PCB\text{-}2]_w$	$[PCB\text{-}11]_o$	$[PCB\text{-}11]_w$
Initial concentration:	x	0	x	0
Change in concentration:	-y	+y	-z	+z
Equilibrium concentration	x - y	y	x - z	z

The equilibrium expressions are the only data we have. We can rearrange each separately to get x:

For PCB-2	$K(2) = \dfrac{x-y}{y}$	$y \times K(2) = x - y$	$[K(2) + 1]y = x$
For PCB-11	$K(11) = \dfrac{x-z}{z}$	$z \times K(11) = x - z$	$[K(11) + 1]z = x$

The expressions in bold give a link between y and z:

$$[K(11) + 1]z = [K(2) + 1]y$$
$$\frac{z}{y} = \frac{K(2)+1}{K(11)+1}$$
$$\frac{[PCB\text{-}11]_w}{[PCB\text{-}2]_w} = \frac{39,801}{126,001} = \mathbf{0.316}$$

This result indicates that the PCB-11 is removed from the water much better than the PCB-2.

7-83. This interesting separation problem is complicated by the fact that there are *unequal volumes of the organic and the water layer*. This means we must keep track of the number of **moles** of I_2 in each layer, not the concentrations. The table of changes in this case is constructed of amounts of moles, and the last line is a return to concentration units.
Moles of I_2 at the start: $0.100 \text{ L} \times 0.002 \text{ mol L}^{-1} = 0.00020 \text{ mol}$

7-83. (continued)

All concentrations in M	$[I_2 (H_2O)]$	$[I_2(CCl_4)]$
Initial No. of moles	0.0002	0
Change in No. of moles in layer	-x	+x
Equilibrium No. of moles	0.0002 - x	x
Volume of layer (liters)	0.100	0.025
Equilibrium concentration	$\dfrac{0.0002 - x}{0.100}$	$\dfrac{x}{0.025}$

The last line contains the values needed for the partition equilibrium expression:

$$K = \frac{[I_2 (CCl_4)]}{[I_2 (aq)]} = \frac{\left(\dfrac{x}{0.025}\right)}{\left(\dfrac{0.0002 - x}{0.100}\right)} = \left(\frac{x}{0.0002 - x}\right)\left(\frac{0.100}{0.025}\right)$$

$$85 = 4\left(\frac{x}{0.0002 - x}\right)$$

Rearranging: $x = 1.9 \times 10^{-4}$; so: $[I_2 (H_2O)] = 9 \times 10^{-5}$ M

This means that 9×10^{-6} moles of I_2 remaining in the aqueous layer.

This is $9 \times 10^{-6} / 2 \times 10^{-4} = \mathbf{0.045 = 4.5\%}$ of the original I_2.

(b) The result from the first part of the problem -- that the 9×10^{-6} mol of I_2 remain in the aqueous phase -- is used in a new separation calculation, with fresh CCl_4. The table is set up as in part (a), except for the initial value for the moles of I_2 in the aqueous phase.

The equilibrium concentrations are then:

$$[I_2 (H_2O)] = \frac{9 \times 10^{-6} - x}{0.100} \qquad \text{and } [I_2 (CCl_4)] = \frac{x}{0.025}.$$

These are put into the expression for K and the result is solved for x:

$$x = 8.6 \times 10^{-6} \text{ mol.}$$

Then the number of moles remaining in the water is

$I_2 (H_2O) = 9.0 \times 10^{-6} - 8.4 \times 10^{-6} = 4 \times 10^{-7}$.

This is a fraction $4 \times 10^{-7} / 2 \times 10^{-4} = \mathbf{0.002 = 0.2\%}$ of the original value.

(c) The single 50.0 mL extraction left 2.3% of the I_2 behind in the aqueous layer. This is more than *ten times* the amount left behind after two 25.0 mL extractions. Smaller separate extractions are much more efficient.

CHAPTER 8 - ACID-BASE EQUILIBRIA

8-1.   (a) Cl^- cannot act as a Brønsted-Lowry acid

         (b) HSO_4^- can act as a Brønsted-Lowry acid. Its conjugate base is SO_4^{2-}.

         (c) NH_4^+ can act as a Brønsted-Lowry acid. Its conjugate base is NH_3.

         (d) NH_3 can act as a Brønsted-Lowry acid. Its conjugate base is NH_2^-.

         (d) H_2O can act as a Brønsted-Lowry acid. Its conjugate base is OH^-.

8-3.   (a) $NH_4Br\ (aq) + H_2O\ (l) \rightarrow H_3O^+\ (aq) + NH_3\ (aq) + Br^-\ (aq)$

         (b) $H_2S\ (aq) + H_2O\ (l) \rightarrow H_3O^+\ (aq) + HS^-(aq)$

         (c) $(NH_4)_2SO_4\ (aq) + H_2O\ (l) \rightarrow 2\ H_3O^+\ (aq) + 2\ NH_3\ (aq) + SO_4^{2-}\ (aq)$

8-5.   For this sample, $pH = -\log_{10}[H_3O^+] = -\log_{10}(2.0 \times 10^{-4}) = +3.70$.

Just as there is a mathematical relationship between $[H_3O^+]$, $[OH^-]$ and K_w, where we know $[H_3O^+] \times [OH^-] = K_w$, so there is a very useful relationship between pH, pOH, and pK_w. As discussed in the chapter: $pH + pOH = pK_w$

At $25^\circ C$, $K_w = 1.0 \times 10^{-14}$, so we then have:

         $pH + pOH\ = 14.00$, which gives: $pOH = 14.00 - pH$

Here: $pOH\ = 14.00 - 3.70 = \mathbf{10.30}$.

An important point concerns **significant figures and the pH (or pOH) function**. Consider a number $x = y \times 10^z$. The digit or digits *before* the decimal point in the pH are determined by the exponent z. The digits *after* the decimal point are determined by the pre-exponential factor y. Any variation in the accuracy or precision of a measurement will be found in the pre-exponential factor y. Therefore, in pH, pOH, or pK_a the number of digits after the decimal point is determined by the number of digits in the pre-exponential part of x. Conversely, when we go from pH or pOH to the number x, the number of digits after the decimal place indicates the number of significant figures that should present in y.

8-7.

	Lower pH level	Upper pH level
pH	5.5	6.5
$pOH = 14.0 - pH$	8.5	7.5
$[H_3O^+] = 10^{(-pH)}$	3×10^{-6}	3×10^{-7}
$[OH^-] = 10^{(-pOH)}$	3×10^{-9}	3×10^{-8}

8-7. (continued)

Note that the pH and pOH values have one digit after the decimal place. Therefore, the pre-exponential factor is determined to *one* significant digit, too.

8-9. Since nitric acid is a strong acid, the major source of H^+ is nitric acid; the pH represents the concentration of all the moles of hydronium ions that could form. In concentration units: $[H_3O^+] = 10^{(-pH)} = 4.79 \times 10^{-3}$ M $= c_{initial}$

The problem is then a dilution problem, where the unknown is c_{final}.

$$c_{final} \times V_{final} = c_{initial} \times V_{initial}$$
$$c_{final} = c_{initial} \times \frac{V_{initial}}{V_{final}} = \left(4.79 \times 10^{-3} \text{ M}\right) \times \frac{1}{8} = 5.99 \times 10^{-5} \text{ M}$$
$$pH_{final} = -\log_{10}(5.99 \times 10^{-5}) = \textbf{3.22}$$
$$pOH = 14.00 - pH = 14.00 - 3.22 = 10.78$$
$$[OH^-] = 10^{(-10.78)} = \textbf{1.7} \times \textbf{10}^{\textbf{-11}} \text{ \textbf{M}}$$

8-11. A neutralization reaction will occur, to the extent of the limiting reactant:
$$H_3O^+ \text{ (aq)} + OH^- \text{ (aq)} \rightarrow 2 H_2O \text{ (l)}$$

Note that the acid volume is 11.74 mL = 0.01174 L and the base volume is 15.78 mL = 0.01578 L.

Number of moles of OH^-: $0.01174 \text{ L} \times 0.071 \text{ mol L}^{-1} = 8.34 \times 10^{-4}$ mol

Number of moles of H_3O^+: $0.01578 \text{ L} \times 0.094 \text{ mol L}^{-1} = 1.48 \times 10^{-3}$ mol

The OH^- is the limiting reactant. The number of moles of excess H_3O^+ is:
1.48×10^{-3} mol $- 8.3 \times 10^{-4}$ mol $= 6.5 \times 10^{-4}$ mol H_3O^+

This amount of hydronium ion is now present in a total water volume of 0.01578 + 0.01174 = 0.02752 L, so the concentration of hydronium ion is

$$[H_3O^+] = 6.5 \times 10^{-4} \text{ mol } H_3O^+ / 0.02752 = 0.0236 \text{ M}$$
$$pH = -\log_{10}(0.0236) = \textbf{1.63}$$

8-13. A smaller acid ionization constant means the concentration of hydronium ion is smaller and an acid is weaker. The information given for ammonium is in pK_a, so we must convert it to the $K_a = 10^{-pK_a} = 5.6 \times 10^{-10}$. In this set the order of the strengths from weakest to strongest is: ammonium < benzoic acid < formic acid.

8-15. (a) We don't need to know anything but that the compound, in acting as a base, can accept a hydrogen ion from water:

$$C_{10}H_{15}ON \ (aq) + H_2O \ (l) \ \rightarrow C_{10}H_{15}ONH^+ \ (aq) + OH^- \ (aq)$$

(b) The general equation is $K_a \times K_b = K_w$, so $K_a = K_w / K_b$

assuming we are working in water at 25°C (where $K_w = 1.0 \times 10^{-14}$), we know:

$$K_a = 1.0 \times 10^{-14} / 1.4 \times 10^{-4} = \mathbf{7.1 \times 10^{-11}}$$

(c) Ammonia has $K_b = 1.8 \times 10^{-5}$, which is smaller than the basicity constant of $C_{10}H_{15}ON$. Therefore, $C_{10}H_{15}ON$ is a **stronger** base than ammonia.

8-17. (a) $\qquad\qquad OCl^- \ (aq) + H_2O \ (l) \ \rightarrow ClOH \ (aq) + H_3O^+ \ (aq)$

(b) The reaction is the hydrolysis of the conjugate base to HOCl.

The acid HOCl has $\qquad\qquad K_a = 3.0 \times 10^{-8}$.

Therefore: $\qquad\qquad\qquad K_b = 1.0 \times 10^{-14} / 3.0 \times 10^{-8} = \mathbf{3.3 \times 10^{-7}}$

8-19. (a) The conjugate base of phenol is $\mathbf{C_6H_5O^-}$. Note that the hydrogen ion leaves the oxygen atom in phenol.

(b) The K_a of phenol is relatively small; phenol is a weak acid. Phenolate ion must therefore be a **weak base**.

(c) $\qquad\qquad\qquad NaC_6H_5O \ (s) \ \rightarrow Na^+ \ (aq) + C_6H_5O^- \ (aq)$

$$C_6H_5O^- \ (aq) + H_2O \ (l) \ \rightleftharpoons C_6H_5OH \ (aq) + OH^- \ (aq)$$

8-21. The new equilibrium reaction must be constructed from two other equilibria. Notice that in the reaction written, nitrite ion, $NO_2^- \ (aq)$ is acting as a base while chlorous acid is acting as an acid. These facts suggest we combine the acid equilibrum equation for $HClO_2$ and another reaction where NO_2^- reacts with H_3O^+ to give H_2O and HNO_2 -- the *reverse* of the acid ionization equilibrium for nitrous acid.

$HClO_2 \ (aq) + H_2O \ (l) \ \rightleftharpoons ClO_2^- \ (aq) + H_3O^+ \ (aq) \qquad K_a(HClO_2) = 1.1 \times 10^{-2}$

$NO_2^- \ (aq) + H_3O^+ \ (aq) \ \rightleftharpoons HNO_2 \ (aq) + H_2O \ (l) \qquad \dfrac{1}{K_a(HNO_2)} = \dfrac{1}{4.6 \times 10^{-4}}$

$HClO_2 \ (aq) + NO_2^- \ (aq) \ \rightleftharpoons HNO_2 \ (aq) + ClO_2^- \ (aq) \qquad K = \dfrac{1.1 \times 10^{-2}}{4.6 \times 10^{-4}} = \mathbf{24}$

8-23. (a) The color change for methyl orange occurs at lower pH (range: 3.2-4.4) than the change for bromocresol green (range: 3.8-5.4). The acid form of methyl orange must lose a hydrogen ion more easily than the acid form of bromocresol green. Therefore, the acid form of **methyl orange** is the stronger acid.

(b) This observation defines a narrow range: around pH 3.8-4.4.

8-25. Although it is not always necessary to use the quadratic formula to solve for the concentration of the hydronium ion in a solution of a weak acid, it isn't all that hard, either. So let's develop a general formula for an acid with acid ionization constant K_a and initial concentration C_0.

(all concentrations in M)	[HA]	[H$_3$O$^+$]	[A$^-$]
Initial concentration :	C_0	≈ 0.00	0.00
Change in concentration:	$-y$	$+y$	$+y$
Equilibrium concentration:	$C_0 - y$	y	y

The solution for the acid ionization equation in terms of y then becomes:

$$K_a = \frac{y^2}{C_0 - y}; \qquad \text{Rearranging:} \quad 0 = y^2 + K_a y - 4K_a C_0$$

This means that whenever we have pH determined by a pure weak acid we can determine the hydronium ion concentration by using $a = 1$, $b = K_a$, and $c = -K_a C_0$.

For **formic acid**, $K_a = 1.8 \times 10^{-4}$, and we get: $a = 1$, $b = 1.8 \times 10^{-4}$, and $c = -3.6 \times 10^{-5}$. We then get $y = 0.00591$, pH = **2.23** and fraction ionized (which is equal to y / C_0) = **0.030**, or 3.0%.

For **benzoic acid**, $K_a = 6.5 \times 10^{-5}$, and we get: $a = 1$, $b = 6.5 \times 10^{-5}$, and $c = -1.3 \times 10^{-5}$. We then get $y = 0.00357$, pH = **2.45** and fraction ionized **0.018**, or 1.8%.

For **ammonium**, $K_a = 5.6 \times 10^{-10}$, and we get: $a = 1$, $b = 5.6 \times 10^{-10}$, and $c = -1.12 \times 10^{-10}$. We then get $y = 1.06 \times 10^{-5}$, pH = **4.98** and fraction ionized (which is equal to y / C_0) = **0.000053**, or 0.0053%.

(b) We see that as K_a decreases, we get a higher pH. Stronger acids give a lower pH; weaker acids give a higher pH.

(c) We see that as K_a decreases the fraction ionized decreases. Stronger acids ionize more; weaker acids ionize less.

8-27. (a) The chemical amount of the aspirin is 0.65 g $/ 180.16$ g mol^{-1} $= 3.6 \times 10^{-3}$ mol. This is used in preparing a 50 mL (0.050 L) solution, so the initial concentration of aspirin is 3.6×10^{-3} mol $/ 0.050$ L $= 7.2 \times 10^{-2}$ M.

The calculation of a weak acid (or a weak base) equilibrium is best done with the appropriate table of changes:

(all concentrations in M)	[HC$_9$H$_7$O$_4$]	[H$_3$O$^+$]	[C$_9$H$_7$O$_4^-$]
Initial concentration :	7.2×10^{-2}	0.00	0.00
Change in concentration:	$-y$	$+y$	$+y$
Equilibrium concentration:	$7.2 \times 10^{-2} - y$	y	y

Note that y is conveniently set equal to the most important chemical quantity, [H$_3$O$^+$]. The equilibrium concentrations are then put into the acid ionization equilibrium expression for HC$_9$H$_7$O$_4$:

$$K_a = \frac{\left[C_9H_7O_4^-\right]\left[H_3O^+\right]}{\left[HC_9H_7O_4\right]} = \frac{y^2}{7.22 \times 10^{-2} - y} = 3.0 \times 10^{-4}$$

Rearranging: $\quad 0 = y^2 + (3.0 \times 10^{-4})y - 2.166 \times 10^{-5}$

Substitution in the quadratic formula gives $y = 4.5 \times 10^{-3}$ as the only root root that has physical meaning (the negative root implies a negative [H$_3$O$^+$]). So [H$_3$O$^+$] $= 4.5 \times 10^{-3}$ M and **pH = 2.35.** The fraction ionized is $4.5 \times 10^{-3} / 7.2 \times 10^{-2} = 0.0625.$

(b) The larger volume will decrease the concentration by a factor of two, to 3.6×10^{-2}. A similar calculation in this case yields for the quadratic formula $a = 1$, $b = 3.0 \times 10^{-4}$, and $c = -1.08 \times 10^{-5}$. From this we get [H$_3$O$^+$] $= 0.0031$, or pH = **2.50.** The fraction ionized is now $0.0031 / 0.036 = $ **0.086.**

(c) The pH is higher in the more dilute solution, but a higher fraction of the aspirin has ionized. What is most important is that the aspirin can still ionize when it is diluted, unlike a strong acid, which is completely ionized in all solutions.

8-29. The K_a for propionic acid (C_2H_5COOH) is 1.3×10^{-5}. This is relatively small, and the initial concentration is relatively large, so we will try to solve the system without the quadratic formula. The table of changes is still set up as before:

(all concentrations in M)	$[C_2H_5COOH]$	$[H_3O^+]$	$[C_2H_5COO^-]$
Initial concentration :	0.35	0.00	0.00
Change in concentration:	-y	+y	+y
Equilibrium concentration:	0.35 - y	y	y

$$K_a = \frac{y^2}{0.35 - y} \approx \frac{y^2}{0.35} = 1.3 \times 10^{-5} \implies y = \sqrt{(0.35)(1.3 \times 10^{-5})} = 0.00213$$

Assumptions to neglect y in pH calculations like this are justified if y ($= [H_3O^+]$) is less than 5% of the final value of $[H_3O^+]$. Here y is much less than 5% of 0.35. so our assumption to neglect y is justified and $[H_3O^+] = 0.00212658$ M; **pH = 2.67.**

(b) Formic acid is only slightly stronger that propionic acid. We can therefore assume that the solution required to generate a pH = 2.67 will be relatively concentrated. In that case, we can take the general equation for $[H_3O^+]$ ("y" in part a) and solve for $[HCOOH]_0$, the initial acid concentration.

$$K_a \approx \frac{[H_3O^+]^2}{[HCOOH]_0} = 1.8 \times 10^{-4}$$

$$[HCOOH]_0 \approx \frac{(0.00212658)^2}{1.8 \times 10^{-4}} = \textbf{0.027 moles per liter}$$

8-31. The wording of the question indicates that the quadratic formula is required.

(all concentrations in M)	$[HIO_3]$	$[H_3O^+]$	$[IO_3^-]$
Initial concentration :	0.100	0.00	0.00
Change in concentration:	-y	+y	+y
Equilibrium concentration:	0.100 - y	y	y

$$K_a = \frac{[IO_3^-][H_3O^+]}{[HIO_3]} = \frac{y^2}{0.100 - y} = 0.16$$

Rearranging: $\qquad 0 = y^2 + (0.16)y - 0.016$

Substitution in the quadratic formula gives only one positive (and therefore physically reasonable) root: $y = 0.0697 = [H_3O^+]$. Then **pH = 1.16.**

Chapter 8

8-33. The problem gives us the pH (3.31, which means $[H_3O^+] = 4.90 \times 10^{-4}$ M) and intitial acid concentration, 0.205 M. A weak acid at relatively high concentration means we can assume :

$$K_a \approx \frac{[H_3O^+]^2}{[papH^+]_o} = \frac{(4.90 \times 10^{-4})^2}{0.205} = \mathbf{1.2 \times 10^{-6}}$$

8-35. The strategy for handling weak bases is similar to weak acids. The starting concentration of the morphine ("morph") is 0.0400 mol / 0.600 mol = 0.0667 M.

(all concentrations in M)	morph	OH⁻	morphH⁺
Initial concentration :	0.0667	0.00	0.00
Change in concentration:	-y	+y	+y
Equilibrium concentration:	0.0667 - y	y	y

8-35. (continued)

$$K_b = \frac{y^2}{0.0667 - y} \approx \frac{y^2}{0.0667} = 8 \times 10^{-7}$$

$$y = \sqrt{(0.0667)(8 \times 10^{-7})} = 0.000230 \ (<< 0.0667)$$

$[OH^-] = 2.3 \times 10^{-4}$ M; pOH = 3.64

pH = 14.00 - pOH = 10.36

8-37. This problem requires that we recognize that the fractional ionization of nitrous acid will indicate the ratio of $[NO_2^-] / [HNO_2] = 4.5\% / 95.5\% = 0.0471$

Then we will have: $K_a = \dfrac{[H_3O^+][NO_2^-]}{[HNO_2]} = 0.0471 \ [H_3O^+] = 4.6 \times 10^{-4}$

$[H_3O^+] = 4.6 \times 10^{-4} / 0.0471 = \mathbf{0.0098 \ M}$

Since the only significant source of hydronium ion in this solution is by the dissociation of nitrous acid into nitrite and hydronium ion, the concentration of nitrite is *also* 0.0098 M. The actual determination of the nitrous acid concentration is done as in the last part of problem 8-29: $[HNO_2] \approx \dfrac{[H_3O^+][NO_2^-]}{K_a} = \dfrac{(0.0098)^2}{4.6 \times 10^{-4}} = \mathbf{0.21 \ M}$

8-39. (a) **basic** - CN⁻, cyanide, is the conjugate base to HCN, a weak acid.

(b) **acidic** - Br⁻ is the conjugate base to HBr, a strong acid; it will have no effect on the pH. The transition metal ion imparts acidity through hydrolysis.

(c) **neutral** - ClO_4^- is the conjugate base to $HClO_4$, a strong acid; it won't affect the pH. Rb^+, an alkali metal cation, won't hydrolyze to any appreciable extent.

(d) **acidic**- NO_3^- is the conjugate base to HNO_3, a strong acid, so it will have no affect on the pH. The ion Mg^{2+} will, however, hydrolyze and form an acidic solution.

8-41. K can be determined from the respective acid ionization equilibria:

$$NH_4^+ \ (aq) + H_2O \ (l) \rightleftharpoons NH_3 \ (aq) + H_3O^+ \ (aq) \qquad K_a(NH_4^+) = 5.6 \times 10^{-10}$$

$$CN^- \ (aq) + H_3O^+ \ (aq) \rightleftharpoons HCN \ (aq) + H_2O \ (l) \qquad \frac{1}{K_a(HCN)} = \frac{1}{6.12 \times 10^{-10}}$$

$$NH_4^+ \ (aq) + CN^- \ (aq) \rightleftharpoons NH_3 \ (aq) + HCN \ (aq) \quad K = \frac{5.6 \times 10^{-10}}{6.2 \times 10^{-10}} = \mathbf{0.90}$$

8-43. The equation for reaction of equal chemical amounts of acetic acid and hydroxide ion is:

$$CH_3COOH \ (aq) + OH^- \ (aq) \rightarrow H_2O \ (l) + CH_3COO^- \ (aq)$$

The resulting solution will have an appreciable amount of acetate, CH_3COO^-, which is itself a weak base. It will hydrolyze to make a basic solution (pH > 7).

8-45. (a) Formic acid is a weak acid, and therefore a mixture of a substantial amount of formic acid and its conjugate base, the formate ion, generates a buffer solution. The total volume is 500 mL, or 0.500 L, so 0.060 mol of formic acid will give [HCOOH] = 0.120 M and 0.045 mol of formate will give [HCOO⁻] = 0.090 M.

This solution *starts* with these concentrations of formic acid and formate. In a table of changes, we allow the formic acid to dissociate to generate hydronium ion and additional formate ion:

(all concentrations in M)	[HCOOH]	[H_3O^+]	[HCOO⁻]
Initial concentration :	0.120	0.000	0.090
Change in concentration:	$-y$	$+y$	$+y$
Equilibrium concentration:	0.120 - y	y	0.090 + y

With a buffer of a weak acid like formic acid the concentration of formic acid and formate will change very little from their initial values; this means y will be small. In this case:

$$[\text{HCOOH}] \approx 0.120 \text{ M} \qquad\qquad [\text{HCOO}^-] \approx 0.090 \text{ M}$$

8-45. (continued)

Then:
$$K_a = \frac{[\text{HCOO}^-][\text{H}_3\text{O}^+]}{[\text{HCOOH}]} \approx \frac{0.090 \times [\text{H}_3\text{O}^+]}{0.120}$$

$$[\text{H}_3\text{O}^+] = K_a \times 1.333 = 1.8 \times 10^{-4} \times 1.333 = \mathbf{2.4 \times 10^{-4}} \text{ M}$$

$$\mathbf{pH} = \mathbf{3.62}.$$

(b) If we add 0.010 mol of NaOH, it will convert 0.010 mol of the formic acid into formate. Then, we will have 0.050 mol of formic acid and 0.055 mol of formate. By the same procedure,

$$[\text{H}_3\text{O}^+] = K_a \times (50/55) = 1.8 \times 10^{-4} \times 1.1 = \mathbf{1.6 \times 10^{-4}} \text{ M}$$

$$\mathbf{pH} = \mathbf{3.79}.$$

8-47. The 0.25 mol of HCl, a strong acid, will react with the 0.50 mol of tris according to the equation: $\qquad$ HCl *(aq)* + tris *(aq)* $\rightarrow$ Cl$^-$ *(aq)* + trisH$^+$ *(aq)*

The HCl is the limiting reactant; after this reaction the solution will contain 0.25 mol of trisH$^+$ and 0.25 mol of tris. This is a mixture of a weak acid and its conjugate base, so a buffer exists. The concentration of each species in the 2.00 L of total solution will be: [tris] = 0.125 M and [trisH$^+$] = 0.125 M
We are given $pK_b = 5.92$. But $pK_b + pK_a = 14.00$, so pK_a (trisH$^+$) = 8.08.
The pH is found most directly through the Henderson-Hasselbalch equation, noting that the [tris] = [trisH$^+$] in this case:

$$\text{pH} = pK_a - \log_{10}\frac{[\text{trisH}^+]}{[\text{tris}]} = 8.08 - \log_{10}(1) = \mathbf{8.08}.$$

8-49. The optimal buffering species will have a K_a close to the desired [H$_3$O$^+$]. In this case, given pH = 3.82, [H$_3$O$^+$] = 1.5×10^{-4} M. The best acid will be ***m*-chlorobenzoic acid**, with $K_a = 1.04 \times 10^{-4}$.

8-51. **i)** At the start: The acid-base properties of the solution are determined by the initial concentration of H$_3$O$^+$, which is determined by the initial concentration of the HBr.

[H$_3$O$^+$] = [HBr] = 0.1439 M; **pH = 0.8419** (4 s.f.)

ii) The equivalence point will occur when the NaOH added is just sufficient to react with all the HBr. We have an initial amount of HBr from the initial concentration and volume:
moles HBr (initial) = 0.1439 mol L^{-1} $\times$ 0.02638 L = 0.003796 mol HBr.

To reach the equivalence point, therefore, requires 0.003796 mol NaOH. This means the volume of NaOH we need for equivalence is 0.003796 mol / (0.1219 mol L^{-1}) = 0.03114 L.

8-51. (continued)

One mL before this we have added 0.03014 L of NaOH. This contains 0.003674 mol NaOH. The reaction mixture at this point contains excess HBr. Since the reaction of NaOH and HBr occurs in a 1:1 ratio, we can find the moles of excess HBr by subtracting the amount of NaOH added from the initial number of moles of HBr:

mol HBr = 0.003976 mol – 0.003674 mol = 0.0001219 mol HBr.

The concentration of HBr must include the total volume of the solution, which is now (assuming volumes are additive) 56.52 mL or 0.05652 L. This means:

$[H_3O^+]$ = [HBr] = 0.0001219 mol / 0.05652 L = 0.002157 M; **pH = 2.6662** (4 s.f.)

iii) At the equivalence point of a strong acid-strong base titration, there is exact neutrality, since there are no acids or base present. Therefore, **pH = 7.00**.

iv) The volume of NaOH added is 0.03214 L, which contains 0.004040 mol NaOH. The NaOH is now in excess, and the amount remaining in the solution is equal to the mol NaOH added minus the initial number of moles of HBr:

mol NaOH = mol NaOH added – mol HBr at start = 0.0002438 mol NaOH.

The concentration of this NaOH (taking account of the total volume, which is now 0.05952 L) is 0.004096 M. This gives a pOH of 2.3876 or a pH of **11.6124**.

8-53. This is a stong acid / strong base titration problem where the only two calculations are just on either side of the equivalence point of 100.00 mL added titrant.

The initial number of moles of $HClO_4$ is 0.1000 L × 0.1000 M = 1.000×10^{-2} mol

At 99.98 mL NaOH, the moles NaOH is 0.09998 L × 0.1000 M = 9.998×10^{-3} mol. The difference in these two amounts is 2×10^{-6} moles. This is the amount of unreacted $HClO_4$. The concentration of $HClO_4$ is now 1.0×10^{-5} M. The pH is 5.00.

Just after the equivalence point we can calculate that we have added 1.0002×10^{-2} mol NaOH. This means we have an excess of 2.00×10^{-6} mol NaOH, with a concentration of 1.0×10^{-5} M. The pOH is 5.00 and the pH is 9.00.

The addition of the 0.04 mL of titrant causes a change of **4.00** pH units.

8-55. Preliminaries: HN_3 is a weak acid with $K_a = 1.9 \times 10^{-5}$ and $pK_a = 4.72$,
The number of moles of HN_3 is $0.05000 \text{ L} \times 0.1000 \text{ mol L}^{-1} = 0.005000 \text{ mol}$
at the start of the titration.

i) Before any OH^- added, all that is needed is a simple weak acid calculation:

$$K_a = \frac{y^2}{0.1000 - y} \approx \frac{y^2}{0.1000} ; y = [H_3O^+]$$

$$y = \sqrt{K_a \times 0.1000} = \sqrt{1.9 \times 10^{-5} \times 0.1000}$$

$$[H_3O^+] = 1.378 \times 10^{-3} \text{ M}; \qquad \textbf{pH} = \textbf{2.86}$$

ii) The number of moles of added OH^- is
$0.02500 \text{ L} \times 0.1000 \text{ mol L}^{-1} = 0.002500 \text{ mol}.$
This will react with an equal amount of HN_3, and the resulting solution will contain
$0.002500 \text{ mol } HN_3$ and $0.002500 \text{ mol } N_3^-$, creating a buffer.
A buffer with an equal amount of a weak acid and its conjugate weak base will have (see
problem No. 45) $pH = pK_a$ and here, **pH = 4.72.**

iii) The number of moles of added OH^- is:

$$0.05000 \text{ L} \times 0.1000 \text{ mol L}^{-1} = 0.005000 \text{ mol}$$

This is equal to the number of moles of HN_3 in the initial solution. Hence, this is the
equivalence point. The HN_3 is all used up, but the product N_3^- is a weak base, and will
hydrolyze somewhat. In this experiment:

$$[N_3^-]_0 = \frac{0.005000 \text{ mol}}{0.10000 \text{ L}} = 0.05000 \text{ M}$$

$$K_b = K_w / K_a = 1.00 \times 10^{-14} / 1.9 \times 10^{-5} = 5.263 \times 10^{-10}.$$

$$K_b = \frac{[HN_3][OH^-]}{[N_3^-]} = \frac{y^2}{[N_3^-]_0 - y} \approx \frac{y^2}{[N_3^-]_0}; \text{ where } y = [OH^-]$$

$$y = \sqrt{K_b \times [N_{3^-}]_0} = \sqrt{5.263 \times 10^{-10} \times 0.05000} = 5.129 \times 10^{-6}$$

$$[OH^-] = 5.129 \times 10^{-6} \text{ M}; \quad pOH = 5.29; \quad \textbf{pH = 8.71.}$$

iv) The total number of moles of hydroxide ion added is

$$0.05100 \text{ L} \times 0.1000 \text{ mol L}^{-1} = 0.005100 \text{ mol}$$

This is a slight excess over the number of moles of HN_3 initially present. The excess
OH^- is $0.005100 - 0.005000 = 1.00 \times 10^{-4}$ mol. This will lead to total OH^- concentration
of: $[OH^-] = 1.00 \times 10^{-4} / 0.10100 \text{ L} = 9.90 \times 10^{-4} \text{ M}.$
This is a much larger $[OH^-]$ than the N_3^- that is present could give, so we can neglect the
hydrolysis of N_3^-. The pOH will be 3.00 and **pH = 11.00.**

8-57. This recalls problem 8-53 above, except it is now a *weak* acid / strong base titration
problem where the only two calculations are just on either side of the equivalence point
of 100.00 mL added titrant. We only care of the point *after* the addition of the acid.
The initial number of moles of HF is $0.1000 \text{ L} \times 0.1000 \text{ M} = 1.000 \times 10^{-2}$ mol
At 99.98 mL NaOH, the moles NaOH is $0.09998 \text{ L} \times 0.1000 \text{ M} = 9.998 \times 10^{-3}$ mol.
The difference in these two amounts is 2×10^{-6} moles. This is the amount of unreacted
HF. The total concentration of HF is now 1.0×10^{-5} M. We use this in the calculation of
the hydronium ion concentration. The quadratic equation is needed here, and we get
$[H_3O^+] = 9.85 \times 10^{-6}$ M and pH = 5.01.

Just after the equivalence point we can calculate that we have added 1.0002×10^{-2} mol
NaOH. This means we have an excess of 2.00×10^{-6} mol NaOH, with a concentration of
1.0×10^{-5} M. The pOH is 5.00 and the pH is 9.00.
The addition of the 0.04 mL of titrant causes a change of **3.99 pH units.**

8-59. The equivalence point is reached when the number of moles of added HCl
($= 0.0750 \text{ mol L}^{-1} \times 0.01590 \text{ L} = 0.0011925$ mol) is equal to the number of moles of
codeine. Thus, there were 0.0011925 mol of codeine in the initial solution. Its mass was
$0.0011925 \text{ mol} \times 299.37 \text{ g mol}^{-1} = \textbf{0.357 g.}$
At the equivalence point, all of the codeine has been converted into its conjugate acid,
$C_{18}H_{21}O_3NH^+$. The pH calculation required is for a weak acid solution.
Note: The total volume of the solution is 115.90 mL (=0.11590 L).

$$K_a(C_{18}H_{21}O_3NH^+) = \quad K_w / K_b = 1.00 \times 10^{-14} / 9.0 \times 10^{-7} = 1.11 \times 10^{-8}$$

$$K_b = \frac{[C_{18}H_{21}O_3N][H_3O^+]}{[C_{18}H_{21}O_3NH^+]} = \frac{y^2}{[C_{18}H_{21}O_3NH^+]_o - y} \approx \frac{y^2}{[C_{18}H_{21}O_3NH^+]_o} ; y = [H_3O^+]$$

$$y = \sqrt{K_a \times [C_{18}H_{21}O_3NH^+]_0} = \sqrt{1.11 \times 10^{-8} \times 0.010289}$$
$$= 1.069 \times 10^{-5} \implies [H_3O^+] = 1.069 \times 10^{-5} \text{ M}; \quad \textbf{pH = 4.97}$$

The ideal indicator changes color near this pH. It would be **bromocresol green**.

8-61. The amount of hydroxide ion added at equivalence is:
$$0.02328 \text{ L} \times 0.2000 \text{ mol L}^{-1} = 0.004656 \text{ mol.}$$
This is the number of moles of cacodylic acid that was present in the initial solution, so
the initial concentration was: $0.004656 \text{ mol} / 0.04000 \text{ L} = \textbf{0.1164 M}$
At the half-equivalence point, half of the cacodylic acid had been converted into its
conjugate base, cacodylate ion. Under these (buffering) conditions:
$$\text{pH} = pK_a = 6.19; \quad K_a = 10^{-6.19} = \textbf{6.5} \times \textbf{10}^{-7}$$

8-63. (a) The pH drops at the equivalence point from above pH = 10 to below pH = 4. This must be the titration of a base by added acid. Such a drop in pH, moreover, could only happen with the titration of **RbOH by HBr.** If NH_3 were present, we would expect a buffer to form around the pK_a of NH_4^+ -- 9.25.

(b) The transition in pH occurs across the pH range 10 to 4. Many indicators would work, including phenophthalein (transition in the range pH = 10 to 8.2), cresol red (pH = 8.8 to 7.0), bromothymol blue (pH = 7.6 to 6.0) and methyl red (6.0 to 4.8).

(c) The transition occurs after an equivalent amount of acid has been added to neutralize the base. The amount of acid added was 0.030 L × 0.065 M = 0.00195 mol acid. This is equal to the mol of NaOH present, so we get ca. **0.020 M** for the concentration of NaOH.

8-65. The data from Table 8-2 are:

$$H_3AsO_4 \ (aq) + H_2O \ (l) \rightleftharpoons H_2AsO_4^- \ (aq) + H_3O^+ \ (aq) \qquad K_{a1} = 5.0 \times 10^{-3}$$
$$H_2AsO_4^- \ (aq) + H_2O \ (l) \rightleftharpoons HAsO_4^{2-} \ (aq) + H_3O^+ \ (aq) \qquad K_{a2} = 9.3 \times 10^{-8}$$
$$HAsO_4^{2-} \ (aq) + H_2O \ (l) \rightleftharpoons AsO_4^{3-} \ (aq) + H_3O^+ \ (aq) \qquad K_{a3} = 3.4 \times 10^{-12}$$

Note that arsenic acid is a weak acid and dihydrogen arsenate ion and hydrogen arsenate ion are very weak acids. Assume that only the first ionization contributes significantly to $[H_3O^+]$, and that $[H_2AsO_4^-] = [H_3O^+]$. Then we calculate the concentration of hydrogen arsenate and arsenate from these values.

(all concentrations in M)	$[H_3AsO_4]$	$[H_3O^+]$	$[H_2AsO_4^-]$
Initial concentration :	0.100	0.000	0.000
Change in concentration:	$-y$	$+y$	$+y$
Equilibrium concentration:	$0.120 - y$	y	y

$$K_a = \frac{[H_2AsO_4^-][H_3O^+]}{[H_3AsO_4]} = \frac{y^2}{0.100 - y} = 5.0 \times 10^{-3}$$

Rearranging: $\qquad 0 = y^2 + (5.0 \times 10^{-3})y - 5.0 \times 10^{-4}$

Substitution in the quadratic formula gives:

$$y = \frac{-5.0 \times 10^{-3} \pm \sqrt{\left(5.0 \times 10^{-3}\right)^2 - 4(1)\left(-5.0 \times 10^{-4}\right)}}{2(1)} = 0.020 \text{ and } -0.025$$

8-65. (continued)

Only the positive root has physical meaning. So:

$[H_3O^+] = 0.020$ M, $\qquad$ $[H_2AsO_4^-] = 0.020$ M, $\qquad$ $[H_3AsO_4] = 0.080$ M

We know from the small value for K_{a2} that very little of the $H_2AsO_4^-$ ionizes, so that the value of $[H_3O^+]$ and $[H_2AsO_4^-]$ are set by the first ionization.

Then: $\qquad K_a = \dfrac{\left[HAsO_4^{2-}\right]\left[H_3O^+\right]}{\left[H_2AsO_4^-\right]} = \dfrac{\left[HAsO_4^-\right] \times 0.020}{0.020} = 9.3 \times 10^{-8}$

$[HAsO_4^{2-}] = 9.3 \times 10^{-8}$ M.

For the third ionization, we proceed as for the second:

$$K_a = \dfrac{\left[AsO_4^{3-}\right]\left[H_3O^+\right]}{\left[HAsO_4^{2-}\right]} = \dfrac{\left[AsO_4^{3-}\right] \times 0.020}{9.3 \times 10^{-8}} = 3.0 \times 10^{-12}$$

$[AsO_4^{3-}] = 1.40 \times 10^{-17}$ M.

8-67. The initial solution is prepared with a base, the phosphate ion. We are much better off working with the K_b's that go with the three species:

PO_4^{3-} *(aq)* + H_2O *(l)* $\rightleftharpoons$ HPO_4^{2-} *(aq)* + OH^- *(aq)* $\qquad K_{b1} = 4.5 \times 10^{-2}$

HPO_4^{2-} *(aq)* + H_2O *(l)* $\rightleftharpoons$ $H_2PO_4^-$ *(aq)* + OH^- *(aq)* $\qquad K_{b2} = 1.605 \times 10^{-7}$

$H_2PO_4^-$ *(aq)* + H_2O *(l)* $\rightleftharpoons$ H_3PO_4 *(aq)* + OH^- *(aq)* $\qquad K_{b3} = 1.33 \times 10^{-12}$

For the calculation, we assume that only the first reaction contributes significantly to the acid-base properties of the solution. Phosphate has such a large K_{b1} that we must determine the concentration of hydroxide using the quadratic formula:

$$K_{b1} = \dfrac{\left[HPO_4^{2-}\right]\left[OH^-\right]}{\left[PO_4^{3-}\right]} = \dfrac{y^2}{0.050 - y} = 0.045$$

$$0 = 0.00225 - 0.045\, y - y^2$$

The only reasonable root is $y = 0.030$ M, so $[OH^-] = 0.030$ M, $[HPO_4^{2-}] = 0.030$ M, and $[PO_4^{3-}] = 0.020$ M.

For the second equilibrium reaction, we use $[OH^-]$ and $[HPO_4^{2-}]$ from the first::

$$K_{b2} = \dfrac{\left[H_2PO_4^-\right]\left[OH^-\right]}{\left[HPO_4^{2-}\right]} = \dfrac{\left[H_2PO_4^-\right](0.030)}{0.030} = 1.61 \times 10^{-7}$$

$[H_2PO_4^-] = 1.61 \times 10^{-7}$

8-67. (continued)

For the third equilibrium reaction:

$$K_{b3} = \frac{[H_3PO_4][OH^-]}{[H_2PO_4^-]} = \frac{[H_3PO_4](0.030)}{1.605 \times 10^{-7}} = 1.33 \times 10^{-12}$$

$$[H_3PO_4] = \mathbf{7.1 \times 10^{-18}}$$

8-69. The layout of this problem is identical to that for example 8-15; only the pH is different and here we are told the total concentration of carbonates. The procedure is again to determine the *fraction* of each form of the carbonate, and then to multiply by the concentration of total carbonate. At pH = 5.6, $[H_3O^+] = 2.5 \times 10^{-6}$ and The ratios of the various forms are determined by $[H_3O^+]$.

$$\frac{[HCO_3^-]}{[H_2CO_3]} = \frac{K_{a1}}{[H_3O^+]} = \frac{4.3 \times 10^{-7}}{2.5 \times 10^{-6}} = 0.172$$

$$\frac{[CO_3^{2-}]}{[HCO_3^-]} = \frac{K_{a2}}{[H_3O^+]} = \frac{4.8 \times 10^{-11}}{2.5 \times 10^{-6}} = 1.92 \times 10^{-5}$$

$$\text{fraction } HCO_3^- = \frac{[HCO_3^-]}{[H_2CO_3] + [HCO_3^-] + [CO_3^{2-}]}$$

This can be solved by dividing through by $[HCO_3^-]$

$$\text{fraction } HCO_3^- = \frac{1}{\dfrac{[H_2CO_3]}{[HCO_3^-]} + 1 + \dfrac{[CO_3^{2-}]}{[HCO_3^-]}} = \frac{1}{\dfrac{1}{.172} + 1 + (1.92 \times 10^{-5})} = 0.14676$$

$$\text{fraction } H_2CO_3 = \frac{\dfrac{[H_2CO_3]}{[HCO_3^-]}}{\dfrac{[H_2CO_3]}{[HCO_3^-]} + 1 + \dfrac{[CO_3^{2-}]}{[HCO_3^-]}} = \frac{\dfrac{1}{.172}}{\dfrac{1}{.172} + 1 + (1.92 \times 10^{-5})} = 0.85324$$

$[H_2CO_3]$ = total carbonate × fraction H_2CO_3

$\qquad = 1.0 \times 10^{-5}$ M × $0.85324 = \mathbf{8.5 \times 10^{-6}}$ **M**

$[HCO_3^-]$ = total carbonate × fraction HCO_3^-

$\qquad = 1.0 \times 10^{-5}$ M × $0.14676 = \mathbf{1.5 \times 10^{-6}}$ **M**

The value for $[CO_3^{2-}]$ will be very small. It is best to determine it from the ratio of hydrogen carbonate to carbonate we calculated above.

8-69. (continued)

$$[CO_3^{2-}] = [HCO_3^-] \times 1.92 \times 10^{-5}$$

$$= 1.47 \times 10^{-6} \times 1.92 \times 10^{-5} = \mathbf{2.8 \times 10^{-11}}\ \mathbf{M.}$$

8-71.

	Lewis Acid	Lewis Base
(a)	Ag^+ *(aq)*	NH_3 *(aq)*
(b)	BF_3 *(g)*	$N(CH_3)_3$ *(g)*
(c)	$B(OH)_3$ *(aq)*	OH^- *(aq)*

8-73. (a) $\qquad CaO\ (s) + H_2O\ (l) \rightarrow Ca(OH)_2\ (s)$

(b) Yes; the calcium oxide acts as a Lewis base, donating electrons to a hydrogen ion on a water molecule, which acts as a Lewis acid.

8-75. (a) The fluoride ion has a closed shell, octet configuration. It can only act as an electron pair donor, so by extension any compound that donates fluoride is acting as an electron pair donor. Therefore, the **base** is the fluoride **donor** and the **acid** is the fluoride **donor**.

(b) Acid: BF_3 $\qquad$ Base: ClF_3O_2

$\quad$ Acid: TiF_4 $\qquad$ Base: KF.

Note that the acids already have some fluorides attached; but they still act as an acid in *accepting* an additional fluoride.

8-77. This problem cannot be solved by averaging the pH values for the two solutions because the pH is a logarithmic function. One can average the hydroxide ion concentrations, but only because equal volumes of the solutions are used. The general way to work such mixing experiments is to determine the total number of moles of OH^-, then the final concentration:

$\quad$ Solution #1: $1.00\ L \times 1.00 \times 10^{-5}\ mol\ L^{-1} = 1.00 \times 10^{-5}\ mol$

$\quad$ Solution #2: $1.00\ L \times 1.00 \times 10^{-6}\ mol\ L^{-1} = 1.00 \times 10^{-6}\ mol$

$\quad$ Total OH^-: $1.10 \times 10^{-5}\ mol$ $\qquad$ Final volume: $2.00\ L$

$\quad$ Final $[OH^-]$: $5.50 \times 10^{-6}\ M$ $\quad$ Final pOH: 5.26 $\quad$ **Final pH: 8.74**

8-79. Hydronium ion, H_3O^+, is present on the same side of the reaction as Cl^-. So increasing $[H_3O^+]$ will drive the reaction away from Cl^- and toward Cl_2. The conditions should be acidic to ensure experimental success.

8-81. As was done at 25 °C, $[H_3O^+] = \sqrt{K_w\left(60\,^\circ C\right)} = \sqrt{9.61 \times 10^{-14}} = 3.1 \times 10^{-7}$

8-83. (a) This problem requires the same methodology as No. 39.

$$H_2SO_3 \ (aq) + H_2O \ (l) \rightleftharpoons HSO_3^- \ (aq) + H_3O^+ \ (aq) \qquad K_a(H_2SO_3) = 1.5 \times 10^{-2}$$

$$CH_3COO^- \ (aq) + H_3O^+ \ (aq) \rightleftharpoons CH_3COOH \ (aq) + H_2O \ (l) \qquad \frac{1}{K_a(CH_3COOH)} = 5.38 \times 10^4$$

$$H_2SO_3(aq) + CH_3COO^-(aq) \rightleftharpoons HSO_3^- \ (aq) + CH_3COOH(aq) \quad K = \frac{1.5 \times 10^{-2}}{1.86 \times 10^{-5}} = 806$$

The reaction written is **spontaneous**.

(b) A more general approach to all of these is to note that there are two reactions -- one an acid ionization reaction of acid 1, the other the opposite of an acid ionization reaction, or acid 2. Therefore, the equilibrium constant for the combined reaction will be :

$$K = K_a(1) \frac{1}{K_a(2)} = \frac{K_a(1)}{K_a(2)}$$

For this part: $K = \dfrac{K_a(HF)}{K_a(CH_2ClCOOH)} = \dfrac{6.6 \times 10^{-4}}{1.4 \times 10^{-3}} = 0.47$ – **not spontaneous**

(c) $\qquad K = \dfrac{K_a(HCN)}{K_a(HNO_2)} = \dfrac{6.2 \times 10^{-10}}{4.6 \times 10^{-4}} = 1.3 \times 10^{-6}$ – **not spontaneous.**

8-85. The K_a of urea acidium ion, $H_2NCONH_3^+$, is given as 0.66 in Table 8-2. This is *not* an acid that only ionizes to a small degree!

(all concentrations in M)	$H_2NCONH_3^+$	H_3O^+	H_2NCONH_2
Initial concentration :	0.15	0.00	0.00
Change in concentration:	-y	+y	+y
Equilibrium concentration:	0.15 - y	y	y

$$K_a = \frac{[H_2NCONH_2][H_3O^+]}{[H_2NCONH_3^+]} = \frac{y^2}{0.15 - y} = 0.66 \implies 0 = y^2 + 0.66y - 0.099$$

The only reasonable root obtained from the quadratic formula is $y = 0.125$, so **[urea] = 0.126 M**, which means $[H_3O^+] = 0.126$ M and **pH = 0.90**.

8-87. (a) Yes, this procedure works. Equal concentrations of CH_3COOH and CH_3COO^- exist in the solution, and with $pKa = 4.74$, this solution buffer solution will have a pH of 4.74.

(b) Yes, this procedure works. The sodium hydroxide is present in one-half the chemical amount of acetic acid, so one-half of the acetic acid will be converted into acetate. Again equal concentrations of CH_3COOH and CH_3COO^- exist and pH = 4.74.

(c) Yes, this procedure works. The strong base will react with some of the acetic acid to generate an acetic acid/acetate buffer. We know that there must be a substantial amount of both acetic acid and acetate if the pH will be 4.74.

(d) No, this procedure does not work. Hydrochloric acid is a strong acid. It cannot form a buffer at pH = 4.74.

(e) Yes, this procedure works. The excess reagent here is acetate; it will react with all the HCl. The final solution will have equal concentrations of CH_3COOH and CH_3COO^- and pH = 4.74.

8-89. The initial concentrations of the substances will be:

$$[C_6F_5COOH]: \quad 0.025 \text{ M} \qquad [C_6F_5COO^-]: \quad 0.030 \text{ M}$$

In this buffer, we cannot assume that only a negligible amount of the acid will ionize. Thus, methods that work for much weaker acids, such as the Henderson-Hasselbalch equation, cannot be used. We must calculate the ionization explicitly.

(all concentrations in M)	$[C_6F_5COOH]$	$[H_3O^+]$	$[C_6F_5COO^-]$
Initial concentration :	0.025	0.000	0.030
Change in concentration:	$-y$	$+y$	$+y$
Equilibrium concentration:	$0.025 - y$	y	$0.030 + y$

Then, in this buffer: $\quad Ka = \dfrac{\left[H_3O^+\right]\left[C_6F_5COO^-\right]}{[C_6F_5COOH]} = \dfrac{y(0.030 + y)}{0.025 - y} = 0.033$

$$0 = y^2 + 0.063\,y - 0.000825$$

The only reasonable root is $y = 0.011$, so $[H_3O^+] = 0.011$ M and **pH = 1.95.**

8-91. In this case we rearrange the K_a expression to get the ratio of $[In^-] / [HIn]$. We note that the pH of 6.50 is the same as $[H_3O^+] = 3.2 \times 10^{-7}$. Then:

$$K_a = \frac{[In^-][H_3O^+]}{[HIn]} \Rightarrow \frac{[In^-]}{[HIn]} = \frac{K_a}{[H_3O^+]} = \frac{9.3 \times 10^{-6}}{3.2 \times 10^{-7}} = \mathbf{29}$$

8-93.
$$\text{Moles of OH}^- = 0.2023 \text{ L} \times 0.451 \text{ mol L}^{-1} = 0.09125 \text{ mol}$$
$$\text{Moles of H}_2\text{SO}_4 = 0.1594 \text{ L} \times 0.251 \text{ mol L}^{-1} = 0.04001 \text{ mol}$$

Since H_2SO_4 is diprotic, it will react with $2 \times 0.4001 \text{ mol} = 0.08001 \text{ mol OH}^-$ and $0.09125 - 0.08001 = 0.01123 \text{ mol}$ of OH^- will remain in the total volume, 0.362 L. Therefore, $[OH^-] = 0.01123/0.362 = 0.031$ M and pOH = 1.51; **pH = 12.49.**

8-95. The experiment involves addition of HCl to a solution of benzoate, to give a solution of unreacted (excess) HCl + benzoic acid. Upon addition of the hydroxide, the *HCl will react first*, and the benzoic acid will react second. Thus, the first equivalence point indicates the complete consumption of the unreacted HCl. At this point, the amount of NaOH added was $0.04650 \text{ L} \times 0.393 \text{ mol L}^{-1} = 0.018275 \text{ mol OH}^-$, so the amount of excess HCl added was 0.018275 mol.

The second equivalence point is reached when all of the benzoic acid was used up. The volume of NaOH used to neutralized the benzoic acid was
$(0.06391 \text{ L} - 0.04650 \text{ L}) = 0.01711 \text{ L}$. The number of moles of NaOH in this volume was $0.01711 \text{ L} \times 0.393 \text{ mol L}^{-1} = 0.006725 \text{ mol OH}^-$. This is equal to the number of moles of benzoic acid, and to the number of moles of benzoate in the original sample. Therefore, the mass of the benzoate in the original sample is
$0.006725 \text{ mol} \times 144.105 \text{ g mol}^{-1}$ = **0.969g sodium benzoate.**

We can check the titration results by noting that the number of moles of benzoic acid and excess HCl must equal the number of moles of HCl added to the salt mixture. In this case, that is:

$$\text{moles HCl} + \text{moles benzoic acid} = \text{moles HCl}_{added}$$
$$0.018275 \text{ mol} + 0.006725 \text{ mol} = 0.500 \text{ mol L}^{-1} \times 0.0500 \text{ L}$$
$$0.025000 \text{ mol} = 0.02500$$

8-97. (a)
$$HCO_3^- \ (aq) + HC_3H_5O_3 \ (aq) \rightarrow H_2CO_3 \ (aq) + C_3H_5O_3^- \ (aq)$$
$$H_2CO_3 \ (aq) \rightarrow CO_2 \ (g) + H_2O \ (l)$$
Overall: $HCO_3^- \ (aq) + HC_3H_5O_3 \ (aq) \rightarrow CO_2 \ (g) + H_2O \ (l) + C_3H_5O_3^- \ (aq)$

(b) $236.6 \text{ mL} / 48 \text{ tsp} = 4.93 \text{ mL} / 1 \text{ tsp} = 4.93 \text{ cm}^3 / 1 \text{ tsp}$

$$\text{g baking soda} = 0.5 \text{ tsp} \times \left(\frac{4.93 \text{ cm}^3}{1 \text{ tsp}}\right) \times \left(\frac{2.16 \text{ g}}{\text{cm}^3}\right) = 5.325 \text{ g}$$

$$\text{mol NaHCO}_3 = \text{mol HC}_3\text{H}_5\text{O}_3 = \frac{5.325 \text{ g}}{84.01 \text{ g mol}^{-1}} = 0.0634 \text{ mol}$$

volume of milk = 1 cup = 0.2366 L

8-97. (continued)

$$[HC_3H_5O_3] = 0.0634 \text{ mol} / 0.2366 \text{ L} = \mathbf{0.27 \text{ M}}$$

(c) We calculated that 0.5 tsp of baking soda contains 0.0634 mol of $NaHCO_3$. This will yield an equivalent number of moles of CO_2, and at $177\ ^\circ C = 450$ K.

The CO_2 will occupy a volume of:

$$V = \frac{nRT}{P} = \frac{(0.0634 \text{ mol}) \times (0.08206 \frac{\text{L - atm}}{\text{mol - K}}) \times (450 \text{ K})}{1 \text{ atm}}$$

$$= \mathbf{2.34 \text{ L}}$$

8-99. For the sake of compactness, phosphonocarboxylic acid will be abbreviated "H_3PC". The three ionization equilbria are:

$$H_3PC\ (aq) + H_2O\ (l) \rightleftharpoons H_2PC^-\ (aq) + H_3O^+ \qquad K_{a1} = \frac{\left[H_3O^+\right]\left[H_2PC^-\right]}{[H_3PC]}$$

$$H_2PC^-\ (aq) + H_2O\ (l) \rightleftharpoons HPC^{2-}\ (aq) + H_3O^+ \qquad K_{a2} = \frac{\left[H_3O^+\right]\left[HPC^{2-}\right]}{\left[H_2PC^-\right]}$$

$$HPC^{2-}\ (aq) + H_2O\ (l) \rightleftharpoons PC^{3-}\ (aq) + H_3O^+ \qquad K_{a3} = \frac{\left[H_3O^+\right]\left[PC^{3-}\right]}{\left[HPC^{2-}\right]}$$

The ratios of the various species is set by the pH of the blood, 7.40 ($[H_3O^+] = 3.981 \times 10^{-8}$ M). This will not vary.

$$\frac{\left[H_2PC^-\right]}{[H_3PC]} = \frac{K_{a1}}{\left[H_3O^+\right]} = \frac{0.01}{3.981 \times 10^{-8}} = 2.512 \times 10^5$$

$$\frac{\left[HPC^{2-}\right]}{\left[H_2PC^-\right]} = \frac{K_{a2}}{\left[H_3O^+\right]} = \frac{7.8 \times 10^{-6}}{3.981 \times 10^{-8}} = 195.9$$

$$\frac{\left[PC^{3-}\right]}{\left[HPC^{2-}\right]} = \frac{K_{a3}}{\left[H_3O^+\right]} = \frac{2.0 \times 10^{-9}}{3.981 \times 10^{-8}} = 0.0502$$

These will all contribute to the fractional amount of each species. For the calculations, though, it helps to have all species as a ratio to just one of them - here chosen to be $[H_3PC]$:

8-99.
$$\frac{\left[HPC^{2-}\right]}{\left[H_3PC\right]} = \frac{\left[HPC^{2-}\right]}{\left[H_2PC^-\right]} \times \frac{\left[H_2PC^-\right]}{\left[H_3PC\right]} = 195.9 \times 2.513 \times 10^5 = 4.921 \times 10^7$$

$$\frac{\left[PC^{3-}\right]}{\left[H_3PC\right]} = \frac{\left[PC^{3-}\right]}{\left[HPC^{2-}\right]} \times \frac{\left[HPC^{2-}\right]}{\left[H_3PC\right]} = 0.0502 \times 4.921 \times 10^7 = 2.47 \times 10^6$$

Before determining the actual concentrations , we should determine the fractional amounts of each of the species.

$$\text{fraction } H_3PC = \frac{[H_3PC]}{[H_3PC] + \left[H_2PC^-\right] + \left[HPC^{2-}\right] + \left[PC^{3-}\right]};$$

Divide top and bottom by [H₃PC]:

$$= \frac{1}{1 + \frac{\left[H_2PC^-\right]}{[H_3PC]} + \frac{\left[HPC^{2-}\right]}{[H_3PC]} + \frac{\left[PC^{3-}\right]}{[H_3PC]}}$$

$$= \frac{1}{1 + 2.512 \times 10^5 + 4.921 \times 10^7 \times 2.47 \times 10^6} = 1.93 \times 10^{-8}$$

$$\text{fraction } H_2PC^- = \frac{\left[H_2PC^-\right]}{[H_3PC] + \left[H_2PC^-\right] + \left[HPC^{2-}\right] + \left[PC^{3-}\right]};$$

Divide top and bottom by [H₃PC]:

$$= \frac{\frac{\left[H_2PC^-\right]}{[H_3PC]}}{1 + \frac{\left[H_2PC^-\right]}{[H_3PC]} + \frac{\left[HPC^{2-}\right]}{[H_3PC]} + \frac{\left[PC^{3-}\right]}{[H_3PC]}}$$

$$\text{fraction } H_2PC^- = \frac{2.513 \times 10^5}{1 + 2.512 \times 10^5 + 4.921 \times 10^7 \times 2.47 \times 10^6} = 4.84 \times 10^{-3}$$

$$\text{fraction } HPC^{2-} = \frac{\left[HPC^{2-}\right]}{[H_3PC] + \left[H_2PC^-\right] + \left[HPC^{2-}\right] + \left[PC^{3-}\right]};$$

8-99. Divide top and bottom by $[H_3PC]$:

$$= \frac{\dfrac{\left[HPC^{2-}\right]}{[H_3PC]}}{1+\dfrac{\left[H_2PC^-\right]}{[H_3PC]}+\dfrac{\left[HPC^{2-}\right]}{[H_3PC]}+\dfrac{\left[PC^{3-}\right]}{[H_3PC]}}$$

$$= \frac{4.923 \times 10^7}{1+2.512 \times 10^5 + 4.921 \times 10^7 \times 2.47 \times 10^6} = 0.9476$$

The fractional amount of PC^{3-} is obtained by difference:

fraction PC^{3-} = 1.00 - 1.92 $\times$ 10^{-8} - 4.84 $\times$ 10^{-3} - 0.9476 = 0.04756

$[H_3PC] = (1.93 \times 10^{-8})(1.0 \times 10^{-5} \text{ M}) = 1.9 \times 10^{-13} \text{ M}$

$[H_2PC^-] = 4.8 \times 10^{-8} \text{ M}; [HP^{2-}] = 9.5 \times 10^{-6} \text{ M}; [P^{3-}] = 4.7 \times 10^{-7} \text{ M}$

8-101. (a) Pure water has a pH of 7 at 25°C, so if we add an acid to pure water, the pH *must* decrease to a value *below* pH = 7.

(b) Since all of the nitric acid dissociates, we know that $[NO_3^-] = 1.0 \times 10^{-8}$. If we set $[H_3O^+] = y$, then we can use the charge balance equation to get $[OH^-]$ in terms of nitrate and hydronium: $[OH^-] = [H_3O^+] - [NO_3^-] = y - 1.0 \times 10^{-8}$. These values are then placed in the K_w equation:

$$K_w = [H_3O^+][OH^-] = y\,(y - 1.0 \times 10^{-8})$$

Rearranging: $\qquad 0 = y^2 - 1.0 \times 10^{-8} - K_w$

This is solved for a single postive root using the quadratic formula, with $a = 1$, $b = -1.0 \times 10^{-8}$, and $c = -K_w = -1.0 \times 10^{-14}$. $y = 1.05 \times 10^{-7}$, and **pH = 6.98.**

CHAPTER 9 - DISSOLUTION AND PRECIPITATION EQUILIBRIA

9-1. The mass of one mole of $CuSeO_4$ is 206.50 g and the mass of five moles of H_2O is 90.08 g. Therefore, the fraction of mass due to $CuSeO_4 \bullet 5H_2O$ is:

$$\frac{206.50 \text{ g } CuSeO_4}{(206.50 + 90.08) \text{ g } CuSeO_4 \bullet 5H_2O} = 0.6963$$

We assume that above $150°$ C all of the water will be driven off, leaving only the mass of the copper (II) selenate that was in the initial sample:

$$64.8 \text{ g x } 0.6963 = \textbf{45.1 g solid}$$

9-3. The mass of the water of cystallization in this sample is 38.4 - 21.2 = 17.2 g H_2O. This corresponds to $\dfrac{17.2 \text{ g}}{18.0153 \text{ g mol}^{-1}} = 0.9547 \text{ mol } H_2O$.

This was combined with 21.2 g $CoSO_4$, which corresponds to

$$\frac{21.2 \text{ g}}{155.00 \text{ g mol}^{-1}} = 0.1368 \text{ mol } CoSO_4.$$

Thus the molar ratio of water to cobalt is $\dfrac{0.9547 \text{ mol}}{0.1368 \text{ mol}} = 6.98 \approx 7.00 \dfrac{\text{mol } H_2O}{\text{mol } CoSO_4}$

The formula for bieberite is $\textbf{CoSO}_4\textbf{•7H}_2\textbf{O}$.

9-5. The temperature where 80 g of KBr will dissolve is at about **47-48 $°$C**.

9-7. Endothermic reactions proceed more to the right with increased temperature. Thus, the solubility of NH_4Cl will **increase**.

9-9. (a) Molarity $SrBr_2 = \dfrac{0.156 \text{ mol } SrBr_2}{4.15 \text{ L solution}} = \textbf{0.0376 M SrBr}_2$

(b) $\left(0.0376 \dfrac{\text{mol } SrBr_2}{\text{L}} \right) \times \left(\dfrac{2 \text{ mol } Br^-}{1 \text{ mol } SrBr_2} \right) = \textbf{0.0752 M Br}^-$

9-11. (a) $SrI_2 \text{ (s)} \rightleftarrows Sr^{2+} \text{ (aq)} + 2 \text{ I}^- \text{ (aq)}$ $K_{sp} = [Sr^{2+}] \, [I^-]^2$

(b) $I_2 \text{ (s)} \rightleftarrows I_2 \text{ (aq)}$ $K_{sp} = [I_2]$

(c) $Ca_3(PO_4)_2 \text{ (s)} \rightleftarrows 3 \text{ Ca}^{2+} \text{ (aq)} + 2 \text{ PO}_4^{3-} \text{ (aq)}$ $K_{sp} = [Ca^{2+}]^3 \, [PO_4^{3-}]^2$

(d) $BaCrO_4 \text{ (s)} \rightleftarrows Ba^{2+} \text{ (aq)} + CrO_4^{2-} \text{ (aq)}$ $K_{sp} = [Ba^{2+}] \, [CrO_4^{2-}]$

9-13. $Fe_2(SO_4)_3 \rightleftarrows 2 \text{ Fe}^{3+} \text{ (aq)} + 3 \text{ SO}_4^{2-} \text{ (aq)}$ $K_{sp} = [Fe^{3+}]^2 \, [SO_4^{2-}]^3$

9-15. For this: $K_{sp} = [Hg_2^{2+}][I^-]^2$; The table of changes is:

(All concentrations in M)	$[Hg_2^{2+}]$	$[I^-]$
Initial Concentration	0.0	0.0
Change in Concentration	$+x$	$+2x$
Equilibrium Concentration	x	$2x$

Here, $[Hg_2^{2+}] = x$ and $[I^-] = 2x$. Therefore:

$$K_{sp} = (x)(2x)^2 = 4x^3$$
$$x = \sqrt[3]{\frac{K_{sp}}{4}} = \sqrt[3]{\frac{1.2 \times 10^{-28}}{4}} = \sqrt[3]{3.0 \times 10^{-29}} = 3.1 \times 10^{-10}$$
$$[Hg_2^{2+}] = \mathbf{3.1 \times 10^{-10}} \text{ M} \qquad [I^-] = \mathbf{6.2 \times 10^{-10}} \text{ M}$$

9-17. It is best to determine solubilities in mol L^{-1}, then convert to the final volume (here 100.0 mL = 0.100 L). For the general case for TlIO$_3$:

(All concentrations in M)	$[Tl^+]$	$[IO_3^-]$
Initial Concentration	0.0	0.0
Change in Concentration	$+x$	$+x$
Equilibrium Concentration	x	x

$$K_{sp} = [Tl^+][IO_3^-] = x^2$$
$$x = \sqrt{K_{sp}} = \sqrt{3.07 \times 10^{-5}} = 1.752 \times 10^{-3}$$

The solubility of TlIO$_3$ is equal to the concentration of Tl$^+$ present at equilibrium, 1.75×10^{-3} M. In 100.0 mL of water, so little TlIO$_3$ will dissolve that we can assume the total solution volume is 100.0 mL (= 0.1000 L). For this solution (note that molar mass of TlIO$_3$ is 379.29 g mol^{-1}):

$$1.752 \times 10^{-3} \frac{\text{mol TlIO}_3}{\text{L}} \times (0.100 \text{ L}) \times \frac{379.29 \text{ g TlIO}_3}{1 \text{ mol TlIO}_3} = \mathbf{0.0665 \text{ g TlIO}_3}$$

9-19.

(a) (All concentrations in M)	$[Pb^{2+}]$	$[Cl^-]$
Initial Concentration	0.0	0.0
Change in Concentration	$+x$	$+2x$
Equilibrium Concentration	x	$2x$

9-19. (continued)

$$K_{sp} = 4x^3 \qquad \Rightarrow \qquad x = \sqrt[3]{K_{sp}/4} = \sqrt[3]{1.6 \times 10^{-5}/4} = 0.0159$$

$$[Pb^{2+}] = 0.016659 \text{ mol L}^{-1} \qquad [Cl^-] = 0.032 \text{ mol L}^{-1}$$

(b) grams Pb $= (0.0159 \text{ mol L}^{-1}) \times (207.2 \text{ g mol}^{-1}) = 3.3 \text{ g L}^{-1}$

If we estimate that one liter of solution weighs 1000 g of water, then this means we have 3.3 g Pb / 1000 g H_2O, or 3300 g Pb / 1,000,000 g H_2O, which is 3300 ppm, sixty times the limit of 50 ppm!

9-21. Another layout used to computer the solubility of salts in **pure water** is to write the relative equilibrium concentrations under the reaction:

$$Ag_2CrO_4 \text{ (s)} \quad \rightleftharpoons \quad 2\ Ag^+ \text{ (aq)} \quad + CrO_4^{2-} \text{ (aq)}$$

at equilibrium: $\qquad\qquad\qquad\qquad\quad 2x \qquad\qquad x$

Given the solubility of 0.027 g in 500 mL of H_2O, the solubility in mol L^{-1}is:

$$\frac{0.027 \text{ g Ag}_2CrO_4}{500 \text{ mL}} \times \frac{500 \text{ mL}}{0.5 \text{ L}} \times \frac{1 \text{ mol Ag}_2CrO_4}{333.73 \text{ g Ag}_2CrO_4} = 1.61 \times 10^{-4} \frac{\text{mol Ag}_2CrO_4}{\text{L}}$$

This value, 1.61×10^{-4} M, is "x" above. Thus:

$$K_{sp} = [Ag^+]^2 [CrO_4^{2-}]$$
$$= (2x)^2 (x) = 4x^3 = 4 (1.61 \times 10^{-4})^3 = \mathbf{1.7 \times 10^{-11}}$$

9-23. The reaction is: $AgCl \text{ (s)} \rightleftharpoons Ag^+ \text{ (aq)} + Cl^- \text{ (aq)}$

At equilibrium: $\qquad\qquad\qquad\qquad x \qquad\quad x$

Solubility in M: $\dfrac{1.8 \times 10^{-2} \text{ g L}^{-1}}{143.32 \text{ g mol}^{-1}} = 1.256 \times 10^{-4}$ M $= x$

$$K_{sp} = x^2 = \mathbf{1.6 \times 10^{-8}}$$

9–25. The key in all three cases is likely to be the hydrolysis of the Pb^{2+} cation formed when these salts dissolve:

$$PbX_2 \rightarrow Pb^{2+} \text{ (aq)} + 2\ X^- \text{ (aq)} \ (X = Cl, Br, I)$$

Pb^{2+} *(aq)* will involved complexation of the Pb^{2+} by water molecules. One of these can hydrolyze to give hydrogen ion and the $Pb(OH)^+$ cation:

$$Pb(OH_2)_x^{2+} \text{ (aq)} \rightarrow Pb(OH)(H_2O)_{x-1}^+ \text{ (aq)} + H^+ \text{ (aq)}$$

The concentration of hydrogen ion, therefore, will affect how much "Pb^{2+}" is present in solution, and this will affect the actual solubility.

9-27. We answer this by determining the value of the reaction quotient Q generated by the high temperature equilibrium:

$$[Ba^{2+}] = [CrO_4^{2-}] = \frac{6.3 \times 10^{-3} \text{ g L}^{-1}}{253.32 \text{ g mol}^{-1}} = 2.49 \times 10^{-5} \text{ M}$$

$$Q = [Ba^{2+}] [CrO_4^{2-}] = (2.49 \times 10^{-5})^2 = \mathbf{6.2 \times 10^{-10}}$$

At 25 °C, $K_{sp} = 2.1 \times 10^{-10}$ (from Table 9-1). Since $Q > K_{sp}$, **precipitation will occur** upon cooling.

9-29. We are only concerned with $[Ce^{3+}]$ and $[IO_3^-]$. They are the species that enter into the solubility product constant. The solubility equilibrium is:

$$Ce(IO_3)_3 \text{ (s)} \rightleftharpoons Ce^{3+} \text{ (aq)} + 3 \, IO_3^- \text{ (aq)} \qquad Q = [Ce^{3+}] [IO_3^-]^3$$

Both will be diluted from their initial concentrations, because the final solution has a larger volume (400.0 mL) due to mixing. The general equation for a concentration C_{final} after dilution is:

$$C_{final} = C_{initial} \times \left(\frac{V_{initial}}{V_{final}} \right)$$

$$[Ce^{3+}] = (2.0 \times 10^{-3} \text{ M}) \times \left(\frac{250.0 \text{ mL}}{400.0 \text{ mL}} \right) = 1.25 \times 10^{-3} \text{ M}$$

$$[IO_3^-] = (1.0 \times 10^{-2} \text{ M}) \times \left(\frac{150 \text{ mL}}{400.0 \text{ mL}} \right) = 3.75 \times 10^{-3} \text{ M}$$

The relevant concentrations are due to the added Ce^{3+} and HIO_3. The concentration of IO_3^- will depend on how much HIO_3 dissociates, since this is a weak acid. At this concentration of HIO_3, however, we can assume that *all* of the HIO_3 dissociates. In particular, in this case we would have $[H^+] = [IO_3^-] = 0.00375$. We can calculate for the acid ionization of HIO_3 that Q in this case will be $(0.00375)^2 = 1.4 \times 10^{-5}$. Since this is less than $K_a(HIO_3)$, all of the HIO_3 will be dissociated. And we can assume $[IO_3^-] = 0.00375$ M.

We put these concentrations into the expression for Q:

$$Q \qquad = (1.25 \times 10^{-3}) (3.75 \times 10^{-3})^3 = 6.6 \times 10^{-11} < K_{sp}$$

No precipitate will form.

9-31. This sort of problem is worked in more detail in No. 27. Note that the volume of NaF solution is 1000 mL - 140 mL = 860 mL. Then:

moles F$^-$ added = 0.860 L × 0.0050 mol L^{-1} = 4.3×10^{-3} moles

$$[Sr^{2+}] = 1.4 \times 10^{-4} \text{ M} \quad [F^-] = 4.3 \times 10^{-3} \text{ M}$$

$$Q = [Sr^{2+}] [F^-]^2 = (1.4 \times 10^{-4})(4.3 \times 10^{-3})^2 = \mathbf{2.6 \times 10^{-9}} < K_{sp}$$

No precipitate will form

9-33. The wording of the problem indicates that a precipitation will occur, so we can skip a calculation of Q. The initial concentrations of Pb^{2+} and IO_3^- *are* required for the calculation of which is the limiting reactant. The volume of the solution after mixing will be 90.0 mL.

$$[Pb^{2+}]_{initial} = [Pb^{2+}]_{stock} \times \left(\frac{V_{stock}}{V_{initial}}\right) = 0.0500 \text{ M} \times \left(\frac{50 \text{ mL}}{90 \text{ mL}}\right) = 0.0278 \text{ M}$$

$$[IO_3^-]_{initial} = [IO_3^-]_{stock} \times \left(\frac{V_{stock}}{V_{initial}}\right) = 0.200 \text{ M} \times \left(\frac{40 \text{ mL}}{90 \text{ mL}}\right) = 0.0889 \text{ M}$$

We need two IO_3^- for every Pb^{2+} ion, so Pb^{2+} is a limiting reactant. This key insight leads to the most convenient way to work the calculation: (i) assume that all the Pb^{2+} will precipitate, along with a stoichiometric amount of IO_3^-. Then (ii) determine how much Pb^{2+} is soluble in the presence of the unreacted IO_3^-.

Here, the moles of IO_3^- that will precipitate is twice the number of moles of Pb^{2+}, 2×0.0278 moles $L^{-1} = 0.0556$ mol L^{-1}. The remaining IO_3^- will be $[IO_3^-] = (0.0889 \text{ mol } L^{-1}) - (0.0556 \text{ mol } L^{-1}) = 0.0333$ M.

If we assume x moles of $Pb(IO_3)_2$ dissolves in the presence of **0.0333 M IO_3^-**, then:

$$K_{sp} = [Pb^{2+}][IO_3^-]^2$$
$$[Pb^{2+}] = \frac{K_{sp}}{[IO_3^-]^2} = \frac{2.6 \times 10^{-13}}{(0.0333)^2} = \mathbf{2.3 \times 10^{-10} \text{ M}}$$

9-35. As in No. 31: Volume of final solution: 80.0 mL = 0.0800 L

moles of Ag^+: $0.0500 \text{ L} \times 0.100 \text{ mol } L^{-1} = 5 \times 10^{-3}$ mol

moles of CrO_4^{2-}: $0.0300 \text{ L} \times (0.0600 \text{ mol } L^{-1}) = 1.8 \times 10^{-3}$ mol

The CrO_4^{2-} is the limiting reactant, and the moles of Ag^+ that remain in solution (note that 2 Ag^+ precipitate for 1 CrO_4^{2-}) is:

$$(5 \times 10^{-3} \text{ mol}) - (2 \times 1.8 \times 10^{-3} \text{ mol}) = 1.4 \times 10^{-3} \text{ mol Ag}^+$$

The silver concentration will be: $[Ag^+] = \dfrac{1.4 \times 10^{-3} \text{ mol Ag}^+}{0.080 \text{ L solution}} = \mathbf{0.018 \text{ M}.}$

In the solubility product equilibrium:

$$K_{sp} = [Ag^+][CrO_4^{2-}]$$
$$[CrO_4^{2-}] = \frac{K_{sp}}{[Ag^+]^2} = \frac{1.9 \times 10^{-12}}{(0.0175)^2} = \mathbf{6.2 \times 10^{-9} \text{ M}}$$

9-37. The solubility of the salt will equal the concentration of Ca^{2+}. K_{sp} is taken from Table 9-1. $K_{sp} = [Ca^{2+}][F^-]^2$

$$[Ca^{2+}] = \frac{K_{sp}}{[F^-]^2} = \frac{3.9 \times 10^{-11}}{(0.040)^2} = \mathbf{2.4 \times 10^{-8} \text{ M}}$$

Chapter 9

9-39. (a) If we set $[Ni^{2+}] = x$, then

$[OH^-] = 2x$ (neglecting the autoionization of water) and:

$$K_{sp} = [Ni^{2+}][OH^-]^2 = 4x^3$$

$$x = \sqrt[3]{\frac{K_{sp}}{4}} = \sqrt[3]{\frac{1.6 \times 10^{-16}}{4}} = 3.4 \times 10^{-6}$$

Solubility $= [Ni^{2+}] = \mathbf{3.4 \times 10^{-6}\ mol\ L^{-1}}$

(b) In this case, $[OH^-]$ is set by $[NaOH]$, just like any common ion problem:

$$K_{sp} = [Ni^{2+}][OH^-]^2$$

$$[Ni^{2+}] = \frac{K_{sp}}{[OH^-]^2} = \frac{1.6 \times 10^{-16}}{(0.100)^2} = 1.6 \times 10^{-14}\ M$$

Solubility $= [Ni^{2+}] = \mathbf{1.6 \times 10^{-14}\ mol\ L^{-1}}$

9-41. If we set $[Ag^+] = x$, then in the *absence of other acid-base effects* such as buffering, $[OH^-] = x$. K_{sp} is taken from Table 9-1. Then:

$$K_{sp} = x^2 \quad \Rightarrow \quad x = \sqrt{K_{sp}} = \sqrt{1.5 \times 10^{-8}} = 1.2 \times 10^{-4}$$

$[Ag^+] = \mathbf{1.2 \times 10^{-4}\ mol\ L^{-1}}$

When buffered at pH = 7, $[OH^-]$ will be a constant 10^{-7} M:

$$K_{sp} = [Ag^+][OH^-] \qquad [Ag^+] = \frac{1.5 \times 10^{-8}}{1 \times 10^{-7}} = \mathbf{0.15\ M}$$

The solubility is over 1000 times higher at a buffered pH = 7.

9-43. If the solubility of the salt is 0.001 mol L^{-1}, then $[Fe^{2+}] = 0.001$ mol L^{-1}, and we can solve for $[OH^-]$. K_{sp} is taken from Table 10-1.

$$K_{sp} = [Fe^{2+}][OH^-]^2 \Rightarrow [OH^-] = \sqrt{\frac{K_{sp}}{[Fe^{2+}]}} = \sqrt{\frac{1.6 \times 10^{-14}}{0.001}} = 4 \times 10^{-6}\ M$$

pOH = 5.4 ; **pH = 8.6**

9-45. (a) From Table 9-1: $K_{sp}(MgC_2O_4) = 8.6 \times 10^{-5}$ $K_{sp}(PbC_2O_4) = 2.7 \times 10^{-11}$

Both are 1 : 1 salts, so we are tell from the smaller K_{sp} for PbC_2O_4 that it is less soluble and Pb^{2+} will be present in the solid. To answer the first part of the question, we must determine the concentration of $C_2O_4^{2-}$ that will just cause the precipitation of magnesium oxalate:

$$[Mg^{2+}][C_2O_4^{2-}] = K_{sp}. \ ; [C_2O_4^{2-}] = \frac{K_{sp}}{[Mg^{2+}]} = \frac{8.6 \times 10^{-5}}{0.10} = \mathbf{8.6 \times 10^{-4}\ M}$$

9-45. (continued)

(b) If we have $[C_2O_4^{2-}] = 8.6 \times 10^{-4}$ M, then what is $[Pb^{2+}]$?

$$[Pb^{2+}] \, [C_2O_4^{2-}] = K_{sp} \, (PbC_2O_4)$$

$$[Pb^{2+}] = \frac{K_{sp}}{[C_2O_4^{2-}]} = \frac{2.7 \times 10^{-11}}{8.6 \times 10^{-4}} = 3.1 \times 10^{-8} \text{ M}$$

Since we started with 0.1 M Pb^{2+}, the remaining fraction is:

$$\frac{3.1 \times 10^{-8}}{0.1} = \textbf{0.00000031}, \text{ or 3.1 parts of every 10 million.}$$

9-47. As in example 9-8, we can manipulate the equilibria to get

$$\frac{[H_3O^+] \, [HS^-]}{[H_2S]} = 9.1 \times 10^{-8}$$

$$[HS^-] = 9.1 \times 10^{-8} \, \frac{[H_2S]}{[H_3O^+]} = 9.1 \times 10^{-8} \times \left(\frac{0.10}{1.0 \times 10^{-5}}\right) = 9.1 \times 10^{-4} \text{ M}$$

Note also that at $[H_3O^+] = 1.0 \times 10^{-5}$ then $[OH^-] = 1.0 \times 10^{-9}$

$$K(ZnS) = [Zn^{2+}] \, [OH^-] \, [HS^-]$$

$$[Zn^{2+}] = \frac{K(ZnS)}{[OH^-][HS^-]} = \frac{2 \times 10^{-25}}{\left(1.0 \times 10^{-9}\right)\left(9.1 \times 10^{-4}\right)} = \textbf{2} \times \textbf{10}^{-13} \text{ M}$$

9-49. The condition that must be met concerns a given concentration of Fe^{2+} (0.10 M) and H_2S (0.10 M). We will calculate the condition when $Q = K$ exactly for the given $[Fe^{2+}]$ and $[H_2S]$. The equilibrium expression is:

$$K(FeS) = [Fe^{2+}] \, [HS^-] \, [OH^-] = 5 \times 10^{-19}$$

We know $[HS^-] = 9.1 \times 10^{-8} \dfrac{[H_2S]}{[H_3O^+]}$ and $[OH^-] = \dfrac{1.0 \times 10^{-14}}{[H_3O^+]}$

Then: $\qquad K(FeS) = [Fe^{2+}] \times \left(9.1 \times 10^{-8} \dfrac{[H_2S]}{[H_3O^+]}\right) \times \dfrac{1.0 \times 10^{-14}}{[H_3O^+]}$

$$K(FeS) = 9.1 \times 10^{-22} \times \frac{\left[Fe^{2+}\right]\left[HS^-\right]}{\left[H_3O^+\right]^2}$$

Rearranging: $[H_3O^+]^2 = \dfrac{9.1 \times 10^{-22} \times [Fe^{2+}] \, [H_2S]}{K(FeS)}$

$$= \frac{9.1 \times 10^{-22} \times 0.10 \times 0.10}{5 \times 10^{-19}} = 1.82 \times 10^{-5}$$

$$[H_3O^+] = \sqrt{1.82 \times 10^{-5}} = 4.3 \times 10^{-3} \text{ M} \qquad \textbf{pH = 2.4}$$

Chapter 9

9-47. (continued)

This represents the maximum pH compatible with all the Fe^{2+} remaining in solution. We must now determine how much Pb^{2+} will dissolve at this pH. We will still have $[H_2S] = 0.1$ M

$$K(PbS) = 9.1 \times 10^{-22} \times \frac{[Pb^{2+}][HS^-]}{[H_3O^+]^2}$$

$$[Pb^{2+}] \quad = \frac{K(PbS) \times [H_3O^+]^2}{[H_2S] \times (9.1 \times 10^{-22})} = \frac{(3 \times 10^{-28}) \times (4.3 \times 10^{-3})^2}{(0.1) \times (9.1 \times 10^{-22})} = \mathbf{6 \times 10^{-11}}\ \textbf{M}$$

9-51. Exercise 9-9 has the following data:

$K_1 = 1 \times 10^4$, $K_2 = 2 \times 10^3$, $K_3 = 5 \times 10^2$, $K_4 = 9 \times 10^1$.

Since we are only interested in the $[Cu^{2+}]$, $[NH_3]$, and $[Cu(NH_3)_4^{2+}]$, we should combine these:

$K_f = K_1 \times K_2 \times K_3 \times K_4 = 9 \times 10^{11}$ where K_f refers to the reaction:

$$Cu^{2+}\ (aq) + 4NH_3\ (aq) \rightleftharpoons Cu(NH_3)_4^{2+}\ (aq)$$

Note that K_f is very large. We may work the problem by assuming the reaction goes all the way to completion and then allowing some $Cu(NH_3)_4^{2+}$ to dissociate. In one liter, before any complexation, there is 0.10 mol of Cu^{2+} and 1.5 mol NH_3. The Cu^{2+} is the limiting reactant for the formation of $Cu(NH_3)_4^{2+}$. It will give rise to:

$$NH_3 \approx 1.5\ \text{mol } NH_3 - \frac{4\ \text{mol } NH_3}{1\ \text{mol } Cu(NH_3)_4{}^{2+}} \times 0.10\ \text{mol } Cu(NH_3)_4{}^{2+}$$

$$\approx 1.1\ \text{mol } NH_3$$

Note that we must use the fact that four moles of NH_3 go into making one mole of $Cu(NH_3)_4^{2+}$. The calculations we have to start the equilibrium calculation are:

$[Cu^{2+}] = 0.00$ M $\qquad [Cu(NH_3)_4^{2+}] = 0.10$ M $\qquad [NH_3] = 1.1$ M

These are used in the table of changes.

(All concentrations in M)	$[Cu(NH_3)_4^{2+}]$	$[Cu^{2+}]$	$[NH_3]$
Initial Concentration	0.10	0.00	1.1
Change in Concentration	-x	+x	+4x
Equilibrium Concentration	0.1 - x	x	1.1 + 4x

9-51. (continued)

The appropriate equilibrium expression is:

$$K = \frac{[Cu(NH_3)_4^{2+}]}{[Cu^{2+}][NH_3]^4} = \frac{0.1 - x}{(x)(1.1 + x)^4}$$

If you multiplied this out, you would get a *fifth order* polynomial, which typically cannot be solved analytically.

A good compromise is to observe that very little $Cu(NH_3)_4^{2+}$ will dissociate, so x will be small. Then: $\quad 1.1 + 4x \approx 1.1 \quad$ and $\quad 0.1 - x \approx 0.1$

We will then have: $\quad K = \dfrac{0.1}{(1.1)^4 x} = \dfrac{0.0683}{x}$

$$x = \frac{0.0683}{K} = \frac{0.0683}{9 \times 10^{11}} = 7 \times 10^{-14}$$

Then $[Cu^{2+}] = 7 \times 10^{-14}$ M; $[Cu(NH_3)_4^{2+}] = 0.10$ M

9-53. The reason for the acidity is the hydrolysis of the aquo complex of Cu^{2+}:

$$Cu^{2+} \text{ (aq)} + 2H_2O \text{ (l)} \rightleftharpoons CuOH^+ \text{ (aq)} + H_3O^+ \text{ (aq)}$$

9-55. We can treat $Co(H_2O)_6^{2+}$ as any other weak acid. Under such conditions --concentration much larger than K; we can avoid a quadratic:

$$[H_3O^+] \approx \sqrt{K[Co(H_2O)_6^{2+}]} = \sqrt{3 \times 10^{-10} \times 0.10} = 5.5 \times 10^{-6}; \textbf{pH} = \textbf{5.3}$$

9-57. We assume the $Pt(NH_3)_4^{2+}$ generates hydronium ion by the reaction:

$$Pt(NH_3)_4^{2+} \text{ (aq)} + H_2O \text{ (l)} \rightleftharpoons Pt(NH_3)_3(NH_2)^+ \text{ (aq)} + H_3O^+ \text{ (aq)}$$

Assuming the acid is very weak (and note $[H_3O^+] = 10^{-4.92} = 1.20 \times 10^{-5}$)

$$K_a \approx \frac{[H_3O^+]^2}{[Pt(NH_3)_4^{2+}]} = \frac{\left(1.20 \times 10^{-5}\right)^2}{0.15} = \textbf{9.6} \times \textbf{10}^{-\textbf{10}}$$

9-59. The "best" separation will have [I⁻] just at the value where $Q(PbI_2) = K_{sp}(PbI_2)$
This requires: $\quad [Pb^{2+}][I^-]^2 = 1.4 \times 10^{-8}$

$$[I^-] = \sqrt{\frac{1.4 \times 10^{-8}}{0.0500}} = \textbf{5.3} \times \textbf{10}^{-\textbf{4}} \textbf{ M}$$

This iodide concentration will strongly suppress the dissolutoin of Hg_2I_2.

9-61. We should perform a qualitative analysis for Ba^{2+}. This ion is in Group 4. As noted in the text, if the pH is set to 9.5 and $(NH_4)_2CO_3$ is added, $BaCO_3$ will precipitate; Cs_2CO_3 will not. A precipitate indicates Ba^{2+} is present. No precipitate (in this experiment) indicates Cs^+. Also barium gives a green flame test, Cesium gives a blue flame test.

Chapter 9

9-63. We can test in the order of the groups:

First, is Hg^{2+}, from Group 2, present? The solution is made to a pH of 0.5 and H_2S is added. A precipitate indicates Hg^{2+}.

Second, is Ni^{2+}, from Group 3, present? Any precipitate is removed. The solution is made basic (pH = 9) by the addition of ammonia, which will react with the acid to give a buffer. A precipitate indicates Ni^{2+} is present.

If these tests are negative, we can conclude only Sr^{2+} is present. But to be sure, and in the event either Hg^{2+} or Ni^{2+} was found, then after removal of any precipitates, and H_2S (by acidification and heating), a test solution at pH = 9.5 is treated with $(NH_4)_2CO_3$. A precipitate indicates Sr^{2+}.

9-65. We can only say Group 1 is absent and ions from Group 2 or 3 are present. The treatment to precipitate Group 3 (basic sulfide) will *also* precipitate Group 2. The test does not tell us anything about Groups 4 and 5.

9-67.
$$Hg_2I_2 \ (s) \rightleftarrows Hg_2^{2+} \ (aq) + 2\ I^- \ (aq)$$
$$K_{sp} = [Hg_2^{2+}][\ I^-]^2$$

9-69. The relative K_a's related to a single reaction, the dissociation of one hydrogen ion from a species. The stoichiometry is always the same: $HA \rightarrow H^+ + A^-$. However, K_{sp}'s can related to one, two, or more species being formed in solution. Therefore, they refer to systems with different stoichiometries and, therefore, different structures for the equilibrium expression.

9-71. Given $K_{sp} (BaSO_4) = 1.1 \times 10^{-10}$ and setting $[Ba^{2+}] = [SO_4^{2-}] = x$
$$x^2 = 1.1 \times 10^{-10}$$
$$x = \sqrt{1.1 \times 10^{-10}} = 1.05 \times 10^{-5}$$
$$[Ba^{2+}] = \mathbf{1.05 \times 10^{-5} \ M}$$

9-73. (a)
$$CaCO_3: \ \frac{14 \times 10^{-3} \ \text{g L}^{-1}}{100.09 \ \text{g mol}^{-1}} = 1.4 \times 10^{-4} \ \text{mol L}^{-1}$$
$$SrCO_3: \ \frac{10 \times 10^{-3} \ \text{g L}^{-1}}{147.63 \ \text{g mol}^{-1}} = 6.8 \times 10^{-5} \ \text{mol L}^{-1}$$
$$BaCO_3: \ \frac{17 \times 10^{-3} \ \text{g L}^{-1}}{197.34 \ \text{g mol}^{-1}} = 8.6 \times 10^{-5} \ \text{mol L}^{-1}$$

The molar solubility decreases then increases down the group. There is not trend.

9-73. (continued)

(b) In pure water with no side-reactions, $[M^{2+}] = [CO_3^{2-}]$ = molar solubility

$K_{sp} = $ (molar solubility)^2 $K_{sp}(CaCO_3) = 2.0 \times 10^{-8}$

$K_{sp}(SrCO_3) = 4.6 \times 10^{-9}$ $K_{sp}(BaCO_3) = 7.4 \times 10^{-9}$

The K_{sp}'s here are for 20 °C. Table 9-1 refers to data at 25 °C. More importantly, there *are* side reactions, especially of carbonate, that make the dissolution reaction more complicated than we have considered.

9-75. (a) We set $[Rb^+] = [ClO_4^-] = x$ = molar solubility

$$x^2 = K_{sp}; \qquad x = \sqrt{K_{sp}} = \sqrt{0.0030} = 0.055 \text{ M}$$

(b) With such a large K_{sp}, we cannot neglect the amount of RbClO$_4$ that will dissolve in the presence of a common ion.

The table of changes becomes (for 0.075 M Rb$_2$SO$_4$, $[Rb^+] = 0.15$ M):

(All concentrations in M)	$[Rb^+]$	$[ClO_4^-]$
Initial Concentration	0.15	0.0
Change in Concentration	+x	+x
Equilibrium Concentration	0.15 + x	x

Then: $(0.15 + x)(x) = K_{sp};$ $x^2 + 0.15x - K_{sp} = 0$

The only reasonable root is $x = 0.018$, so the solubility is **0.018 M**.

(c) Here, the common ion is ClO$_4^-$, at $[ClO_4^-] = 0.075$ M

(All concentrations in M)	$[Rb^+]$	$[ClO_4^-]$
Initial Concentration	0.0	0.075
Change in Concentration	+x	+x
Equilibrium Concentration	x	0.075 + x

Then: $(x)(0.075 + x) = K_{sp};$ $x^2 + 0.075x - K_{sp} = 0$

The only reasonable root is $x = 0.14$, so the solubility is **0.14 M.**

(d) The salt NaCl contains no ion in common with RbClO$_4$. It should have no effect on the solubility.

9-77. We need a ratio of concentrations, not an absolute value. We take the ratio of the K_{sp}'s:

$$\frac{[Cu^+][Br^-]}{[Ag^+][Br^-]} = \frac{K_{sp}(CuBr)}{K_{sp}(AgBr)}$$

The $[Br^-]$ is unknown and, in this case, *irrelevant*, because it is in the numerator and the denominator.

$$\frac{[Cu^+]}{[Ag^+]} = \frac{K_{sp}(CuBr)}{K_{sp}(AgBr)} = \frac{4.2 \times 10^{-8}}{7.7 \times 10^{-13}} = \mathbf{55,000.}$$

9-79. For this problem, we determine the number of moles of chloride that is left when the Ag_2CrO_4 first appears. However, we do not know any volumes. So we can assume that the initial solution had a volume of V. The number of moles of Cl^- present was $0.100\ V$ mol. This will require V liters of $AgNO_3$ solution, so the total solution volume is $2.00\ V$. The concentration of chromate present to complete the titration is now 0.00125 M CrO_4^{2-}, half of its initial concentration.

The Ag^+ concentration that will cause precipitation is, if we set $[Ag^+] = x$:

$$x^2[CrO_4^{2-}] = K_{sp}(Ag_2CrO_4); \qquad x = \sqrt{\frac{K_{sp}}{[CrO_4^{2-}]}} = \sqrt{\frac{1.9 \times 10^{-12}}{0.00125}} = 3.9 \times 10^{-5}$$

The $[Cl^-]$ in equilibrium with this is, if $[Cl^-] = x$:

$$[Ag^+]\,x = K_{sp}(AgCl) \qquad x = \frac{K_{sp}}{[Ag^+]} = \frac{1.6 \times 10^{-10}}{3.9 \times 10^{-5}} = 4.1 \times 10^{-6}\ M$$

The number of *moles* of Cl^- in *this* $2.00\ V$ L solution is:

4.1×10^{-6} mol $L^{-1} \times 2.00\ V$ L $= 8.2 \times 10^{-6}\ V$ mol Cl^-

This compares to the initial 0.100 mol:

$$\text{fraction } Cl^- = \frac{8.2 \times 10^{-6}\ V}{0.100\ V} = 0.00008 = \mathbf{0.008\ \%.}$$

9-81. (a) The "tiny" change in the solution is quite large from the point of view of a silver ion. The 0.10 mol of ammonia is a massive excess for complexation of silver:

$$Ag^+\ (aq) + 2NH_3\ (aq) \rightleftharpoons Ag(NH_3)_2^+\ (aq) \qquad\qquad K = 1.7 \times 10^7$$

We can combine this with the solubility equilibrium for AgBr:

$$AgBr\ (s) \rightleftharpoons Ag^+\ (aq) + Br^-\ (aq) \qquad\qquad K_{sp} = 7.7 \times 10^{-13}$$

Together these give us a solubility/complexation equilibrium:

$$AgBr\ (s) + 2NH_3\ (aq) \rightleftharpoons Ag(NH_3)_2^+ + Br^- \qquad\qquad K = 1.3 \times 10^{-5}$$

9-81. (continued)

This is still small, but it is *much* larger then K_{sp} (AgBr). The ammonia complex of silver ion "pulls" the silver into solution.

(b) If we calculate K's for the dissolution of AgCl and AgI in ammonia solution we obtain:

$$AgCl\ (s) + 2NH_3\ (aq) \rightleftharpoons Ag(NH_3)_2^+ + Cl^- \qquad K = 3 \times 10^{-3}$$
$$AgI\ (s) + 2NH_3\ (aq) \rightleftharpoons Ag(NH_3)_2^+ + I^- \qquad K = 3 \times 10^{-9}$$

We can see that AgI is a million times less soluble than AgCl, whether in pure water or in ammonia solution.

9-83. The equilibrium for the potassium complexation by 18-crown-6 (abbreviated "18-C-6" here) :

$$K^+\ (aq) + 18\text{-}C\text{-}6\ (aq) \rightleftharpoons K(18\text{-}C\text{-}6)^+\ (aq) \qquad K = 6.6$$

This is *not* a large equilibrium constant, and we must avoid any assumptions if possible. One operation that is *not* an assumption is to set up the table of changes so that the all of the potassium and 18-crown-6 are initially complexed.

(All concentrations in M)	$[K^+]$	$[18\text{-}C\text{-}6]$	$[K(18\text{-}C\text{-}6)^+]$
Initial Concentration	0.0	0.0	0.0080
Change in Concentration	$+x$	$+x$	$-x$
Equilibrium Concentration	x	x	$0.0080 - x$

The appropriate equilibrium expression is:

$$K = \frac{[K(18\text{-}C\text{-}6)^+]}{[K^+][18\text{-}C\text{-}6]} = \frac{0.00800 - x}{x^2}$$

Rearranging: $0 = Kx^2 + x - 0.00800 = 6.6\,x^2 + x - 0.00800$

Use of the quadratic formula for this yields two roots: $x = +0.00762$ and $x = -0.159$

Only the positive root is reasonable, so

$[18\text{-}C\text{-}6] = [K^+] = 0.0076$ M and $[K(18\text{-}C\text{-}6)^+] = 0.0004$ M

For sodium, we have the same conditions, except there is a different equilibrium constant. Then: $0 = Kx^2 + x - 0.00800 = 111.6\,x^2 + x - 0.00800$

Use of the quadratic formula for this yields two roots:

$$x = +0.00510 \text{ and } x = -0.014$$

Only the positive root is reasonable, so

$[18\text{-}C\text{-}6] = [Na^+] = 0.0051$ M and $[K(18\text{-}C\text{-}6)^+] = 0.0029$ M

9-85. $Pt(OH)_6^{2-}\ (aq) + H_2O\ (l) \rightleftharpoons Pt(OH)_5(OH_2)^-\ (aq) + OH^-\ (aq)$

CHAPTER 10 - THERMOCHEMISTRY

10-1. The heat comes from the friction of the drill as it grinds against, and tears apart, the wood. The energy for the heat comes from the mechanical energy of the motor, which in turn comes from electrical energy.

10–3. There are actual two things wrong with this. The first is the "heat content" phrase, since this suggests that heat is a fluid that can be contained. The second is that the heat that can be transferred to or from the engine block depends on the magnitude of the temperature change.

10-5. The specific heat of water is over five times that of iron (Table 10-1) and the heat capacity of two samples of the same substance is larger for the larger mass of that substance. So the ranking is:

1 kg iron ball < 2 kg iron plate < 1 kg water < 2 kg water.

10-7. The molar heat capacity is equal to the specific heat multiplied by the mass of one mole of the element.

Element	Specific heat $(J\,K^{-1}\,g^{-1})$	Molar mass $(g\,mol^{-1})$	Molar heat capacity $(J\,K^{-1}\,mol^{-1})$
Lithium	3.57	6.941	**24.8**
Sodium	1.23	22.990	**28.3**
Potassium	0.756	39.098	**29.6**
Rubidium	0.363	85.468	**31.0**
Cesium	0.242	132.905	**32.2**

The molar heat capacity of all of the elements is a gradually increasing -- about $1\,J\,K^{-1}\,mol^{-1}$ per row of the periodic table (the jump from Li to Na is slightly more). Thus, we expect the heat capacity for Fr to be around **33.5 J K^{-1} mol^{-1}**.

10-9.

Element	Specific heat $(J\,K^{-1}\,g^{-1})$	Molar mass $(g\,mol^{-1})$	Molar heat capacity $(J\,K^{-1}\,mol^{-1})$	Defiation from Dulong-Petit
Nickel	0.444	58.6934	**26.1**	4.4%
Zinc	0.388	65.39	**25.4**	1.6%
Rhodium	0.243	102.906	**25.0**	0.0%

10-9. (continued)

Element	Specific heat $(J\ K^{-1}\ g^{-1})$	Molar mass $(g\ mol^{-1})$	Molar heat capacity $(J\ K^{-1}\ mol^{-1})$	Defiation from Dulong-Petit
Tungsten	0.132	183.84	**24.3**	−2.8%
Gold	0.129	196.965	**25.4**	1.6%

10-11. The molar mass of P_4 is $4 \times 30.974 = 123.896$ g mol^{-1}, so the molar heat capacity of P_4 is: 0.757 J K^{-1} g^{-1} $\times 123.896$ g mol^{-1} = **93.8 J K^{-1} mol^{-1}**.

10-13. The equation for equilibrating two masses at different temperatures is:

$$m_1\ c_{s1}\ \Delta t_1 = -\ m_2\ c_{s2}\ \Delta t_2$$

For this problem, substance 1 will be the water ("w") and substance 2 will be the zinc ("Zn"). The final temperature for each substance is the same. Then:

$$m_w\ c_{s,w}\ (t_f - t_{w,i}) = -\ m_{Zn}\ c_{s,Zn}\ (t_f - t_{Zn,i})$$

We have

$$m_w = 200.0\ g \qquad m_{Zn} = 60.0\ g$$
$$c_{s,w} = 4.22\ J\ (^oC)^{-1}\ g^{-1} \qquad c_{s,Zn} = 0.389\ J\ (^oC)^{-1}\ g^{-1}$$
$$t_{w,i} = 100.0\ ^oC \qquad t_{Zn,i} = 20^oC$$

This set-up uses temperature as small "t," and units of degrees Celsius. This is permissible because we are only concerned with *changes* or *differences* in temperature, and the Kelvin and the degree Celsius are the same size. Then the equation becomes:

$$200.0\ g \times 4.22\ J\ (^oC)^{-1}\ g^{-1}\ (t_f - 100.0^oC) = -\ 60.0\ g \times 0.389\ J\ (^oC)^{-1}\ g^{-1}\ (t_f - 20.0^oC)$$
$$844\ t_f - 84,400 = -23.34\ t_f + 466.8$$
$$(844 + 23.34)\ t_f = 84,400 + 466.8$$
$$t_f = 84,867\ /\ 867.3 = \textbf{97.8}^o\textbf{C}$$

10–15. (a) We need to account for the heat that will be absorbed by the water in this temperature change. $q_{total} = q_{water} + q_{calorimeter}$. The remaining heat is absorbed by the calorimeter and will be used to determine the calorimeter constant.

For 200.0 g of water:
$$q_{water} = c_s m \Delta T = 4.184\ J\ K^{-1}\ g^{-1} \times 200.0\ g \times 1.67\ K$$
$$= 1397\ J$$
$$q_{calorimeter} = q_{total} - q_{water} = 1770\ J - 1397\ J = 373\ J$$
$$C = q_{calorimeter}\ /\ \Delta t$$
$$= 373\ J\ /\ 1.67\ K = \textbf{223 J K}^{-1}$$

10-15. (continued)

(b) We need to calculate the heat change for both the caloromieter and for the water.

Calorimeter: $q_{calorimeter} = C\,\Delta t = 223\ \text{J K}^{-1} \times 2.64\ \text{K} = 589\ \text{J}$

Water: $q_{water} = c_s m\,\Delta t = 4.184\ \text{J g}^{-1}\,\text{K}^{-1} \times 225.0\ \text{g} \times 2.64\ \text{K} = 2485\ \text{J}$

Total: $q_{total} = q_{calorimeter} + q_{water} = 589 + 2485\ \text{J} = \mathbf{3070\ J}$.

10-17. (a) For a heat transfer q and a temperature change Δt:

$$q = C\,\Delta t \qquad \Rightarrow \qquad C = q / \Delta t = 5682\ \text{J} / 2.31^\circ\text{C}$$

$$= 2460\ \text{J}\ {^\circ}\text{C}^{-1} = \mathbf{2460\ J\ K^{-1}}$$

(b) $\qquad q = C\,\Delta t = 2460\ \text{J K}^{-1} \times 4.40\ \text{K} = 10820\ \text{K} = \mathbf{10.8\ kJ}$

This is the heat evolved by the reaction and absorbed by the calorimeter. Thus, q is positive for the calorimeter, and negative for this reaction.

10-19. The increase in temperature is an indication that the reacting chemical system is releasing energy as heat. The reaction is exothermic and $\Delta \mathbf{H} < \mathbf{0}$ **(is negative).**

10-21. The heat due to *just* the reaction of the bromine with the excess NaOH is the difference between the value for the bromine experiment and the "blank" without the bromine. Specifically: $\Delta H_{reac} = 121.3\ \text{J} - 2.34\ \text{J} = 118.9\ \text{J}$

This is the enthalpy change for 2.88×10^{-3} mol of Br_2 *(l)*. To convert this to the enthalpy change per mole of bromine:

$$\frac{118.9\ \text{J}}{2.88 \times 10^{-3}\ \text{mol Br}_2} = 41{,}300\ \text{J mol}^{-1}\ \text{Br}_2 = \mathbf{41.3\ kJ\ (mol\ Br_2)^{-1}}$$

10-23. (a) There are two moles of Na_2O and the evolution of 828 kJ of heat. Therefore:

$$\Delta H = \left(\frac{-828\ \text{kJ}}{2\ \text{mol Na}_2\text{O}} \right) \times \left(\frac{1\ \text{mol Na}_2\text{O}}{61.979\ \text{g Na}_2\text{O}} \right) = \mathbf{-6.68\ kJ\ (g\ Na_2O)^{-1}}.$$

(b) There is one mole of MgO indicated, and the uptake of 303 kJ of heat.

Therefore : $\qquad \Delta H = \left(\frac{+302\ \text{kJ}}{1\ \text{mol MgO}} \right) \times \left(\frac{1\ \text{mol MgO}}{40.3044\ \text{g MgO}} \right) = \mathbf{+7.49\ kJ\ (g\ MgO)^{-1}}$

(c) There are two moles of CO and the uptake of 33.3 kJ of heat.

Therefore: $\qquad \Delta H = \left(\frac{+33.3\ \text{kJ}}{2\ \text{mol CO}} \right) \times \left(\frac{1\ \text{mol CO}}{28.010\ \text{g CO}} \right) = \mathbf{+\ 0.594\ kJ\ (g\ CO)^{-1}}$

10-25. (a) This is a vaporization reaction : $\Delta \mathbf{H} > \mathbf{0}$ (heat is needed to drive reaction)

(b) This is a condensation reaction to a solid: $\Delta \mathbf{H} < \mathbf{0}$ (heat is given off)

(c) This is the freezing of a liquid: $\Delta \mathbf{H} < \mathbf{0}$ (heat is given off)

10-27. The molar mass of CBr_4 is 331.63 g mol^{-1}, so 10.00 g of the compound is 0.03015 mol CBr_4. We then have: $\dfrac{1.359 \text{ kJ}}{0.03015 \text{ mol } CBr_4} = \dfrac{45.07 \text{ kJ}}{\text{mol } CBr_4}$.

For 2.500 mole, we have: $2.500 \text{ mol } CBr_4 \times \dfrac{45.07 \text{ kJ}}{1 \text{ mol } CBr_4} = \textbf{112.7 kJ}$

10-29. $\left(\dfrac{6.04 \text{ kJ}}{1 \text{ mol CO}}\right) \times \left(\dfrac{1 \text{ mol CO}}{28.0104 \text{ g CO}}\right) \times 2.38 \text{ g CO} = \textbf{+0.513 kJ = 513 J}$

10-31.
$Cs\ (l) + O_2\ (g) \rightarrow CsO_2\ (s)$ $\qquad\qquad \Delta H = -268.8$ kJ
$CsO_2\ (s) \rightarrow Cs\ (s) + O_2\ (g)$ $\qquad\qquad \Delta H = +266.1$ kJ

$Cs\ (l) \rightarrow Cs\ (s)$ $\qquad \Delta H = +266.1 \text{ kJ} + (-268.8 \text{ kJ}) = \textbf{- 2.7 kJ}$

10-33.
$2\ CO_2\ (g) + H_2O\ (g) \rightarrow CH_2CO\ (g) + 2\ O_2\ (g)$ $\qquad -\Delta H_1 = +981.1$
$2\ CH_4\ (g) + 4\ O_2\ (g) \rightarrow 2\ CO_2\ (g) + 4\ H_2O\ (g)$ $\qquad 2 \times \Delta H_2 = -1604.6$

$2\ CH_4\ (g) + 2\ O_2\ (g) \rightarrow CH_2CO\ (g) + 3\ H_2O\ (g)$ $\qquad \textbf{ΔH = - 623.5 kJ}$

10-35. A mole of diamonds. They lie at higher enthalpy than graphite, so burning releases more enthalpy (heat in this case) on the path to the very low enthalpy product, carbon dioxide.

10-37. Tabulations of enthalpies of formation and similar thermodynamic quantities, such as those we will see in Chapter 12, are almost always done on a per mole basis. Enthalpy data and other thermodynamic data for reactions are given *for the reaction as written*. To determine the enthalpy change for a reaction we multiply molar enthalpies of formation by the number of moles of the substances *in the reaction as written*. So when we write "$2 \times \Delta H_f^o(NO_2)$" we are really saying "the heat of formation of two moles of NO_2."

$\Delta H^o = 2 \times \Delta H_f^o(NO_2\ (g)) + 2 \times \Delta H_f^o(H_2O\ (l)) - 3 \times \Delta H_f^o(O_2\ (g)) - \Delta H_f^o(N_2H_4\ (l))$

$= 2 \text{ mol} \times (+33.18 \dfrac{\text{kJ}}{\text{mol}}) + 2 \text{ mol} \times (-285.83 \dfrac{\text{kJ}}{\text{mol}})$

$- 3 \text{ mol} \times (0.0\dfrac{\text{kJ}}{\text{mol}}) - 1 \text{ mol} \times (50.63\dfrac{\text{kJ}}{\text{mol}}) = \textbf{-555.93 kJ}$

10-39. (a) $\Delta H^o = 2 \times \Delta H_f^o(ZnO\ (s)) + 2 \times \Delta H_f^o(SO_2\ (g)) - 3 \times \Delta H_f^o(O_2\ (g)) - 2 \times \Delta H_f^o(ZnS\ (s))$

$= 2 \text{ mol} \times (-348.28\dfrac{\text{kJ}}{\text{mol}}) + 2 \text{ mol} \times (-296.83\dfrac{\text{kJ}}{\text{mol}})$

$- 3 \text{ mol} \times (0.0\dfrac{\text{kJ}}{\text{mol}}) - 2 \text{ mol} \times (-205.98\dfrac{\text{kJ}}{\text{mol}}) = \textbf{-878.26 kJ}$

10-39. (continued)

(b) The molar mass of sphalerite is 97.46 g mol^{-1}, so 3.00 metric tons (= 3.00×10^6 g) will contain 3.078×10^4 mol. The value determined in (a) is for the reaction as written -- with *two* moles of ZnS. So the heat absorbed from the 3.00 metric ton sample is: $\Delta H = 3.078 \times 10^4$ mol ZnS $\times \dfrac{-878.26 \text{ kJ}}{2 \text{ mol ZnS}} =$ **-1.35 $\times$ 10^7 kJ**

(c) The difference results from the use of thermodynamic data at 298 K for the calculations in (a) and (b), whereas the correct answer will have to use thermodynamic data at the higher temperatures.

10-41. (a) $\Delta H^o = \Delta H^o_f(\text{Ca}^{2+} \text{ (aq)}) + 2 \times \Delta H^o_f(\text{Cl}^- \text{ (aq)}) - \Delta H^o_f(\text{CaCl}_2 \text{ (s)})$

$$= 1 \text{ mol} \times (-542.83 \frac{\text{kJ}}{\text{mol}}) + 2 \text{ mol} \times (-167.16 \frac{\text{kJ}}{\text{mol}}) - 1 \text{ mol } (-795.8 \frac{\text{kJ}}{\text{mol}})$$

$$= \textbf{-81.4 kJ}$$

(b) The value obtained is for the dissolution of one mole of $CaCl_2$. The molar mass of $CaCl_2$ is 110.98 g mol^{-1}, so 11.1 g $CaCl_2$ contains 0.1000 mol $CaCl_2$. The enthalpy change in this experiment is just:

$$q - \Delta H = 0.1000 \text{ mol CaCl}_2 \times \dfrac{-81.4 \text{ kJ}}{1 \text{ mol CaCl}_2} = -8.14 \text{ kJ} = -8140 \text{ J}$$

This is the amount of heat given off to the surroundings by the reacting chemical system. In this case, the surroundings are the water.

$$q_{\text{water}} = - q_{\text{reaction}} = C \, \Delta t, \qquad \text{where } C = 418 \text{ J K}^{-1}$$

$$\Delta t = \dfrac{-q_{\text{reaction}}}{C} = \dfrac{-(8,140)}{418 \text{ J K}^{-1}} = 19.5 \,^{\circ}\text{C}$$

The final temperature will be 20.0 $^{\circ}$C + 19.5 $^{\circ}$C = **39.5°C**

10-43. For the combustion reaction ($\Delta H^o_f(\text{O}_2) = 0.0$ and is omitted):

$$C_6H_{12} \text{ (l)} + 9 \text{ O}_2 \text{ (g)} \rightarrow 6 \text{ CO}_2 \text{ (g)} + 6 \text{ H}_2\text{O} \text{ (l)}$$

$$\Delta H^o_{comb.} = 6 \times \Delta H^o_f(\text{CO}_2 \text{ (g)}) + 6 \times \Delta H^o_f(\text{H}_2\text{O (l)}) - \Delta H^o_f(\text{C}_6\text{H}_{12} \text{ (l)})$$

$$1 \text{ mol} \times \Delta H^o_f(\text{C}_6\text{H}_{12} \text{ (l)}) = 6 \times \Delta H^o_f(\text{CO}_2 \text{ (g)}) + 6 \times \Delta H^o_f(\text{H}_2\text{O (l)}) - \Delta H^o_{comb.}$$

$$= 6 \text{ mol} \times (-393.51 \frac{\text{kJ}}{\text{mol}}) + 6 \text{ mol} \times (-285.83 \frac{\text{kJ}}{\text{mol}})$$

$$- 1 \text{ mol} \times (-3923.7 \frac{\text{kJ}}{\text{mol}}) = -152.3 \text{ kJ}$$

$$\Delta H^o_f(\text{C}_6\text{H}_{12} \text{ (l)}) = \textbf{--152.3 kJ mol}^{-1}$$

10-45. The condensation reaction is:

$$CHCl_3 \text{ (g)} \rightarrow CHCl_3 \text{ (l)}$$

For this:

$$\Delta H^o = \Delta H_f^o(CHCl_3 \text{ (l)}) - \Delta H_f^o(CHCl_3 \text{ (g)})$$

$$= -134.5 \frac{kJ}{mol} - (-103.1 \frac{kJ}{mol}) = -31.4 \text{ kJ mol}^{-1}$$

For 2.5 mol: $2.5 \text{ mol} \times (-31.4 \frac{kJ}{mol}) = \textbf{-78 kJ}$

10-47. For the reaction:

$$C_3H_8 \text{ (g)} + 5 \text{ } O_2 \text{ (g)} \rightarrow 3 \text{ } CO_2 \text{ (g)} + 4 \text{ } H_2O \text{ (g)}$$

$$\Delta H_{comb.}^o = 3 \times \Delta H_f^o(CO_2 \text{ (g)}) + 4 \times \Delta H_f^o(H_2O \text{ (l)}) - \Delta H_f^o(C_3H_8 \text{ (l)})$$

$$= 3 \text{ mol} \times (-393.51 \frac{kJ}{mol}) + 4 \text{ mol} \times (-285.83 \frac{kJ}{mol})$$

$$- 1 \text{ mol} \times (-103.85 \frac{kJ}{mol})$$

$$= -2220 \text{ kJ}$$

This is for one mole of C_3H_8. Since the molar mass of C_3H_8 is 44.096 g mol^{-1}, the enthalpy change for one gram is:

$$\frac{-2220 \text{ kJ}}{1 \text{ mol } C_3H_8} \times \frac{1 \text{ mol}}{44.096 \text{ g}} = -50.3 \text{ kJ g}^{-1}$$

For the reaction of C_8H_{18} (note that to make the reaction reflect only one mole of C_8H_{18}, we keep the stoichiometry of the O_2 fractional):

$$C_8H_{18} \text{ (g)} + 12 \text{ } 1/2 \text{ } O_2 \text{ (g)} \rightarrow 8 \text{ } CO_2 \text{ (g)} + 9 \text{ } H_2O \text{ (g)}$$

$$\Delta H_{comb.}^o = 8 \times \Delta H_f^o(CO_2 \text{ (g)}) + 9 \times \Delta H_f^o(H_2O \text{ (g)}) - \Delta H_f^o(C_8H_{18} \text{ (l)})$$

$$= 8 \text{ mol} \times (-393.51 \frac{kJ}{mol}) + 9 \text{ mol} \times (-285.83 \frac{kJ}{mol})$$

$$- 1 \text{ mol} \times (-269.1 \frac{kJ}{mol})$$

$$= -5451 \text{ kJ}$$

This is for one mole of C_8H_{18}. Since the molar mass of C_8H_{18} is 114.231 g mol^{-1}, the enthalpy change for one gram is:

$$\frac{-5451 \text{ kJ}}{1 \text{ mol } C_8H_{18}} \times \frac{1 \text{ mol}}{114.23 \text{ g}} = -47.7 \text{ kJ g}^{-1}$$

The enthalpy change for the propane is larger. We can take this as confirmation that more energy is released per gram, also.

10-47. (continued)

(b) The usefulness of a fuel includes the energy released when it is burned. In this regard, propane is a superior fuel. However, propane is a gas at normal pressures, which means it is difficult to transfer into a tank without leakage. The liquid octane is also much denser than even compressed propane. So octane and other liquid hydrocarbons are favored from a storage and transport standpoint.

10-49. The atomization reaction here is: $CCl_3F\ (g) \rightarrow C\ (g) + 3\ Cl\ (g) + F\ (g)$

Breaking three moles of C-Cl bonds:	$+3 \times 338$ kJ
Breaking one mole of C-F bonds:	+484 kJ
TOTAL:	**+1498** kJ

10-51. (a)

Breaking one mole of C-H bonds:	+412 kJ
Breaking one mole of C≡N bonds:	+890 kJ
TOTAL:	**+1302** kJ

(b)

Breaking one mole of C-H bonds:	+412 kJ
Breaking one mole of C≡N bonds:	+890 kJ
Making $^1/_2$ mole of H-H bonds:	$-^1/_2 \times 436$ kJ
Making $^1/_2$ mole of N≡N bonds:	$-^1/_2 \times 946$ kJ
TOTAL:	**+611** kJ

(c)

Breaking one mole of C-H bonds:	+412 kJ
Breaking one mole of C≡N bonds:	+890 kJ
Making mole of N-H bonds:	-388 kJ
Making one mole of C≡N bonds:	+890 kJ
TOTAL:	**+24** kJ

10-53. (a)

```
    H   H   H
    |   |   |
H — C — C — C — H
    |   |   |
    H   H   H
```

(b) Note that CO_2 has two C=O double bonds and O_2 has one O=O double bond.

Breaking eight moles of C-H bonds:	+8(412) kJ
Breaking two moles of C-C bonds:	+2(348) kJ
Breaking five moles of O=O bonds:	+5(497) kJ
Making six moles of C=O bonds:	-6(743) kJ
Making eight moles of O-H bonds:	-8(463) kJ
TOTAL:	**-1685** kJ.

10-55.

The reaction is completely neutral with respect to the type and number of the different bonds - three B-Br and three B-Cl bonds on the right and left. Thus, from the standpoint of average bond enthalpies, $\Delta H = 0$.

10–57. For the burning of methane:

$$CH_4\ (g) + 2\ O_2(g) \rightarrow CO_2\ (g) + 2\ H_2O\ (g)$$

Breaking four moles of C-H bonds:	+4(412) kJ
Breaking two moles of O=O bonds:	+2(497) kJ
Making two moles of C=O bonds:	-2(743) kJ
Making four moles of O-H bonds:	-4(463) kJ
TOTAL:	−696 kJ

For the burning of carbon tetrafluoride:

$$CF_4\ (g) + 2\ O_2(g) \rightarrow CO_2\ (g) + 2\ F_2O\ (g)$$

Breaking four moles of C-F bonds:	+4(484) kJ
Breaking two moles of O=O bonds:	+2(497) kJ
Making two moles of C=O bonds:	-2(743) kJ
Making four moles of O-H bonds:	-4(185) kJ
TOTAL:	+704 kJ

The larges difference is in the relative strength of the O–H and the O–F bond. Since the bond enthalpy of the latter is so much smaller than the bond enthalpy of the former, the O–F bonds "give back" far less energy than the O–H bonds.

10-59. The constant pressure input of heat into the nitrogen gas causes the gas to expand and perform pressure-volume work.

$$w = -P_{ext}\Delta V = -P_{ext}\ (V_{final} - V_{initial})$$
$$= -\ 50.0\ atm \times (974\ L - 542\ L)$$
$$= \mathbf{-\ 21,600\ L\text{-}atm}$$

To convert this to the conventional energy unit the joule

$-21,600\ L\text{-}atm \times 101.325\ J\ L^{-1}\ atm^{-1} = \mathbf{-\ 2.19 \times 10^6\ J}$

10-61.

$$w = -P_{ext}\Delta V = -\ 1.14\ atm \times (0.40\ L - 4.00\ L) = \mathbf{+\ 4.10\ L\text{-}atm}$$
$$= +\ 4.10\ L\text{-}atm \times 101.325\ J\ L^{-1}\ atm^{-1} = \mathbf{+416\ J}$$

Chapter 10

10-63. Because the gas is expanding against an external pressure of zero atmospheres, the work done is zero.

10-65. (a) The ΔE and q of the system are both positive as the thermal energy enters the system; w is zero because in the absence of a volume change work is not possible.
(b) The ΔE and q for the system are both negative as the thermal energy leaves the system. w is zero again.

 (c) Because the system returns to its original temperature, the overall energy change is equal to the overall heat change, because we also know w is zero.
Therefore: $q_1 + q_2 = \Delta E_1 + \Delta E_2.$

10-67.
$$H = E + PV$$
$$H - E = PV = nRT = 2.00 \text{ mol} \times 8.3145 \text{ J mol}^{-1} \text{ K}^{-1} \times 400 \text{ K} = \textbf{6650 J}$$

10-69. (a)
$$C_{10}H_8 \ (s) + 12 \ O_2 \ (g) \rightarrow 10 \ CO_2 \ (g) + 4 \ H_2O \ (l)$$
(b) The experiment used 0.6410 g/128.17 g mol^{-1}= 0.005001 mol and evolved 25.79 kJ. The heat absorbed at constant volume for the combustion of one mole of naphthalene is then -25.79 kJ / 0.005001 mol = -5157 kJ mol$^{-1.}$
Since the reaction is done at constant volume, this heat transfer is equal to the energy change for the reaction, and since the reaction is done at 25 °C, (=298.15 K), this is equal to the standard energy change: $\Delta E^o = \textbf{-5,157 kJ}$.
(c) For this reaction, there are 10 moles of product gases and 12 moles of reactant gases, so for the gas phase $\Delta n = -2$ mol.

$$\Delta H^o = \Delta E^o + RT\Delta n = -5,157,000 \text{ J} + (8.3145 \text{ J mol}^{-1} \text{ K}^{-1})(298.15 \text{ K})(-2.00 \text{ mol})$$
$$= -5,162,000 \text{ J} = \textbf{-5,162 kJ}$$

(d) $\Delta H^o = 4 \times \Delta H^o_f(H_2O \ (l)) + 10 \times \Delta H^o_f(CO_2 \ (g)) - 12 \times \Delta H^o_f(O_2 \ (g)) - \Delta H^o_f(C_{10}H_8 \ (s))$

$$-5162 \text{ kJ} = \quad 4 \text{ mol} \times (-285.83\frac{\text{kJ}}{\text{mol}}) + 10 \times (-393.51\frac{\text{kJ}}{\text{mol}})$$
$$- 12 \times (-0.00\frac{\text{kJ}}{\text{mol}}) - 1 \text{ mol} \times \Delta H^o_f(C_{10}H_8)$$

1 mol $\times \Delta H^o_f(C_{10}H_8) = + 84$ kJ $\quad \Rightarrow \quad \Delta H^o_f(C_{10}H_8) = \textbf{+ 84 kJ mol}^{-1}$

10-71. (a) molar heat capacity $= \left(0.215\dfrac{\text{cal}}{\text{g K}}\right) \times \left(\dfrac{4.184 \text{ J}}{\text{cal}}\right) \times \left(\dfrac{26.9815 \text{ g Al}}{1 \text{ mol Al}}\right)$
$$= \textbf{24.3} \ \dfrac{\textbf{J}}{\textbf{mol K}^{-1}}$$

152

10-71. (continued)

(b)
$$C_p = \left(0.215\,\frac{cal}{g\ K}\right) \times \left(\frac{4.184\ J}{cal}\right) \times \left(\frac{1000\ g}{1\ kg}\right) \times \left(\frac{1000\ kg}{1\ metric\ ton}\right)$$
$$= 9.0 \times 10^5\ \textbf{J K}^{-1}$$

10-73. The equation for equilibrating two masses at different temperatures is:
$$m_1\,c_{s1}\,\Delta t_1 = -\,m_2\,c_{s2}\,\Delta t_2$$
For this problem, substance 1 will be the water ("w") and substance 2 will be the unknown ("u"). The final temperature for each substance is the same.

Then: $\qquad\qquad m_w\,c_{s,w}\,(t_f - t_{w,i}) = -\,m_u\,c_{s,u}\,(t_f - t_{u,i})$

We have
$$m_w = 100.0\ g \qquad\qquad m_u = 61.0\ g$$
$$c_{s,w} = 4.18\ J\ {}^\circ C^{-1}\ g^{-1} \qquad t_f = 26.39\ {}^\circ C$$
$$t_{w,i} = 20.00 {}^\circ C \qquad\qquad t_{u,i} = 120.0 {}^\circ C$$

$$100.0\ g \times 4.18\ J\ ({}^\circ C)^{-1}\ g^{-1}\ (26.39 - 20.00 {}^\circ C) = -\,c_{s,u} \times 61.0\ g \times (26.39 - 120.0 {}^\circ C)$$
$$2671\ J = c_{s,u} \times 5710\ g\ {}^\circ C$$
$$c_{s,u} = \left(\frac{2671\ J}{5710\ g\ {}^\circ C}\right) = \textbf{0.468 J g}^{-1}\ \textbf{(}^\circ\textbf{C)}^{-1}$$

10-75. The heat released in the combustion is:
$$q = 14.3\ g\ glucose \times \left(\frac{1\ mol\ glucose}{180.16\ g\ glucose}\right) \times \left(\frac{-2{,}820{,}000\ J}{1\ mol\ glucose}\right) = -2.24 \times 10^5\ J.$$

If this is absorbed by the water in the body, then $q_{water} = +2.24 \times 10^5\ J.$

Then: $\qquad m_w\,c_w\,\Delta t = 2.24 \times 10^5\ J$; We have a mass of 50 kg = 50,000 g.
$$\Delta t = \frac{2.24 \times 10^5\ J}{(50{,}000g) \times \left(4.18\ J\ K^{-1}\ g^{-1}\right)} = 1.07\ K = \textbf{1.07} {}^\circ\textbf{C}.$$

10-77. This problem is best done by a careful budget of the heat changes, knowing that the total heat change for all components of the system is zero:

$0 = q\ (ice\ melting) + q(water\ from\ ice\ warming) + q\ (water\ in\ glass\ cooling)$

The terms are listed separately for clarity.

$$q\ (ice\ melting) = 36.0\ g\ water \times \left(\frac{+\ 333J}{1\ g\ water}\right) = 11{,}988\ J$$

$$q(water\ from\ ice\ warming) = 36.0\ g\ water \times \left(\frac{4.18\ J}{1^\circ C\ g^{-1}}\right) \times \left(t_{final} - 0.0\ {}^\circ C\right)$$
$$= 150.48\ J\ {}^\circ C^{-1}\ t_{final}$$

10-77.
$$q \text{ (water in glass cooling)} = 360 \text{ g water} \times \left(\frac{4.18 \text{ J}}{1^{\circ}\text{C g}^{-1}}\right) \times \left(t_{\text{final}} - 20.0 \text{ }^{\circ}\text{C}\right)$$

$$= 1504.8 \text{ J }^{\circ}\text{C}^{-1} t_{\text{final}} - 30{,}096 \text{ J}$$

Then in our original heat budget equation:

$$0 = 11{,}988 \text{ J} + 150.48 \text{ J } (^{\circ}\text{C})^{-1} t_{\text{final}} + 1504.8 \text{ J } (^{\circ}\text{C})^{-1} t_{\text{final}} - 30{,}096 \text{ J}$$

$$1655 \text{ J } (^{\circ}\text{C})^{-1} t_{\text{final}} = 30{,}096 \text{ J} - 11{,}988 \text{ J} = 18{,}108 \text{ J}$$

$$t_{\text{final}} = \frac{18{,}108 \text{ J}}{1{,}655 \text{ J } (^{\circ}\text{C})^{-1}} = \textbf{10.9 }^{\circ}\textbf{C}.$$

10-79. This is true. If the substance exists as a liquid in equilibrium with a gas at 1 atm and 298 K, then it can have standard states associated with both phases.

10-81. The cooling ability *"near its boiling point"* is connected to the specific heat. The cooling ability *"at its boiling point"* is connected to the enthalpy of vaporization. Helium can absorb more heat per gram near the boiling point, so helium is a better coolant **near** the boiling point. However, nitrogen requires more heat per gram for vaporization. It is a better coolant **at** its boiling point.

10-83. (a) We assume that the specific heats of the initial and product solutions are the same as that of the water. We approximate that the water has a density of 1.00 g mL^{-1}. We have a total mass of water of 200.0 g and a temperature change of +1.31 K. For the heat change of the calorimeter, the water, and the overall system:

Calorimeter: $q_{\text{calorimeter}} = C \Delta T = 192 \text{ J K}^{-1} \times 1.31 \text{ K} = 251 \text{ J}$

Water: $q_{\text{water}} = c_s m \Delta t = 4.184 \text{ J g}^{-1} \text{ K}^{-1} \times 225.0 \text{ g} \times 2.64 \text{ K} = 1096 \text{ J}$

Total: $q_{\text{total}} = q_{\text{calorimeter}} + q_{\text{water}} = 251 + 1096 \text{ J} = 1347 \text{ J}.$

This is the opposite of the heat change for the chemical reaction, $q_r = -1347 \text{ J}$. The reaction in this case is: $Ag^+ + Cl^- \rightarrow AgCl$. We react equal mole amounts of both silver and chloride ions. So the mole amount for the reaction is 0.0200 mol. We use this and the heat change for the reaction to get the reaction enthalpy:

$$\Delta H_r = \frac{-1347 \text{ J}}{0.0200} = -67350 \text{ J} = \textbf{- 67.4 kJ}$$

(b) These values yield a result of **–65.49 kJ**

(c) Of course the answers would differ slightly, because the tabulated values are for a temperature of exactly 298.15 K and the reaction is studies at a lower temperature. ΔH will be expected to vary slightly by temperature.

10-85. (a) $\Delta H^o = 2 \Delta H_f^o(CO_2) + \Delta H_f^o(H_2O \ (g)) - \Delta H_f^o(C_2H_2) - \frac{5}{2} \Delta H_f^o(O_2)$

$= 2 \ mol \times (-393.51 \ kJ \ mol^{-1}) + 1 \ mol \ (-241.82 \ kJ \ mol^{-1})$

$- 1 \ mol \ (+226.73 \ kJ \ mol^{-1}) - \frac{5}{2} \ mol \ (0 \ kJ \ mol^{-1})$

$= - 1255.57 \ kJ = \mathbf{-1,255,570 \ J}$

(b) $c_p = c_{p,carbon \ dioxide} + c_{p,water}$

$= 2.00 \ mol \ CO_2 \times \left(\dfrac{37 \ J}{K - mol \ CO_2} \right) + 1.00 \ mol \ H_2O \times \left(\dfrac{36 \ J}{K - mol \ H_2O} \right)$

$= \mathbf{110 \ J \ K^{-1}}.$

(c) The heat released in the combustion of one mol of acetylene (1,255,570 J) goes into warming the two mol of CO_2 and one mol of H_2O produced.

The combined heat capacity of the gases is $110 \ J \ K^{-1}$.

$$q_{gases} = -q_{reaction} = +1,255,570 \ J$$
$$c_{p,gases} \Delta T = 1,255,570 \ J$$
$$\Delta T = {}^{1,255,570 \ J}/_{110 \ J \ K^{-1}} = 11,400 \ K \ (= 11,100^oC)$$

Given an initial temperature of 20^oC, the final temperature will be $\approx \mathbf{11,100^oC}$.

10-87. We only need $\Delta H_f^o(CO_2)$ and $\Delta H_f^o(SiO_2)$, because we know

$\Delta H_f^o(N_2) = \Delta H_f^o(C) = 0 \ kJ \ mol^{-1}$.

10-89. The hint reminds us to go entirely with one system of units - SI in this case. Then the heat capacity of the ball becomes $850 \ J \ K^{-1} \ kg^{-1}$. We also know that the change in temperature must be $1.00^oC = 1.00 \ K$

The heat transfers must equal zero in the end:

$0 = q$(generated in falling ball) $+ q$ (absorbed in warming the ball)

where q(generated in falling ball) is the ball's decrease in potential energy.

$0 = m_{ball}g\Delta h + m_{ball} \ c_{s,ball} \ \Delta t = m_{ball} \ (g\Delta h + c_{s,ball} \ \Delta t)$

The common factor m_{ball} cancels out, so:

$0 = g\Delta h + c_{s,ball} \ \Delta t$

Or: $\Delta h = \dfrac{-c_{s,ball} \ \Delta t}{g} = \dfrac{\left(850 \ J \ K^{-1} \ kg^{-1} \right) \times (1.00 \ K)}{9.81 \ m \ s^{-2}}$

$= 86.6 \ J \left(\dfrac{s^2}{kg - m} \right) = 86.6 \left(\dfrac{kg - m^2}{s^2} \right) \times \left(\dfrac{s^2}{kg - m} \right) = \mathbf{86.6 \ m}$

10-91. The work done by the expanding gas is equal to the work done by the paddle on the water. Hence:

$$w_{gas} = q_{water}$$

$$P_{ext} \Delta V = m \, c_s \, \Delta T$$

$$1.00 \text{ atm } (13.00 - 5.00 \text{ L}) = (1{,}000 \text{ g}) \times (4.18 \text{ J K}^{-1} \text{ g}^{-1}) \times \Delta T$$

$$(8.00 \text{ L-atm}) \times (101.325 \text{ J/ L-atm}) = 4180 \text{ J K}^{-1} \times \Delta T$$

$$(810.5 \text{ J}/4180 \text{ J K}^{-1}) = \Delta T = 0.194 \text{ K} = \mathbf{0.194^oC}$$

10-93. (a) The energy consumed in running a 26 mile race at 418 kJ per mile is 10,900 kJ (that's almost 11 megajoules!). If this is converted to heat, and the heat is dissipated by evaporation of water, then the number of grams of water evaporated is:

$$10{,}900 \text{ kJ } \times \frac{1 \text{ mol g H}_2\text{O}}{44 \text{ kJ}} \times \frac{18.015 \text{ g H}_2\text{O}}{1 \text{ mol H}_2\text{O}} = 4500 \text{ g H}_2\text{O}.$$

This corresponds to 4,500 mL of water, or 4.5 L.

(b) If you have ever run a marathon (I have not, but text author Wade Freeman has) or seen someone who has (I have), then you can easily imagine 4.5 L of water being consumed during the race. So the estimate seems low for the amount of water lost.

In fact, much of the water excreted as perspiration is lost, not as water vapor, but as liquid water. Thus, perspiration is an imperfect cooling mechanism.

Heat can also be lost by direct cooling by air. But unlike some other animals, human beings have a limited ability to deliberately increase direct cooling (as when an elephant flaps its ears or a dog pants) beyond wearing very little clothing and shaving the head.

Regarding other energy dissipation mechanisms, this problem posits energy loss only by heat. Energy can also be lost through work, including the considerable amount of energy converted into mechanical energy of motion of the legs and arms.

CHAPTER 11 - SPONTANEOUS CHANGE AND EQUILIBRIUM

11-1. (a) $$4\ Fe\ (s) + 3\ O_2\ (g) \rightarrow 2\ Fe_2O_3\ (s)$$
This reaction is exothermic and spontaneous at room temperature.

(b) $$NH_4NO_3\ (s) \rightarrow NH_4^+\ (aq) + NO_3^-\ (aq)$$
This reaction is endothermic and spontaneous at room temperature.

11-3. Good examples are given by procedures used for cooling:

(i) vaporization of isopropyl alcohol ("rubbing alcohol") from skin.

(ii) melting ice in warm water.

(iii) use of a "cold pack," possibly based on dissolving NH_4NO_3 in water

11-5. (a) This is best done by enumeration. There are six states for die 1. Each can be paired with six states of die 2. Thus: No. microstates = $6 \times 6 = $ **36**

(b) This combination represents one of thirty-six equally probable states, so the probability of two sixes is $1/36 = $ **0.02778**.

11-7. (a) *Any* melting process of a pure substance has $\Delta S > 0$.

(b) Any reduction of structure (literally in this case) is a reduction in order and therefore has $\Delta S > 0$.

(c) Any separation of a mixture represents an increase in order: $\Delta S < 0$.

11-9. Appendix D has the data we could use for this, but we can also answer these questions conceptually.

(a) Notice that there are *two* atoms of F in F_2, and that this particle has the ability to tumble. This increases the number of states, so F_2 (*g*, 300 K) has the higher standard entropy.

(b) The vaporization of any substance involves a massive increase in the space available for it to occupy, and a similar increase in the number of positions the molecules can be found at. So H_2O (*g*, 300 K) has the higher standard entropy.

(c) Any substance increases its standard entropy when it warms. So H_2O (*g*, 400 K) has the higher standard entropy.

(d) The ozone molecule has three atoms and is bent. This increases the number of positions each molecule can occupy. So O_3 (*g*, 300 K) has the higher standard entropy.

11-11. (a) This reaction has two moles of gas reacting to form one mole of gas so $\Delta S < 0$.

(b) This reaction two moles of gas condense to a solid and $\Delta S < 0$.

(c) Simple breaking of one molecule into two atoms increases the available positions for the particles to occupy and always has $\Delta S > 0$.

(d) This involves freezing of a pure substance so $\Delta S < 0$.

11-13. (a)
$$\Delta S_r^o = 2 \times S^o (H_2O\ (l)) + 2S^o (NO_2\ (g)) - S^o (N_2H_4\ (l)) - 3S^o (O_2\ (g))$$
$$= 2\ mol\ (69.91\ J\ K^{-1}\ mol^{-1}) + 2\ mol\ (239.95\ J\ K^{-1}\ mol^{-1})$$
$$-\ 1\ mol\ (121.21\ J\ K^{-1}\ mol^{-1}) - 3\ mol\ (205.03\ J\ K^{-1}\ mol^{-1})$$
$$= \mathbf{-116.58\ J\ K^{-1}}$$

(b) The 120.00 g of N_2H_4 is the same as 3.7447 moles. We have the entropy change when one mole of N_2H_4 reacts. So when 120.00 g reacts, the entropy change is:

$$\Delta S = 3.7447\ mol\ N_2H_4 \times \frac{-116.58\ J\ K^{-1}}{1\ mol\ N_2H_4} = \mathbf{436.56\ J\ K^{-1}}$$

11-15. All reactions involve $Cl_2\ (g)$ as a reactant. The reaction in each case (using "M" for the metals) is: $2\ M + Cl_2 \rightarrow 2\ MCl$. We use $S^o\ (Cl_2\ (g)) = 222.96$. The other data and results may be presented in a table:

M	$S^o(M)$ $J\ K^{-1}\ mol^{-1}$	$S^o(MCl)$ $J\ K^{-1}\ mol^{-1}$	$\Delta S_r^o = 2S^o(MCl) - 2S^o(M) - S^o(Cl_2)$ $J\ K^{-1}$
Li	29.12	59.33	-162.54
Na	51.21	72.13	-181.12
K	64.18	82.59	-186.14
Rb	76.78	95.90	-184.72
Cs	85.23	101.17	-191.08

Except for Rb, ΔS_r^o becomes **increasingly negative** down the group.

11-17.
$$\Delta S_r^o = S^o(BaCl_2\ (s)) - S^o(Ba^{2+}(aq)) - 2S^o\ (Cl^-(aq))$$
$$= 1\ mol\ (123.68\ J\ K^{-1}\ mol^{-1}) - 1\ mol\ (9.6\ J\ K^{-1}\ mol^{-1})$$
$$-\ 2\ mol\ (56.5\ J\ K^{-1}\ mol^{-1})$$
$$= \mathbf{+1.1\ J\ K^{-1}}$$

11-19. The reaction is largely driven by an increase in entropy, for there is now a new chemical species -- HOD -- in the mixture.

11-21. Spontaneity in a chemical change is determined by the sign of the change in the entropy of the system plus its surroundings. An increase of the entropy of the surroundings can offset decreases of entropy (= increases in order) within a system.

11-23. The entropy of the surroundings must increase by *at least* **44.7 J K⁻¹**.

11-25. The entropy of a substance has an absolute reference--0 J K⁻¹ at 0 K. Therefore, we can define and use an absolute entropy, equal to the entropy change in bringing that substance from 0 K to some other conditions. It is therefore a property of the substance. This differs from enthalpy, where there is no absolute reference and it is necessary to define a "zero", in fact the elements in their standard state. The enthalpies of formation that we read in tables refer to the only meaning of enthalpy, as a *change* from some defined zero.

11-27. $\Delta S_{fus} = \dfrac{\Delta H_{fus}}{T_{fus}} = \dfrac{35,400 \text{J mol}^{-1}}{(3410+273)\text{K}} = \textbf{+9.61 J K}^{-1}\textbf{ mol}^{-1}$

11-29. (a) $\Delta G = \Delta H - T\Delta S = +5,650 \text{ J K}^{-1} - (170 \text{ K})(28.9 \text{ J K}^{-1}\text{ mol}^{-1}) = \textbf{+737 J mol}^{-1}$

(b) $(737 \text{ J mol}^{-1}) \times (3.60 \text{ mol}) = 2650 \text{ J} = \textbf{2.65 kJ}$

(c) The answer is **no**. The melting of ammonia is non-spontaneous at 170 K

(d) $\Delta G = 0 \quad = \Delta H = T\Delta S \quad \Rightarrow \quad T = \dfrac{\Delta H}{\Delta S} = \dfrac{5,650 \text{J mol}^{-1}}{28.9 \text{ Jmol K}^{-1}} = \textbf{196 K}$

11-31. Trouton's rule: $\Delta S_{vap} \approx 88 \text{ J K}^{-1}\text{ mol}^{-1}$

$\dfrac{\Delta H_{vap}}{T_{vap}} \approx 88 \text{ J K}^{-1}\text{ mol}^{-1} \rightarrow \Delta H_{vap} \approx (T_{vap})(88 \text{ J K mol}^{-1})$

$\approx (56.2 + 273.15 \text{ K})(88 \text{ J K}^{-1}\text{ mol}^{-1}) = 29,000 \text{ J K}^{-1} = \textbf{29 kJ mol}^{-1}$

11-33. In each case we must consider the S^{o} and ΔH_r^{o} for the reaction, then determine

$\Delta G_r^{o} = \Delta H_r^{o} - T\Delta S_r^{o}$

Note that the value of ΔS_r^{o} has to be calculated for the formation of the substance *from its element* form.

(a) Reaction for ΔS_r^{o} calculation: Li (s) + 1/2 Cl₂ (g) → LiCl (s)

$\Delta S_r^{o} = S^{o}\text{ (LiCl)} + 1/2 \, S^{o}\text{ (Cl}_2) - S^{o}\text{(Li)}$

$= 59.33 \text{ J K}^{-1} - \frac{1}{2}(222.96 \text{ J K}^{-1}) - (29.12 \text{ J K}^{-1})$

$= -81.27 \text{ J K}^{-1} = -0.08127 \text{ kJ K}^{-1}$

$\Delta G_r^{o} = \Delta H_f^{o} - T \Delta S_r^{o} = -408.61 \text{ kJ} - (298 \text{ K})(-0.08127 \text{ kJ K}^{-1})$

$= \textbf{-384.38 kJ}$

11-33. (continued)

(b) Reaction: Na $(s) \rightarrow$ Na (g)

$\Delta S_r^O = S^o$ (Na, g) $- S^o$(Na, s) $= 153.6 \text{ J K}^{-1} - 51.21 \text{ J K}^{-1}$

$= +102.39 \text{ J K}^{-1} = 0.10239 \text{ kJ K}^{-1}$

$\Delta G_r^o = \Delta H_f^o - T \Delta S_r^o = -107.32 \text{ kJ} - (298 \text{ K}) (0.10239 \text{ kJ K}^{-1})$

$= \textbf{76.79 kJ}$

(c) Reaction: C + 2 H$_2$ + 1/2 O$_2 \rightarrow$ CH$_3$OH (l)

$\Delta S_r^o = S^o$ (CH$_3$OH, l) $- S^o$(C, s) $- 2\, S^o$(H$_2$, g) $- \frac{1}{2} S^o$(O$_2$, g)

$= 160.7 \text{ J K}^{-1} - 51.21 \text{ J K}^{-1}$

$= +102.39 \text{ J K}^{-1} = 0.10239 \text{ kJ K}^{-1}$

$\Delta G_r^o = \Delta H_f^o - T \Delta S_r^o = -107.32 \text{ kJ} - (298 \text{ K}) (0.10239 \text{ kJ K}^{-1})$

$= 76.79 \text{ kJ}$

11-35. (a) $\Delta G_r^o = \Delta G_f^o$ (Sr (s)) $+ \Delta G_f^o$ (CaCl$_2$ (s)) $- \Delta G_f^o$ (Ca (s)) $- \Delta G_f^o$ (SrCl$_2$ (s))

$= 1 \text{ mol } (0 \text{ kJ mol}^{-1}) + 1 \text{ mol } (-748.1 \text{ kJ mol}^{-1})$

$- 1 \text{ mol } (0 \text{ kJ mol}^{-1}) - 1 \text{ mol } (-781.1 \text{ kJ mol}^{-1}) = \textbf{+33.0 kJ}$

(b) $\Delta G_r^o = \Delta G_f^o$ (Al$_2$O$_3$ (s)) $+ 2 \times \Delta G_f^o$ (Fe (s))

$- 2 \times \Delta G_f^o$ (Al (s)) $- \Delta G_f^o$ (Fe$_2$O$_3$ (s))

$= 1 \text{ mol } (-1582.3 \text{ kJ mol}^{-1}) + 2 \text{ mol } (0 \text{ kJ mol}^{-1})$

$- 2 \text{ mol } (0 \text{ kJ mol}^{-1}) - 1 \text{ mol } (-742.2 \text{ kJ mol}^{-1}) = \textbf{-840.1 kJ}$

11-37. $\Delta H_r^o = \Delta H_f^o$ (CoCl$_2 \bullet$6H$_2$O (s)) $- \Delta H_f^o$ (CoCl$_2$ (s)) $- 6 \times \Delta H_f^o$ (H$_2$O (l))

$= 1 \text{ mol } (-2115.4 \text{ kJ mol}^{-1}) - 1 \text{ mol } (-312.5 \text{ kJ mol}^{-1})$

$- 6 \text{ mol } (-285.83 \text{ kJ mol}^{-1}) = \textbf{-87.9 kJ}$

$\Delta S_r^o = S^o$ (CoCl$_2 \bullet$6 H$_2$O (s)) $- S^o$ (CoCl$_2$ (s)) $- 6 \times S^o$ (H$_2$O (l))

$= 1 \text{ mol } (343 \text{ J K}^{-1} \text{ mol}^{-1}) - 1 \text{ mol } (109.16 \text{ J K}^{-1} \text{ mol}^{-1})$

$- 6 \text{ mol } (69.91 \text{ J K}^{-1} \text{ mol}^{-1}) = \textbf{-186 J K}^{-1}$

$\Delta G_r^o = \Delta G_f^o$ (CoCl$_2 \bullet$6 H$_2$O (s)) $- \Delta G_f^o$ (CoCl$_2$ (s)) $- 6 \times \Delta G_f^o$ (H$_2$O (l))

$= 1 \text{ mol } (-1725.2 \text{ kJ mol}^{-1}) - 1 \text{ mol } (-269.8 \text{ kJ mol}^{-1})$

$- 6 \text{ mol } (-237.18 \text{ kJ mol}^{-1}) = \textbf{-32.3 kJ}$

11-39. At *low* temperature, $T\Delta S$ becomes small, so the enthalpy change is most important in determining ΔG.   (a) $\Delta G < 0$ (b) $\Delta G < 0$ (c) $\Delta G > 0$ (d) $\Delta G > 0$

11-41. (a) $\Delta H_r^o = 2 \times \Delta H_f^o$ (Fe$_2$O$_3$ (s)) - 4 $\times \Delta H_f^o$ (Fe (s)) - 3 $\times \Delta H_f^o$ (O$_2$ (g))

$\qquad$ = 2 mol (-824.2 kJ mol^{-1}) - 4 mol (0 kJ mol^{-1}) - 3 mol (0 kJ mol^{-1}) = **−1648.4 kJ**

$\Delta S_r^o = 2 \times S^o$ (Fe$_2$O$_3$ (s)) - 4 $\times S^o$ (Fe (s)) - 3 S^o (O$_2$ (g))

$\qquad$ = 2 mol (87.40 J K^{-1} mol^{-1}) - 4 mol (27.28 J K^{-1} mol^{-1}) - 3 mol (205.03 J K^{-1} mol^{-1})

$\qquad$ = **- 549 J K^{-1}**

$$\Delta H^o - T\Delta S^o = \Delta G^o$$

$$\Delta H^o - T\Delta S^o = 0 \text{ at equilibrium}$$

$$T = \frac{\Delta H}{\Delta S} = \frac{-1648.400 J}{-549 J K^{-1}} = 3000 \text{ K}$$

Since both ΔH^o and ΔS^o are negative, the reaction will be spontaneous at temperatures *lower* than 3000 K: **0 < T < 3000 K.**

(b)

$\Delta H_r^o = \Delta H_f^o$ (SO$_3$ (g)) - ΔH_f^o (SO$_2$ (g)) - $\frac{1}{2}$ ΔH_f^o (O$_2$ (g))

$\qquad$ = 1 mol (-395.72 kJ mol^{-1}) - 1 mol (-296.83 kJ mol^{-1}) - $\frac{1}{2}$ mol (0 kJ mol^{-1})

$\qquad$ = -98.89 kJ = -98,890 J

$\Delta S_r^o = S^o$ (SO$_3$) - S^o (SO$_2$) - $\frac{1}{2}$ S^o (O$_2$)

$\qquad$ = 1 mol (256.65 J K^{-1} mol^{-1}) - 1 mol (248.11 J K^{-1} mol^{-1}) - $\frac{1}{2}$ mol(205.03 J K^{-1} mol^{-1})

$\qquad$ = -93.97 J K^{-1}

$$\Delta H - T\Delta S = 0 \text{ at equilibrium}$$

$$\frac{\Delta H}{\Delta S} = T = \frac{-98,890 J}{-93.97 J K^{-1}} = 1052K$$

Since both ΔH^o and ΔS^o are negative, the reaction will be spontaneous at temperatures *lower* than 1050 K: **0 < T < 1050 K.**

(c)

$\Delta H_r^o = \Delta H_f^o$ (N$_2$O (g)) + 2 $\times \Delta H_f^o$ (H$_2$O (g)) - ΔH_f^o (NH$_4$NO$_3$ (s))

$\qquad$ = 1 mol (82.05 kJ mol^{-1}) + 2 mol (-241.82 kJ mol^{-1}) - 1 mol (-365.56 kJ mol^{-1})

$\qquad$ = -36.03 kJ = -36,030 J

$\Delta S_r^o = S^o$ (N$_2$O (g)) + 2 $\times S^o$ (H$_2$O (g)) - S^o (NH$_4$NO$_3$ (s))

$\qquad$ = 1 mol (219.74 J K^{-1} mol^{-1}) + 2 mol (118.72 J K^{-1} mol^{-1}) - 1 mol (151.08 J K^{-1} mol^{-1})

$\qquad$ = +306.1 J K^{-1}

With $\Delta H^o < 0$ and $\Delta S^o > 0$, the reaction will be **spontaneous at all T.**

11-43. The reaction is: WO_3 *(s)* + 3 H_2 *(g)* $\rightarrow$ W *(s)* + 3 H_2O *(g)*

$$\Delta H_r^o = \Delta H_f^o \text{ (W (s))} + 3\ \Delta H_f^o \text{ (}H_2O\text{ (g)) - } \Delta H_f^o \text{ }WO_3 \text{ (s)) - 3 } \Delta H_f^o \text{ (}H_2 \text{ (g))}$$

$$= 1 \text{ mol } (0.0 \text{ kJ mol}^{-1}) + 3 \text{ mol } (-241.82 \text{ kJ mol}^{-1})$$

$$- 1 \text{ mol } (-842.87 \text{ kJ mol}^{-1}) - 3 \text{ mol } (-0.0 \text{ kJ mol}^{-1}) = +\ \mathbf{117.41\ kJ}$$

$$\Delta S_r^o = S^o \text{(W (s))} + 3\ S^o \text{(}H_2O\text{ (g)) - } S^o\text{(}WO_3\text{ (s)) - 3 } S^o \text{(}H_2 \text{ (g))}$$

$$= 1 \text{ mol } (32.64 \text{ J K}^{-1} \text{ mol}^{-1}) + 3 \text{ mol } (\ 188.72 \text{ J K}^{-1} \text{ mol}^{-1})$$

$$- 1 \text{ mol } (75.90 \text{ J mol}^{-1} \text{ K}^{-1}) - 3 \text{ K } (130.57 \text{ J mol}^{-1} \text{ mol}^{-1}) = +\ \mathbf{131.19\ J\ K^{-1}}$$

The significant positive standard entropy change indicates that, at high enough temperatures, the $-T\Delta S^o$ term in the expression for the Gibbs' energy will become large and negative -- large enough to overcome the positive ΔH^o.

$$\Delta G < 0 \text{ for } T > \frac{\Delta H^o}{\Delta S^o} = \frac{117,410 \text{ J}}{131.19 \text{ J K}^{-1}} = 895 \text{ K}$$

11-45. We must carry out a Hess' law calculation using the first reaction and the *reverse* of the

second reaction:		ΔH^o	ΔG^o
	H_2 *(g)* + CO_2 *(g)* $\rightarrow$ HCOOH *(g)*	+30.9 kJ	+58.6 kJ
	HCOOH *(l)* $\rightarrow$ H_2 *(g)* + CO_2 *(g)*	+31.2 kJ	-32.9 kJ
combined:	HCOOH *(l)* $\rightarrow$ HCOOH *(g)*	30.9 +31.2	58.6 $-$ 32.9
		= + 62.1 kJ	**= + 25.7 kJ**

We can determine ΔS^o by solving:

$$\Delta G^o = \Delta H^o - T\Delta S^o$$

$$\Delta S^o = -(\Delta G^o - \Delta H^o) / T$$

$$= -(25,700 \text{ J} - 62,100 \text{ J}) / 298 \text{ K} = \mathbf{-122\ J\ K^{-1}}$$

11-47. $\Delta G_r^o = 3 \times \Delta G_f^o \text{ (}H_2O\text{ (g)) + } 2 \times \Delta G_f^o \text{ (}NO_2\text{ (g)) - } \Delta G_f^o \text{ (}O_2\text{ (g)) - } 2 \times \Delta G_f^o \text{ (}NH_3 \text{ (g))}$

$$= 3 \text{ mol } (-228.59 \text{ kJ mol}^{-1}) + 2 \text{ mol } (51.29 \text{ kJ mol}^{-1})$$

$$- \frac{7}{2} \text{ mol } (0 \text{ kJ mol}^{-1}) - 2 \text{ mol } (-16.48 \text{ kJ mol}^{-1})$$

$$= -550.23 \text{ kJ} = \mathbf{-550,230\ J} = -550,230 \text{ J per mol of reaction}$$

$$K = e^{-\left(\Delta G^o / RT\right)} = e^{-\left(-550,230 \text{ J mol}^{-1} / \left(8.3145 \text{J mol}^{-1} \text{ K}^{-1} \times 298.15 \text{ K}\right)\right)} = \mathbf{2.5 \times 10^{+96}}$$

11-49. (a) $\Delta G_r^o = \Delta G_f^o \, (SO_3 \; (g)) - \Delta G_f^o \, (SO_2 \; (g)) - \dfrac{1}{2} \Delta G_f^o \, (O_2 \; (g))$

$= 1\text{mol} \, (-371.08 \text{ kJ mol}^{-1}) - 1\text{mol} \, (-300.19 \text{ kJ mol}^{-1}) - \dfrac{1}{2}\text{mol} \, (0 \text{ kJ mol}^{-1})$

$= -70.89 \text{ kJ} = -70{,}890 \text{ J} = -70.890 \text{ J per mol of reaction}$

$K = e^{-\left(-70{,}890 \text{ Jmol}^{-1}/\left(8.3145 \text{J mol}^{-1} \text{ K}^{-1}\right)\left(298.15 \text{ K}\right)\right)} = e^{+28.60} = \mathbf{2.6 \times 10^{12}}$

$$K = \dfrac{P_{SO_3}}{\left(P_{SO_2}\right)\left(P_{O_2}\right)^{1/2}}$$

(b) $\Delta G_r^o = 2 \times \Delta G_f^o \, (Fe_3O_4 \; (s)) + \dfrac{1}{2} \Delta G_f^o \, (O_2 \; (g)) - 3 \times \Delta G_f^o \, (Fe_2O_3 \; (s))$

$= 2 \text{ mol} \, (-1015.5 \text{ kJ mol}^{-1}) + \dfrac{1}{2} \text{ mol} \, (0 \text{ kJ mol}^{-1}) - 3 \text{ mol} \, (-742.2 \text{ kJ mol}^{-1})$

$= +195.6 \text{ kJ} = +195{,}600 \text{ J} = +195{,}600 \text{ J per mol of reaction}$

$K = e^{-\left(+195{,}600 \text{ J mol}^{-1}/\left(8.3145 \text{J mol}^{-1} \text{ K}^{-1}\right)\left(298.15 \text{ K}\right)\right)} = e^{-78.90} = \mathbf{5.4 \times 10^{-35}}$

$$K = \left(P_{O_2}\right)^{1/2}$$

(c) $\Delta G^o = \Delta G_f^o \, (Cu^{2+} \; (aq)) + 2 \times \Delta G_f^o \, (Cl^- \; (aq)) - \Delta G_f^o \, (CuCl_2 \; (s))$

$= 1 \text{ mol} \, (+65.49 \text{ kJ mol}^{-1}) + 2 \text{ mol} \, (-131.23 \text{ kJ mol}^{-1})$

$- 1 \text{ mol} \, (-175.7 \text{ kJ mol}^{-1})$

$= -21.27 \text{ kJ} = \mathbf{-21{,}270 \text{ J}} = -21{,}270 \text{ J per mol of reaction}$

$K = e^{-\left(-21{,}270 \text{ J mol}^{-1} /\left(8.3145 \text{J mol}^{-1} \text{ K}^{-1}\right)\left(298.15 \text{ K}\right)\right)} = e^{+8.58} = \mathbf{5300}$

$$K = [Cu^{2+}] \, [Cl^-]^2$$

11-51. $\Delta G^o = \Delta G_f^o \, (HPO_4^{2-} \; (aq)) + \Delta G_f^o \, (H_3O^+ \; (aq))$

$- \Delta G_f^o \, (H_2PO_4^- \; (aq)) - \Delta G_f^o \, (H_2O \; (l))$

$= 1 \text{ mol} \, (-1089.15 \text{ kJ mol}^{-1}) + 1 \text{ mol} \, (-237.18 \text{ kJ mol}^{-1})$

$- 1 \text{ mol} \, (-1130.28 \text{ kJ mol}^{-1}) - 1 \text{ mol} \, (-237.18 \text{ kJ mol}^{-1})$

$= +41.13 \text{ kJ} = +41{,}130 \text{ J} = 41{,}130 \text{ J per mol of reaction}$

$K = e^{-\left(+41{,}130 \text{ J mol}^{-1} /\left(8.3145 \text{J mol}^{-1} \text{ K}^{-1}\right)\left(298.15 \text{ K}\right)\right)} = e^{-16.59} = \mathbf{6.2 \times 10^{-8}}$

11-53. All that is required is that ΔS be sufficiently negative to "overwhelm" ΔH. It makes sense that ΔS is very negative, because the reaction involves trapping (condensing) a gas in a liquid.

11-55. $\Delta G_{590}^o = 29{,}400 \text{ J} = 29{,}400 \text{ per mol of reaction}$

$K = e^{-\left(+29{,}400 \text{ J mol}^{-1}/\left(8.3145 \text{J mol}^{-1} \text{ K}^{-1}\right)\left(590 \text{K}\right)\right)} = e^{-5.99} = \mathbf{0.0025}$

Chapter 11

11-57. The information in the problem indicates that heat drives O_2 out of water, so that process must be endothermic. The reaction written in the problem is for dissolving O_2 *into* water. This must be exothermic, or $\Delta H < 0$.

11-59. Since the reaction gives fewer moles of gaseous products than gaseous reactants, **increasing pressure** will favor the products . The reaction is exothermic and **decreasing temperatures** will likewise drive the reaction to the right.

11-61.
$$\Delta H^o = 2\,\Delta H^o_f\ (NH_3\ (g)) - 3\,\Delta H^o_f\ (H_2\ (g)) - \Delta H^o_f\ (N_2\ (g))$$

$$= 2\ \text{mol}\ (-46.11\ \text{kJ mol}^{-1}) - 3\ \text{mol}\ (0\ \text{kJ mol}^{-1}) - 1\ \text{mol}\ (0\ \text{kJ mol}^{-1})$$

$$= -92.22\ \text{kJ} = -92,220\ \text{J}$$

$$\Delta S^o = 2\,S^o\ (NH_3\ (g)) - 3\,S^o\ (H_2\ (g)) - S^o\ (N_2\ (g))$$

$$= 2\ \text{mol}\ (+192.34\ \text{J mol}^{-1}\ \text{K}^{-1}) - 3\ \text{mol}\ (+130.57\ \text{J mol}^{-1}\ \text{K}^{-1})$$

$$\qquad - 1\ \text{mol}\ (191.50\ \text{J mol}^{-1}\ \text{K}^{-1})$$

$$= -198.53\ \text{J K}^{-1}$$

$$\Delta G\ (600\ \text{K}) = \Delta H^o - 600\,\Delta S^o$$

$$= -92,220\ \text{J} - (600\ \text{K})\ (-198.53\ \text{J K}^{-1}) = 26,898\ \text{J}$$

$$= 26,898\ \text{J per mol of the reaction}$$

$$K = e^{-\left(26,898\ \text{J mol}^{-1}/\left(8.3145\text{J mol}^{-1}\ \text{K}^{-1}\right)\left(600\text{K}\right)\right)} = e^{-5.39} = \mathbf{0.0046}$$

11-63. The assumption that ΔH and ΔS are independent of temperature permits the use of the van't Hoff equation.

$$\ln\left(\frac{K_2}{K_1}\right) = \frac{\Delta H^o}{R}\left(\frac{1}{T_1} - \frac{1}{T_2}\right)$$

$K_2 = 1.21 \times 10^{-3}$; $T_2 = 200^o\text{C} = 473.15\ \text{K}$

$K_1 = 6.8$; $T_1 = 25^o\text{C} = 298.15\ \text{K}$

$$\ln\left(\frac{1.21\times10^{-3}}{6.8}\right) = \frac{\Delta H^o}{8.3145\text{J K}^{-1}\ \text{mol}^{-1}}\left(1.2405\times10^{-3}\text{K}^{-1}\right)$$

$$\frac{8.3145\text{J mol}^{-1}\ \text{K}^{-1}}{1.2405\times10^{-3}\text{K}^{-1}}\left(-8.634\right) = \Delta H^o = -58,000\ \text{J mol}^{-1} = -58\ \text{kJ mol}^{-1}$$

This answer gives the enthalpy change for "one mole of the reaction" For the reaction as written, we will have $\Delta H^o = \mathbf{-58\ kJ}$.

11-65. $K_2 = 2900$; $T_2 = 28\ ^oC = 301.15\ K$

$K_1 = 40$ $T_1 = 48\ ^oC = 321.15\ K$

$$\ln\left(\frac{2900}{40}\right) = \frac{\Delta H^o}{8.3145 J K^{-1} mol^{-1}}\left(\frac{1}{321.15 K} - \frac{1}{301.15 K}\right)$$

This yields $\Delta H^o = -170,000\ J\ mol^{-1}$

ΔH for the reaction as written is **-170,000 J = -170 kJ.**

At 28 ^oC:

$$\Delta G^o = -RT \ln K = -(8.3145\ \frac{J}{K\ mol^{-1}})\ (301.15\ K)\ (\ln 2900) = -19,960\ J\ mol^{-1}$$

For the reaction as written, $\Delta G^o = -19,960\ J.$

$$\Delta S^o = -\frac{\Delta G^o - \Delta H^o}{T} = -\frac{(-19,960\ J) - (-172,200\ J)}{301.15\ K} = \textbf{-500 J K}^{-1}$$

11-67. (a) $\Delta G^o = -RT \ln K = -(8.3145\ \frac{J}{K\ mol^{-1}})\ (298\ K)\ \ln (9.3 \times 10^9) = -56,872\ J\ mol^{-1}$

For the reaction as written, $\Delta G^o = -56,900\ J = \textbf{-56.9 kJ}$

(b) For ΔH^o, we employ the van't Hoff equation:

$$\ln\left(\frac{K_2}{K_1}\right) = \frac{\Delta H^o}{R}\left(\frac{1}{T_1} - \frac{1}{T_2}\right)$$

$$\ln\left(\frac{3.3 \times 10^7}{9.3 \times 10^9}\right) = \frac{\Delta H^o}{8.3145 J K^{-1} mol^{-1}}\left(\frac{1}{298\ K} - \frac{1}{398\ K}\right)$$

$$-5.64 = \Delta H^o\ \frac{\left(8.43 \times 10^{-4}\ K^{-1}\right)}{8.3145 J K^{-1} mol^{-1}}$$

$$\Delta H^o = -5.64 \times \left(\frac{8.3145 J\ mol^{-1}}{8.43 \times 10^{-4}}\right) = -55,630\ J\ mol^{-1}$$

For the reaction as written, $\Delta H^o = -55,630\ J = \textbf{-55.6 kJ}$

$$\Delta S^o = -\frac{\Delta G^o - \Delta H^o}{T} = -\frac{(-56,872\ J) - (-55,630\ J)}{298.15\ K} = \textbf{+4.2 J K}^{-1}$$

11-69. The Clausius-Clapeyron equation is used here:

$P_1 = 0.4034\ atm\ T_1 = -50\ ^oC = 223.15\ K$ $P_2 = 4.2380\ atm\ T_2 = 0\ ^oC = 273.15\ K$

$$\ln\left(\frac{P_2}{P_1}\right) = \frac{\Delta H_{vap}}{R}\left(\frac{1}{T_1} - \frac{1}{T_2}\right)$$

$$\ln\left(\frac{4.2380}{0.4034}\right) = \frac{\Delta H_{vap}}{8.3145 J K^{-1} mol^{-1}}\left(\frac{1}{223.15 K} - \frac{1}{273.15 K}\right)$$

11-69. (continued)

$$2.3519 = \frac{\Delta H_{vap}}{8.3145\text{J K}^{-1}\text{ mol}^{-1}}\left(-8.203\times10^{-4}\text{ K}^{-1}\right)$$

$$\frac{\left(8.3145\text{J mol}^{-1}\right)\left(2.3519\right)}{\left(8.203\times10^{-4}\right)} = \Delta H_{vap} = +23{,}800 \text{ J mol}^{-1} = \mathbf{+23.8\ kJ\ mol^{-1}}$$

(b) The normal boiling point is when P_{vap} = 1 atm. If we set this equal to P_1 and $T_1 = T_b$,

then:
$$\ln(P_2) = \frac{\Delta H_{vap}}{R}\left(\frac{1}{T_b} - \frac{1}{T_2}\right)$$

$$\frac{R\ln(P_2)}{\Delta H_{vap}} = \left(\frac{1}{T_b} - \frac{1}{T_2}\right)$$

$$\frac{1}{T_b} = \frac{R\ln(P_2)}{\Delta H_{vap}} + \frac{1}{T_2}$$

$$= \frac{8.3145 \text{ J mol}^{-1}\text{ K}^{-1}\ \ln(0.4034)}{23{,}800 \text{ J mol}^{-1}} + \frac{1}{223.15 \text{ K}}$$

$$= 0.004164 \text{ K}^{-1}$$

$$\mathbf{T_b = 240\ K}$$

11-71. These data imply: P_2 = 1.00 atm T_2 = 124°C = 397.15 K

P_1 = 0.63 atm T_1 = 107°C = 380.15 K

$$\ln\left(\frac{1.00}{0.63}\right) = \frac{\Delta H_{vap}}{8.3145\text{J K}^{-1}\text{ mol}^{-1}}\left(\frac{1}{380.15\text{K}} - \frac{1}{397.15\text{K}}\right)$$

$$\frac{\left(8.3145\text{J mol}^{-1}\right)\left(0.462\right)}{\left(1.126\times10^{-4}\right)} = \Delta H_{vap} = +34{,}000 \text{ J mol}^{-1} = \mathbf{+34\ kJ\ mol^{-1}}$$

11-73. The data are, for $6\ CO_2\ (g) + 12\ H_2O\ (l) \rightarrow C_6H_{12}O_6 + 6\ O_2\ (g)$:

$$\Delta G^o = +2872 \text{ kJ} \qquad \Delta S^o = -207 \text{ J K}^{-1}$$

$$\Delta H^o - T\Delta S^o = \Delta G^o$$

$$\Delta H^o = \Delta G^o + T\Delta S^o$$

$$= (2{,}872{,}000 \text{ J}) + (298.15 \text{ K})\,(-207 \text{ J K}^{-1})$$

$$= 2{,}810{,}000 \text{ J} = \mathbf{2{,}810\ kJ}$$

11-75. (a)$\Delta G_r^o = \Delta G_f^o\ (Pb^{2+},\ aq) + 2\ \Delta G_f^o\ (I^-,\ aq) - \Delta G_f^o\ (PbI_2\ (s))$

$$= 1 \text{ mol } (-24.43 \text{ kJ mol}^{-1}) + 2 \text{ mol } (-51.57 \text{ kJ mol}^{-1}) - 1 \text{ mol } (-173.64 \text{ kJ mol}^{-1})$$

$$= \mathbf{+46.07\ kJ}$$

$$K = e^{-\left(\Delta G^o/RT\right)} = e^{-\left(46070 \text{ J mol}^{-1}/(8.31451 \text{ J mol}^{-1}\text{ K}^{-1}\times298.15\text{ K})\right)} = \mathbf{1.1 \times 10^{+8}}$$

11-75. (continued)

(b) For this we need to recalculate ΔG_r^o for 373 K, using $\Delta G_r^o = \Delta H_r^o - T\Delta S_r^o$

We approximate that neither ΔH_r^o nor ΔS_r^o will change with temperature. Using the

values at 298.15 K, calculated using Hess' law:

$\Delta G_r^o = \Delta H_r^o - T\Delta S_r^o = 63{,}400 \text{ kJ} - (373.15 \text{ K} \times 58.25 \text{ J K}^{-1}) = +41{,}600 \text{ J}$

From this we calculate $\Delta G_r^o = \mathbf{1.5 \times 10^{-6}}$

(c) $\Delta G < 0$ for $T > \dfrac{\Delta H^o}{\Delta S^o} = \dfrac{67{,}400 \text{ J}}{58.25 \text{ J K}^{-1}} = 1100 \text{ K}$. This will not be possible for water at

standard pressures.

11-77. (a) Each individual keystroke has a $\dfrac{1}{2}$ chance of producing the value specified, since

there are 50 even and 50 odd numbers in the range between 1 and 100.

The question gives 10 results in *specific* order, so the probability is:

$\dfrac{1}{2} \times \dfrac{1}{2} \times \dfrac{1}{2} \times \dfrac{1}{2} \times \dfrac{1}{2} \times \dfrac{1}{2} \times \dfrac{1}{2} \times \dfrac{1}{2} \times \dfrac{1}{2} \times \dfrac{1}{2} = \left(\dfrac{1}{2}\right)^{10} = \mathbf{9.77 \times 10^{-4}}$

(b) Here the pattern involves 9 even results and 1 odd result in *random* order. Picture the

case where all ten numbers are even. This has a probability of 9.76×10^{-4}, just as in part

(a). If we allow *any* individual keystroke to yield an odd number, then this can happen

ten different ways--the first keystroke, the second, etc. Thus the probability is

$10 \times 9.77 \times 10^{-4} = \mathbf{9.77 \times 10^{-3}}$

11-79. (a) The **random**, less ordered structure will have **higher** entropy.

(b) With each site having 2 possible orientations, "up" or "down," the number of possible

states is $(2)^{6.022 \times 10^{23}}$.

11-81. (a) The area under the curve from 0 to 200 K is obviously larger for gold. **Gold** has

higher absolute entropy.

(b) Each square is 50 K wide and 0.05 J mol^{-1} K^{-2} high, for an area of 2.5 J mol^{-1} K^{-1}

There are approximately 12.8 squares under the curve for Cu, so we estimate the entropy

to be 32 J mol^{-1} K^{-1}. For Au there are approximately 18 squares, so we estimate an

entropy of 45 J mol^{-1} K^{-1}. These estimates compare well with the tabulated values of 33

and 48, respectively.

11-83. (a) This is **false**. Energy is *transformed* in power plants.

(b) This is **false**. Scientists have found *no* exceptions to the second law of

thermodynamics.

(c) This is **true**.

11-83. (continued)

(d) This is **false**. The second law of thermodynamics allows scientists to predict the *tendency*, in terms of spontaneity vs. non-spontaneity, of reactions.

(e) This is **false**. Classical thermodynamics is derived from the *observed laws of thermodynamics. These laws are independent of any theory of matter.*

11-85. (a)

$$\Delta H^o = 3 \Delta H^o_f \ (S_8 \ (g)) - 4 \Delta H^o_f \ (S_6 \ (g))$$

$$= 3 \ \text{mol} \ (101.55 \ \text{kJ mol}^{-1}) - 4 \ \text{mol} \ (103.72 \ \text{kJ mol}^{-1})$$

$$= \textbf{-110.23 kJ}$$

The energetics favor $S_8 \ (g)$.

(b) The reaction has fewer gaseous molecules in the products, so $\Delta S^o < 0$. Thus the entropy change favors the reactant $S_6 \ (g)$

11-87. $\Delta S^o = \dfrac{\Delta H_{vap}}{T_{vap}} = \dfrac{38,740 \text{J mol}^{-1}}{(273.15 \ \text{K} + 78.4 \ \text{K})} = \textbf{+110 J K}^{-1} \textbf{mol}^{-1}$

Trouton's rule predicts $\Delta S_{vap} = 88 \pm 5 \ \text{kJ mol}^{-1}$

The vaporization of ethanol has a larger entropy change than expected. This suggests ethanol is more ordered in the liquid phase than most compounds. One reason may be extensive hydrogen bonding networks.

11-89. (a) $\Delta G < 0$ for $T = \dfrac{\Delta H^o}{\Delta S^o} = \dfrac{33,900 \text{ J}}{96.4 \text{ J K}^{-1}} = 351 \text{ K} = 79^\circ C$

(b) $\Delta S = \dfrac{\Delta H_{vap}}{T_b} = \dfrac{30800 \text{ J}}{353.25 \text{ K}} = 87.2 \text{ J K}^{-1}$

(c) Yes, both the enthalpy change and the entropy change at the boiling point are about 9% different from those at 298 K.

11-91.

$$\Delta G^o = 3 \Delta G^o_f \ (SiO_2, \text{quartz}) + 2 \Delta G^o_f \ (N_2 \ (g)) + 3 \Delta G^o_f \ (C, \text{gr})$$

$$- 3 \Delta G^o_f \ (CO_2 \ (g)) - \Delta G^o_f \ (Si_3N_4 \ (s))$$

$$= 3 \ \text{mol} \ (-856.67 \ \frac{\text{kJ}}{\text{mol}}) + 2 \ \text{mol} \ (0 \ \frac{\text{kJ}}{\text{mol}}) + 3 \ \text{mol} \ (0 \ \frac{\text{kJ}}{\text{mol}})$$

$$- 3 \ \text{mol} \ (-394.36 \ \frac{\text{kJ}}{\text{mol}}) - 1 \ \text{mol} \ (-642.6 \ \frac{\text{kJ}}{\text{mol}}) = \textbf{- 744.3 kJ}$$

11-93. (a) $\Delta H^o_r = 2\,\Delta H^o_f\,(CuCl) + \Delta H^o_f\,(Cl_2) - 2\,\Delta H^o_f\,(CuCl_2)$

$= 2\ mol\ (-137.2\ kJ/mol) + 1\ mol\ (0\ kJ/mol) - 2\ mol\ (-220.1\ kJ/mol)$

$= \mathbf{+165.8\ kJ}$

$\Delta S^o_r = 2S^o\,(CuCl) + S^o\,(Cl_2) - 2S^o\,(CuCl_2)$

$= 2\ mol\ (86.2\ J\ K^{-1}\ mol^{-1}) + 1\ mol\ (222.96\ J\ K^{-1}\ mol^{-1})$

$- 2\ mol\ (108.07\ J\ K^{-1}\ mol^{-1}) = \mathbf{+179.2\ J\ K^{-1}}$

(b) $\Delta G^o_{r,590} \approx \Delta H^o - T\Delta S^o$

$\approx 165,800\ J - (590\ K)(179.2\ J\ K^{-1}) = 60,072\ J = \mathbf{60.07\ kJ}$

(c) $\Delta G^o_{r,590} = \Delta H^o_{r,590} - T\Delta S^o_{r,590}$

$= 158,360\ J - (590\ K)(177.74\ J\ K^{-1}) = 53,493\ J = \mathbf{53.49\ kJ}$

(d) $\%\ error = \dfrac{\Delta G^o_{incorrect} - \Delta G^o_{correct}}{\Delta G^o_{correct}} \times 100\% = \dfrac{60.07\ kJ - 53.49\ kJ}{53.49\ kJ} \times 100\% = \mathbf{12.3\%}$

11-95. (a)

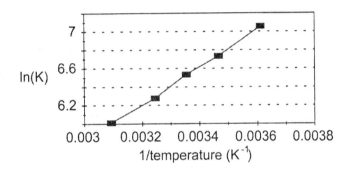

(b) The slope of the graph is equal to $-\dfrac{\Delta H^0}{R}$, and here the slope has a value, determined by linear regression (it can also be done by hand) of +2020 K, so:

$$-\frac{\Delta H^0}{R} = slope$$

$$\Delta H^o = -\ slope \times R = -\ 2020\ K \times 8.3145\ J\ mol^{-1}\ K^{-1}$$

$$= -\ 16,800\ J\ mol^{-1}$$

For the reaction as written, $\Delta H^o = \mathbf{-16.8\ kJ}$. The reaction is exothermic. Higher temperatures reduce K, shifting the reaction to the **right**.

11-97. The table of changes must be arranged to include the variable x, which indicates the fraction of N_2 that undergoes dissociation, and the variable y, which indicates the initial pressure of N_2:

ALL P IN ATM	N_2	N
INITIAL PRESSURE	y	0
CHANGE IN PRESSURE	$-yx$	$+2yx$
FINAL PRESSURE	$y(1-x)$	$2yx$

We know:

$$P_{TOTAL} = P_{N_2} + P_N$$

$$1.00 \text{ atm} = y(1-x) + (2yx) = y(1+x)$$

Therefore:

$$y = 1.00 / (1+x)$$

$$P_N = {}^{1-x}/_{1+x} \qquad P_{N_2} = {}^{2x}/_{1+x}$$

Therefore:

$$K = \frac{\dfrac{(2x)^2}{(1.00+x)^2}}{\dfrac{(1.00-x)}{(1.00+x)}} = \frac{4x^2}{1-x^2}$$

At 5000 K, $x = 0.65\% = 0.0065$

$$K_1 = \frac{4(0.0065)^2}{1-(0.0065)^2} = 1.690 \times 10^{-4}$$

At 6000 K, $x = 11.6\% = 0.116$

$$K_2 = \frac{4(0.116)^2}{1-(0.116)^2} = 5.456 \times 10^{-2}$$

From van't Hoff:

$$\ln\left(\frac{K_2}{K_1}\right) = \frac{\Delta H^o}{R}\left(\frac{1}{T_1} - \frac{1}{T_2}\right)$$

$$\ln\left(\frac{5.456 \times 10^{-2}}{1.690 \times 10^{-4}}\right) = \frac{\Delta H^o}{\left(8.3145 \text{J mol}^{-1} \text{ K}^{-1}\right)}\left(\frac{1}{5000\text{K}} - \frac{1}{6000\text{K}}\right)$$

$$\frac{(8.3145)(5.777)}{3.33\times10^{-5}} = \Delta H^o = \textbf{1442 kJ}$$

11-99. To an outside observer, a system at equilibrium never changes -- like a dead object. But the fact that equilibria are dynamic on a microscopic level -- constantly going backwards and forwards--means the system is in fact very active.

11-101.

(a) $\Delta H_r^o - T\Delta S_r^o = \Delta G_r^o$

$$\Delta S_r^o = -\frac{\Delta G^o - \Delta H^o}{T} = -\frac{(-34,500J) - (-19,700J)}{298.15K}$$

$$= +49.7 \text{ J K}^{-1}$$

(b) Formation of 38 mol of ATP from ADP requires $38 \times (34.5 \text{ kJ}) = 1311 \text{ kJ}$. Conversion of glucose and O_2 to carbon dioxide and water yields 2872 kJ. The efficiency is then 0.456 = **45.6%**

11-103.(a) Since the F's lie across from one or on the same side as one another, let us call the structure on the left "trans" and the one on the right "cis," from the Latin roots meaning "across from" and "same side."

At equilibrium, the ratio of [cis] / [trans] is 95 / 5, or **19**. This is the equilibrium constant. The change in the Gibbs' energy is then:

$$\Delta G = -RT \ln K = -(8.314 \text{ J mol}^{-1} \text{ K}^{-1})(295 \text{ K}) \ln 19$$
$$= 0.88 \text{ kJ mol}^{-1}.$$

(b) Since this is an equilibrium, it does not matter how it established. We expect the [cis] / [trans] ratio to be 19 also.

CHAPTER 12 - REDOX REACTIONS AND ELECTROCHEMISTRY

12-1. (a) Reduction: $\qquad$ $2\,H^+\,(aq) + H_2O_2\,(aq) + 2\,e^- \rightarrow 2\,H_2O\,(l)$

Oxidation: $\qquad\qquad\qquad\qquad\qquad$ $Fe^{2+}(aq) \rightarrow Fe^{3+}\,(aq) + e^-$

(b) In this problem H^+ and H_2O appear in *both* the reduction and the oxidation reaction. The half-equations must be reconstructed by considering the reduced and oxidized species.

Reduction:	$MnO_4^- \rightarrow Mn^{2+}$
Add H_2O to balance O:	$MnO_4^- \rightarrow Mn^{2+} + 4\,H_2O$
Add H^+ to balance H:	$8\,H^+ + MnO_4^- \rightarrow Mn^{2+} + 4\,H_2O$
Add e^- to balance charge:	$5\,e^- + 8\,H^+ + MnO_4^- \rightarrow Mn^{2+} + 4\,H_2O$
Add phase descriptions:	

$$5\,e^- + 8\,H^+\,(aq) + MnO_4^-\,(aq) \rightarrow Mn^{2+}\,(aq) + 4\,H_2O\,(l)$$

Oxidation:	$SO_2 \rightarrow HSO_4^-$
Add H_2O to balance O:	$2\,H_2O + SO_2 \rightarrow HSO_4^-$
Add H^+ to balance H:	$2\,H_2O + SO_2 \rightarrow HSO_4^- + 3\,H^+$
Add e^- to balance charge:	$2\,H_2O + SO_2 \rightarrow HSO_4^- + 3\,H^+ + 2\,e^-$
Add phase descriptions:	$2\,H_2O\,(l) + SO_2\,(aq) \rightarrow HSO_4^-\,(aq) + 3\,H^+\,(aq) + 2\,e^-$

(c) This is a disproportionation reaction, where a single substance (here ClO_2^-) is both reduced (to Cl^-) *and* oxidized (to ClO_2).

Reduction:	$ClO_2^- \rightarrow Cl^-$
Add H_2O to balance O:	$ClO_2^- \rightarrow Cl^- + 2\,H_2O$
Add H^+ to balance H:	$4\,H^+ + ClO_2^- \rightarrow Cl^- + 2\,H_2O$
Add e^- to balance charge:	$4\,e^- + 4\,H^+ + ClO_2^- \rightarrow Cl^- + 2\,H_2O$
Add phase descriptions:	

$$4\,e^- + 4\,H^+\,(aq) + ClO_2^-\,(aq) \rightarrow Cl^-\,(aq) + 2\,H_2O\,(l)$$

Oxidation:	$ClO_2^- \rightarrow ClO_2$

Add H_2O to balance O and add H^+ to balance H (not needed):

Add e^- to balance charge:	$ClO_2^- \rightarrow ClO_2 + e^-$

Add appropriate phase descriptions: $\quad ClO_2^-\,(aq) \rightarrow ClO_2\,(g) + e^-$

12-3. Alternate reactions will be balanced in full detail. For the others the balanced half-reactions and the final equation are presented.

(a) Reduction: $$VO_2^+ \rightarrow VO^{2+}$$

Add H_2O to balance O: $$VO_2^+ \rightarrow VO^{2+} + H_2O$$

Add H^+ to balance H: $$2\,H^+ + VO_2^+ \rightarrow VO^{2+} + H_2O$$

Add e^- to balance charge: $$e^- + 2\,H^+ + VO_2^+ \rightarrow VO^{2+} + H_2O$$

Add appropriate phase descriptions:
$$e^- + 2\,H^+\ (aq) + VO_2^+\ (aq) \rightarrow VO^{2+}\ (aq) + H_2O\ (l)$$

Oxidation: $$SO_2 \rightarrow SO_4^{2-}$$

Add H_2O to balance O: $$2\,H_2O + SO_2 \rightarrow SO_4^{2-}$$

Add H^+ to balance H : $$2\,H_2O + SO_2 \rightarrow SO_4^{2-} + 4\,H^+$$

Add e^- to balance charge: $$2\,H_2O + SO_2 \rightarrow SO_4^{2-} + 4\,H^+ + 2\,e^-$$

Add appropriate phase descriptions:
$$2\,H_2O\ (l) + SO_2\ (g) \rightarrow SO_4^{2-}\ (aq) + 4\,H^+\ (aq) + 2\,e^-$$

Combine: The reduction reaction has 1 e^- transferred; the oxidation reaction has 2 e^-. So the reduction reaction must be multiplied by 2 and the reduction reaction must be multiplied by 1 to cancel electrons. The final balanced equation is:
$$\mathbf{SO_2\ (g) + 2\,VO_2^+\ (aq) \rightarrow 2\,VO^{2+}\ (aq) + SO_4^{2-}\ (aq)}$$

(b) Oxidation: $$2\,H_2O\ (l) + SO_2\ (g) \rightarrow SO_4^{2-}\ (aq) + 4\,H^+\ (aq) + 2\,e^-$$

Reduction: $$Br_2\ (l) + 2\,e^- \rightarrow 2\,Br^-\ (aq)$$

Combine: $$\mathbf{Br_2\ (l) + 2\,H_2O\ (l) + SO_2\ (g) \rightarrow SO_4^{2-}\ (aq) + 4\,H^+\ (aq)}$$
$$\mathbf{+\ 2\,Br^-\ (aq)}$$

(c) Oxidation: $$HCOOH\ (aq) \rightarrow CO_2 + 2\,H^+\ (aq) + 2\,e^-$$

Reduction: $$5\,e^- + 8\,H^+\ (aq) + MnO_4^-\ (aq) \rightarrow Mn^{2+}\ (aq) + 4\,H_2O\ (l)$$

Combine: $$\mathbf{5\,HCOOH\ (aq) + 2\,MnO_4^-\ (aq) + 6\,H^+\ (aq)}$$
$$\mathbf{\rightarrow 5\,CO_2\ (g) + 2\,Mn^{2+}\ (aq) + 8\,H_2O\ (l)}$$

(e) In this disproportionation Pb_3O_4 is reduced to Pb^{2+} and oxidized to PbO_2

Reduction: $$Pb_3O_4 \rightarrow Pb^{2+}$$

Balance all but H and O: $$Pb_3O_4 \rightarrow 3\,Pb^{2+}$$

Add H_2O to balance O: $$Pb_3O_4 \rightarrow 3\,Pb^{2+} + 4\,H_2O$$

Add H^+ to balance H: $$8\,H^+ + Pb_3O_4 \rightarrow 3\,Pb^{2+} + 4\,H_2O$$

Add e^- to balance charge: $$2e^- + 8\,H^+ + Pb_3O_4 \rightarrow 3\,Pb^{2+} + 4\,H_2O$$

Add phase descriptions: $$2e^- + 8\,H^+\ (aq) + Pb_3O_4\ (s) \rightarrow 3\,Pb^{2+} + 4\,H_2O\ (l)$$

12-3. (continued)

Oxidation: $\qquad\qquad\qquad\qquad\qquad$ $Pb_3O_4 \rightarrow PbO_2$

Balance all but H and O: $\qquad\qquad\qquad$ $Pb_3O_4 \rightarrow 3\,PbO_2$

Add H_2O to balance O: $\qquad\quad$ $2\,H_2O + Pb_3O_4 \rightarrow 3\,PbO_2$

Add H^+ to balance H : $\qquad\quad$ $2\,H_2O + Pb_3O_4 \rightarrow 3\,PbO_2 + 4\,H^+$

Add e^- to balance charge: $\qquad$ $2\,H_2O + Pb_3O_4 \rightarrow 3\,PbO_2 + 4\,H^+ + 4\,e^-$

Add phase descriptions: $\quad$ $2\,H_2O\ (l) + Pb_3O_4\ (s) \rightarrow 3\,PbO_2\ (s) + 4\,H^+\ (aq) + 4\,e^-$

Combine: The oxidation reaction involves four electrons, the reduction two. We must multiply the reduction reaction by two before adding it to the oxidation reaction.

$$12\,H^+\ (aq) + 3\,Pb_3O_4\ (s) \rightarrow 6\,Pb^{2+} + 3\,PbO_2\ (s) + 6\,H_2O\ (l)$$

This can be simplified by dividing through by 3:

$$\mathbf{4\,H^+\ (aq) + Pb_3O_4\ (s) \rightarrow 2\,Pb^{2+} + PbO_2\ (s) + 2\,H_2O\ (l)}$$

(f) There are two points to make about this problem. First, *three* elements might change: Hg, P, and Au. But P appears on both sides in phosphate, so a change oxidation numbers occurs only with Hg and Au. Secondly, in combining the oxidation reaction (three electrons transferred) and the reduction reaction (two electrons transferred), we must multiply *both* by an integer to make the electrons equal.

Oxidation: $\qquad\qquad$ $Au\ (s) + 4\,Cl^-\ (aq) \rightarrow AuCl_4^-\ (aq) + 3\,e^-$

Reduction: $\qquad$ $2\,e^- + Hg_2HPO_4\ (s) + H^+\ (aq) \rightarrow 2\,Hg\ (l) + H_2PO_4^-\ (aq)$

Combine: $\mathbf{3\,Hg_2HPO_4\ (s) + 3\,H^+\ (aq) + 2\,Au\ (s) + 8\,Cl^-\ (aq)}$

$$\mathbf{\rightarrow 6\,Hg\ (l) + 3\,H_2PO_4^-\ (aq) + 2\,AuCl_4^-\ (aq)}$$

12-5. (a) This particular problem requires solution algebraically, using the coefficients listed (as in Chapter 2–1)

$$a\,HMnO_4\ (aq) + b\,H_2S\ (aq) + c\,H_2SO_4\ (aq) \rightarrow d\,MnSO_4\ (aq) + e\,H_2O\ (l)$$

$\qquad\qquad$ For H: $\qquad\quad a + 2b + 2c = 2e$

$\qquad\qquad$ For Mn: $\qquad\qquad\quad a = d$

$\qquad\qquad$ For O: $\qquad\quad 4a + 4c = 4d + e$

$\qquad\qquad$ For S: $\qquad\qquad b + c = d$

This solves for: $a = 8$, $b = 5$, $c = 3$, $d = 8$, and $e = 12$.

$$8\,HMnO_4\ (aq) + 5\,H_2S\ (aq) + 3\,H_2SO_4\ (aq) \rightarrow 8\,MnSO_4\ (aq) + 12\,H_2O\ (l)$$

(b) This is best done by noting that the ratio of Zn (oxidation number changes from 0 to +2, or +2) and N (oxidation number changes from +5 to –3, or –8) must be 4:1. This lets us "lock in" Zn and $ZnCl_2$ with coefficients of 4 and HNO_3 and NH_4Cl with coefficients of 1.

12-5. (continued)

Working by inspection from that point, we obtain:

$$4\ Zn\ (s) + HNO_3\ (aq) + 9\ HCl\ (aq) \rightarrow 4\ ZnCl_2\ (aq) + NH_4Cl\ (aq) + 3\ H_2O\ (l)$$

(c) This is best done by noting that the ratio of Sn (oxidation number changes from 0 to +4, or +4) and N (oxidation number changes from +5 to +1, or –4) must be 1:1. This lets us "lock in" Sn and $SnCl_4$ with coefficients of 2, HNO_3 with a coefficient of 2, and N2O with a coefficient of 1. Working by inspection from that point, we obtain:

$$2\ Sn\ (s) + 2\ HNO_3\ (aq) + 4\ HCl\ (aq) \rightarrow 2\ SnCl_4\ (aq) + N_2O\ (g) + 3\ H_2O\ (l)$$

(d) This is another problem whose complexity means the algebraic method is best.

$$a\ H_2O + b\ KMnO_4\ (aq) + c\ H_2SO_4\ (aq) \rightarrow d\ MnSO_4\ (aq) + e\ K_2SO_4\ (aq) + f\ H_2O_2\ (l)$$

For H:	$2a + 2c = 2f$ or $a + c = f$
For O:	$a + 4b + 4c = 4d + 4e + 2f$
For Mn:	$b = d$
For K:	$b = 2e$
For S:	$c = d + e$

This solves for: $a = 2, b = 2, c = 3, d = 2, e = 1$ and $f = 5$

$$2\ H_2O + 2\ KMnO_4\ (aq) + 3\ H_2SO_4\ (aq) \rightarrow 2\ MnSO_4\ (aq) + 1K_2SO_4\ (aq) + 5\ H_2O_2\ (l)$$

12-7. As with No. 3 above, only some parts (here, a and d) are worked in detail.

(a) <u>Reduction:</u> $Br_2 \rightarrow Br^-$

Balance all but H and O: $Br_2 \rightarrow 2\ Br^-$

Add H_2O to balance O, Add H^+ to balance H: *neither* needed here.

Add e^- to balance charge: $2\ e^- + Br_2 \rightarrow 2\ Br^-$

Add phase descriptions: $2\ e^- + Br_2\ (aq) \rightarrow 2\ Br^-\ (aq)$

<u>Oxidation:</u> $Cr(OH)_3 \rightarrow CrO_4^{2-}$

Add H_2O to balance O: $H_2O + Cr(OH)_3 \rightarrow CrO_4^{2-}$

Add H^+ to balance H : $H_2O + Cr(OH)_3 \rightarrow CrO_4^{2-} + 5\ H^+$

"Neutralize" H^+ with OH^- added to both sides; make $H^+ + OH^-$ into H_2O:
$$H_2O + 5\ OH^- + Cr(OH)_3 \rightarrow CrO_4^{2-} + 5\ H_2O$$

Simplify: $5\ OH^- + Cr(OH)_3 \rightarrow CrO_4^{2-} + 4\ H_2O$

Add e^- to balance charge: $5\ OH^- + Cr(OH)_3 \rightarrow CrO_4^{2-} + 4\ H_2O + 3\ e^-$

Add appropriate phase descriptions:
$$5\ OH^-\ (aq) + Cr(OH)_3\ (s) \rightarrow CrO_4^{2-}\ (aq) + 4\ H_2O\ (l) + 3\ e^-$$

Chapter 12

12-7. (continued)

<u>Combine:</u> The reduction reaction has two e^- transferred; the oxidation reaction has 3 e^-. So the reduction reaction must be multiplied by **3** and the reduction reaction must be multiplied by **2** to cancel electrons. The final equation is:

$$10\ OH^-(aq) + 2\ Cr(OH)_3(s) + 3\ Br_2(aq) \rightarrow 6\ Br^-(aq) + 2\ CrO_4^{2-}(aq) + 8H_2O(l)$$

(b) Briefly:

<u>Reduction:</u> $\quad\quad\quad 4\ e^- + H_2O\ (l) + ZrO(OH)_2\ (s)\ \rightarrow Zr\ (aq) + 4\ OH^-\ (aq)$

<u>Oxidation:</u> $\quad\quad\quad\quad\ 2\ OH^-\ (aq) + SO_3^{2-}\ (aq)\ \rightarrow SO_4^{2-}\ (aq) + H_2O\ (l) + 2\ e^-$

<u>Combined:</u> $\quad\quad\ ZrO(OH)_2\ (s) + 2\ SO_3^{2-}\ (aq)\ \rightarrow 2\ SO_4^{2-}\ (aq) + Zr\ (aq) + H_2O\ (l)$

(c) Briefly:

<u>Reduction:</u> $\quad\quad\quad 2\ e^- + H_2O\ (l) + HPbO_2^-\ (aq)\ \rightarrow Pb\ (s) + 3\ OH^-\ (aq)$

<u>Oxidation:</u> $\quad\quad\quad\quad\ 8\ OH^-\ (aq) + Re\ (s)\ \rightarrow ReO_4^-\ (aq) + 4\ H_2O\ (l) + 7\ e^-$

<u>Combined:</u> $\quad\quad\ 7\ HPbO_2^-\ (aq) + 2\ Re\ (s)\ \rightarrow$

$$2\ ReO_4^-\ (aq) + 7\ Pb\ (s) + 5\ OH^-\ (aq) + H_2O\ (l)$$

(d) <u>Reduction:</u> $\quad\quad\quad\quad\quad\quad HXeO_4^-\ \rightarrow Xe$

Add H_2O to balance O: $\quad\quad\ HXeO_4^-\ \rightarrow Xe + 4\ H_2O$

Add H^+ to balance H: $\quad\ 7\ H^+ + HXeO_4^-\ \rightarrow Xe + 4\ H_2O$

"Neutralize" H^+ with OH^- added to both sides; make $H^+ + OH^-$ into H_2O:

$$7\ H_2O + HXeO_4^-\ \rightarrow Xe + 4\ H_2O + 7\ OH^-$$

Simplify: $\quad\quad\quad\quad 3\ H_2O + HXeO_4^-\ \rightarrow Xe + 7\ OH^-$

Add e^- to balance charge: $\ 6\ e^- + 3\ H_2O + HXeO_4^-\ \rightarrow Xe + 7\ OH^-$

Add phase descriptions:

$$6\ e^- + 3\ H_2O\ (l) + HXeO_4^-\ (aq)\ \rightarrow Xe\ (g) + 7\ OH^-\ (aq)$$

<u>Oxidation:</u> $\quad\quad\quad\quad\quad\quad HXeO_4^-\ \rightarrow XeO_6^{4-}$

Add H_2O to balance O: $\ 2\ H_2O + HXeO_4^-\ \rightarrow XeO_6^{4-}$

Add H^+ to balance H : $\ 2\ H_2O + HXeO_4^-\ \rightarrow XeO_6^{4-} + 5\ H^+$

"Neutralize" H^+ with OH^- added to both sides; make $H^+ + OH^-$ into H_2O:

$$2\ H_2O + 5\ OH^- + HXeO_4^-\ \rightarrow XeO_6^{4-} + 5\ H_2O$$

Simplify: $\quad\quad\quad\quad 5\ OH^- + HXeO_4^-\ \rightarrow XeO_6^{4-} + 3\ H_2O$

Insert e^- to balance charge: $5\ OH^- + HXeO_4^-\ \rightarrow XeO_6^{4-} + 3\ H_2O + 2\ e^-$

Add phase descriptions:

$$5\ OH^-\ (aq) + HXeO_4^-\ (aq) \rightarrow XeO_6^{4-}\ (aq) + 3\ H_2O\ (l) + 2\ e^-$$

12-7. (continued)

Combine: The reduction reaction has six e⁻ transferred; the oxidation reaction has 2 e⁻. So the oxidation reaction must be multiplied by **3** to cancel electrons. The final equation is:

$$8\ OH^-\ (aq) + 4\ HXeO_4^-\ (aq) \rightarrow Xe\ (g) + 3\ XeO_6^{4-}\ (aq) + 6\ H_2O\ (l)$$

(e) Briefly:

Reduction: $\quad 2\ e^- + H_2O\ (l) + Ag_2S\ (s) \rightarrow 2\ Ag\ (s) + HS^-\ (aq) + OH^-\ (aq)$

Oxidation: $\quad 5\ OH^-(aq) + Cr(OH)_3(s) \rightarrow CrO_4^{2-}\ (aq) + 4\ H_2O\ (l) + 3\ e^-$

Combined: $7\ OH^-\ (aq) + 2\ Cr(OH)_3\ (s) + 3\ Ag_2S\ (s)$
$$\rightarrow 6\ Ag\ (s) + 3\ HS^-\ (aq) + 2\ CrO_4^{2-}\ (aq) + 5\ H_2O\ (l)$$

(f) Briefly:

Reduction: $\quad 2\ e^- + 2\ H_2O\ (l) + CO_3^{2-}\ (aq) \rightarrow CO\ (g) + 4\ OH^-\ (aq)$

Oxidation: $\quad 4\ OH^-\ (aq) + N_2H_4\ (aq) \rightarrow N_2\ (g) + 4\ H_2O\ (l) + 4\ e^-$

Combined: $\quad N_2H_4\ (aq) + 2\ CO_3^{2-}\ (aq) \rightarrow 2\ CO\ (g) + 4\ OH^-\ (aq) + N_2\ (g)$

12-9. This problem may seem unusual because the one element -- oxygen with a 2- oxidation number -- is present in *four* different substances: PbO_2^{2-}, SO_4^{2-}, H_2O, and OH^-. But systematic use of the half-reaction method described in the text will lead to the correct answer without a problem. We develop a half-reaction for oxygen involving the only species containing oxygen and hydrogen -- water, hydroxide and oxygen gas. Generally, *whenever oxygen changes its oxidation number to or from 2-, its half reaction should be balanced against **water**.*

Reduction: $\quad O_2 \rightarrow H_2O$

Add H_2O to balance O: $\quad O_2 \rightarrow 2\ H_2O$

Add H^+ to balance H: $4\ H^+ + O_2 \rightarrow 2\ H_2O$

Add e^- to balance charge: $\quad 4\ e^- + 4\ H^+ + O_2 \rightarrow 2\ H_2O$

"Neutralize" H^+ with OH^- added to both sides; make $H^+ + OH^-$ into H_2O:

$$4\ e^- + 4\ H_2O + O_2 \rightarrow 2\ H_2O + 4\ OH^-$$

Simplify: $\quad 4\ e^- + 2\ H_2O + O_2 \rightarrow 4\ OH^-$

Add phase descriptions: $4\ e^- + 2\ H_2O\ (l) + O_2\ (g) \rightarrow 4\ OH^-\ (aq)$

Oxidation: $\quad PbS \rightarrow PbO_2^{2-} + SO_4^{2-}$

(note: both the Pb and the S are balanced already)

Add H_2O to balance O: $\quad 6\ H_2O + PbS \rightarrow PbO_2^{2-} + SO_4^{2-}$

Add H^+ to balance H : $\quad 6\ H_2O + PbS \rightarrow PbO_2^{2-} + SO_4^{2-} + 12\ H^+$

12-9. (continued)

"Neutralize H^+ with OH^- added to both sides; make $H^+ + OH^-$ into H_2O:

$$12\ OH^- + 6\ H_2O + PbS \rightarrow PbO_2^{2-} + SO_4^{2-} + 12\ H_2O$$

Simplify: $\qquad 12\ OH^- + PbS \rightarrow PbO_2^{2-} + SO_4^{2-} + 6\ H_2O$

Add e^- to balance charge: $\qquad 12\ OH^- + PbS \rightarrow PbO_2^{2-} + SO_4^{2-} + 6\ H_2O + 8\ e^-$

Add phase descriptions:

$$12\ OH^-\ (aq) + PbS\ (s) \rightarrow PbO_2^{2-}\ (aq) + SO_4^{2-}\ (aq) + 6\ H_2O\ (l) + 8\ e^-$$

Combine: The reduction reaction has four e^- transferred; the oxidation reaction has eight e^- transferred. So the reduction reaction must be multiplied by 2 to cancel electrons. The final equation is:

$$4\ OH^-\ (aq) + PbS\ (s) + 2\ O_2\ (g) \rightarrow PbO_2^{2-}\ (aq) + SO_4^{2-}\ (aq) + 2\ H_2O\ (l)$$

12-11. (a) $\qquad Ca\ (s) + Cl_2\ (g) \rightarrow CaCl_2\ (s)$

(b) $\qquad 2\ Fe^{3+}\ (aq) + Sn^{2+}\ (aq) \rightarrow 2\ Fe^{2+}\ (aq) + Sn^{4+}\ (aq)$

(c) $\qquad Fe_2O_3\ (s) + 3\ H_2\ (g) \rightarrow 2\ Fe\ (s) + 3\ H_2O\ (g)$

(d) $\qquad 2\ K\ (s) + H_2O_2\ (aq) \rightarrow 2\ KOH\ (aq)$

12-13. Half reactions aren't required here, but they make the solution much clearer.

Reduction: $e^- + H^+\ (aq) + HNO_2\ (aq) \rightarrow NO\ (g) + H_2O\ (l)$

Oxidation: $H_2O\ (l) + HNO_2\ (aq) \rightarrow NO_3^-\ (aq) + 3\ H^+\ (aq) + 2\ e^-$

Combined: $3\ HNO_2\ (aq) \rightarrow 2\ NO\ (g) + NO_3^-\ (aq) + H^+\ (aq) + H_2O\ (l)$

12-15. 1 amp = $1\ C\ sec^{-1}$, so 45 A = $45\ C\ sec^{-1}$, and the charge is:

$$Q = (45\ C\ s^{-1}) \times 1.2\ s = \textbf{54 C}$$

12-17. $\left(\dfrac{6.5 \times 10^3\ C}{18\ min}\right) \times \left(\dfrac{1\ min}{60\ s}\right) = 6.0\ C\ s^{-1} = \textbf{6.0 A}$

12-19. Cathode: $\qquad 2e^- + 2\ H^+\ (aq) \rightarrow H_2\ (g)$

Anode: $\qquad 2\ H_2O\ (l) \rightarrow 4\ H^+\ (aq) + O_2\ (g) + 4\ e^-$

Total: $\qquad 2\ H_2O\ (l) \rightarrow 2\ H_2\ (g) + O_2\ (g)$

12-21.

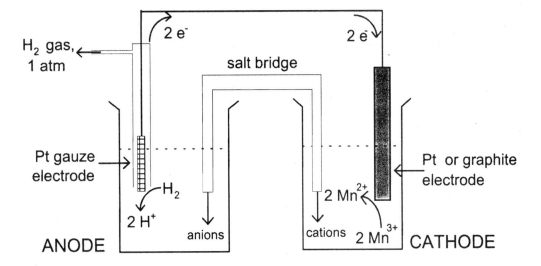

During the reaction electrons are released at the anode, where hydrogen gas becomes hydrogen ion. The electrons flow through an external circuit to the cathode, where manganese (III) ions are reduced to manganese (II) ions.

12-23.

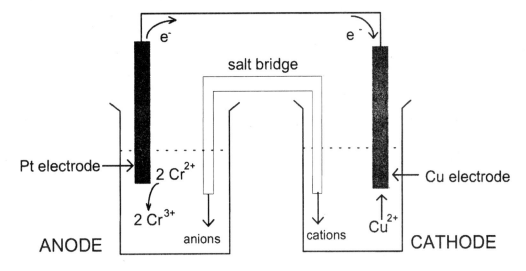

During the reaction electrons are released at the anode, where chromium (II) ion becomes chromium (III) ions. The electrons flow through an external circuit to the cathode, where copper (II) ions are reduced to copper metal, which deposits on the copper electrode. The overall reaction is:

$$2 \, Cr^{2+} \, (aq) + Cu^{2+} \, (aq) \quad \rightarrow Cu \, (s) + 2 \, Cr^{3+} \, (aq)$$

12-25. (a)

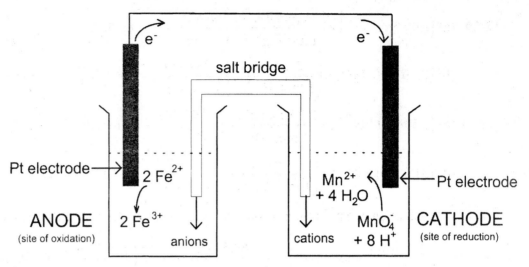

Anode (oxidation) reaction: $\qquad$ Fe^{2+} *(aq)* $\rightarrow Fe^{3+}$ *(aq)* $+ e^-$

Cathode (reduction) reaction:

$$5\,e^- + 8\,H^+\ (aq) + MnO_4^-\ (aq) \rightarrow Mn^{2+}\ (aq) + 4\,H_2O\ (l)$$

Balanced reaction:

$$10\,Fe^{2+}\ (aq) + 16\,H^+\ (aq) + 2\,MnO_4^-\ (aq) \rightarrow 2\,Mn^{2+}\ (aq) + 8\,H_2O\ (l) + 10\,Fe^{3+}\ (aq)$$

(b)

graphite
electrode

$2\,I^-$

ANODE
(site of oxidation)

I_2

anions

salt bridge

cations

Ag^+

Ag
electrode

CATHODE
(site of reduction)

e^- e^-

Anode (oxidation) reaction: $\qquad$ $2\,I^-$ *(aq)* $\rightarrow I_2$ *(aq)* $+ 2\,e^-$

Cathode (reduction) reaction: $\qquad$ $e^- + Ag^+$ *(aq)* $\rightarrow Ag$ *(s)*

Balanced reaction: $\qquad$ $2\,Ag^+$ *(aq)* $+ 2\,I^-$ *(aq)* $\rightarrow I_2$ *(aq)* $+ 2\,Ag$ *(s)*

12-27. $2.5\ A \times \left(\dfrac{60\ s}{min} \right) \times \left(\dfrac{1\ C\ s^{-1}}{1\ A} \right) \times \left(\dfrac{6.0221 \times 10^{23}\ e^-\ mol^{-1}}{96{,}485\ C\ mol^{-1}} \right) = \mathbf{9.36 \times 10^{20}\ e^-\ min^{-1}}$

12-29. (a) $\left(1.00 \text{ mol Ag}^+\right) \times \left(\dfrac{1 \text{ mol e}^-}{1 \text{ mol Ag}^+}\right) \times \left(\dfrac{96,485 \text{ C}}{1 \text{ mol e}^-}\right) = \textbf{96,500 C}$

(b) $(2.00 \text{ mol H}_2\text{O}) \times \left(\dfrac{2 \text{ mol e}^-}{1 \text{ mol H}_2\text{O}}\right) \times \left(\dfrac{96,485 \text{ C}}{1 \text{ mol e}^-}\right) = \textbf{386,000 C}$

(c) $\left(1.50 \text{ mol Mn}^{3+}\right) \times \left(\dfrac{1 \text{ mol e}^-}{1 \text{ mol Mn}^{3+}}\right) \times \left(\dfrac{96,485 \text{ C}}{1 \text{ mol e}^-}\right) = \textbf{145,000 C}$

(d) $\left(3.00 \text{ mol ClO}_4^-\right) \times \left(\dfrac{8 \text{ mol e}^-}{1 \text{ mol ClO}_4^-}\right) \times \left(\dfrac{96,485 \text{ C}}{1 \text{ mol e}^-}\right) = \textbf{2.32} \times \textbf{10}^\textbf{6} \textbf{ C}$

12-31. The reduction of Sn^{4+} requires four moles of electrons per mol of Sn. Therefore:

$$\text{mol Sn} = \left(\dfrac{69,500 \text{ C}}{96,485 \text{ C }\left(\text{mol e}^-\right)^{-1}}\right) \times \left(\dfrac{1 \text{ mol Sn}}{4 \text{ mol e}^-}\right) = \textbf{0.180 mol Sn}$$

This corresponds to $0.180 \text{ mol} \times 118.71 \text{ g mol}^{-1} = \textbf{21.4 g Sn.}$

12-33. (a) $\qquad\qquad\qquad \text{Zn } (s) + \text{Cl}_2 \ (g) \ \rightarrow \text{Zn}^{2+} \ (aq) + 2 \text{ Cl}^- \ (aq)$

(b) $Q = 0.800 \text{ C s}^{-1} \times 25 \text{ min} \times 60 \text{ s min}^{-1} = \textbf{1200 C}$

$\text{mol e}^- = 1200 \text{ C} / 96,485 \text{ C }(\text{mol e}^-)^{-1} = 0.01244 = \textbf{0.0124 mol e}^-$

(c) Two electrons must pass through the circuit for each zinc atom oxidized. So the number of moles of zinc deposited is one-half the number of mole of e^-:

$$\text{mol Zn} = .01244 / 2 = 0.00622 \text{ mol Zn}$$

$$\text{g Zn } = 0.00622 \text{ mol} \times 65.39 \text{ g mol}^{-1} = \textbf{0.407 g Zn}$$

(d) The number of mol of Cl_2 consumed is equal the number of mol of Zn, or 0.00622 mol Cl_2. This corresponds to a volume of:

$$V = \dfrac{nRT}{P} = \dfrac{(0.00622 \text{ mol})\left(0.08206 \dfrac{\text{L} - \text{atm}}{\text{mol} - \text{K}}\right)(298.15 \text{ K})}{1 \text{ atm}} = \textbf{0.152 L Cl}_\textbf{2}$$

12-35. The experimental setup requires the electrons that cause the reduction of Ag^+ also to cause the reduction of Cd^{2+}. The production of Ag requires one mole of electrons per mole of Ag and the production of Cd requires two moles of electrons per mole of Cd. The mole ratio of Cd to Ag is:

$$\left(\dfrac{1 \text{ mol e}^-}{1 \text{ mol Ag}}\right) \times \left(\dfrac{1 \text{ mol Cd}}{2 \text{ mol e}^-}\right) = \dfrac{1 \text{ mol Cd}}{2 \text{ mol Ag}}$$

And we then have:

$$0.475 \text{ g Ag } \times \left(\dfrac{1 \text{ mol Ag}}{107.87 \text{ g Ag}}\right) \times \left(\dfrac{1 \text{ mol Cd}}{2 \text{ mol Ag}}\right) \times \left(\dfrac{112.41 \text{ g Cd}}{1 \text{ mol Cd}}\right) = \textbf{0.248 g Cd}$$

12-37. The number of electrons used in oxidation of O in H_2O to O_2 (4 mol e^- per mol O_2) equals the number of electrons used in the reduction of dissolved copper. The number of moles of O_2 is 16.0 g / 31.9988 g mol^{-1} = 0.500 mol O_2.

Therefore, the number of moles of electrons passed is 2.00 mol e^-. The mass of Cu produced is

63.5 g / 63.546 g mol^{-1} = 1.00 mol. Therefore, each Cu gains 2 electrons, so the initial oxidation state must be **+2**.

12-39.
$$3 \, MnO_2 \, (s) + 4 \, Al \, (s) \rightarrow 3 \, Mn \, (s) + 2 \, Al_2O_3 \, (s)$$

$$\Delta H^o_r = 3 \, \Delta H^o_f(Mn \, (s)) + 2 \, \Delta H^o_f(Al_2O_3 \, (s)) - 3\Delta H^o_f(MnO_2 \, (s)) - 4 \, \Delta H^o_f(Al \, (s))$$

$$= 3 \text{ mol } (0.0 \, \frac{kJ}{mol}) + 2 \text{ mol } (-1675.7 \, \frac{kJ}{mol})$$

$$- 3 \text{ mol } (-520.03 \, \frac{kJ}{mol}) - 4 \text{ mol } (0.0 \, \frac{kJ}{mol}) = \textbf{-1791.3 kJ}$$

$$\Delta G^o_r = 3 \, \Delta G^o_f(Mn \, (s)) + 2 \, \Delta G^o_f(Al_2O_3 \, (s)) - 3 \, \Delta G^o_f(MnO_2 \, (s)) - 4 \, \Delta G^o_f(Al \, (s))$$

$$= 3 \text{ mol } (0.0 \, \frac{kJ}{mol}) + 2 \text{ mol } (-1582.3 \, \frac{kJ}{mol})$$

$$- 3 \text{ mol } (-465.17 \, \frac{kJ}{mol}) - 4 \text{ mol } (0.0 \, \frac{kJ}{mol}) = \textbf{-1769.1 kJ}$$

12-41. The reaction is:
$$2 \, HgO \, (s) \rightarrow 2 \, Hg \, (l) + O_2 \, (g)$$

$$\Delta H^o_r = 2 \, \Delta H^o_f(Hg \, (l)) + \Delta H^o_f(O_2 \, (g)) - 2 \, \Delta H^o_f(HgO \, (s))$$

$$= 2 \text{ mol}(0.0 \, \frac{kJ}{mol}) + 1 \text{ mol } (0.0 \, \frac{kJ}{mol}) - 2 \text{ mol } (-90.83 \, \frac{kJ}{mol})$$

$$= \textbf{181.66 kJ}$$

$$\Delta S^o_r = 2 \, S^o(Hg \, (l)) + S^o(O_2 \, (g)) - 2 \, S^o(HgO \, (s))$$

$$= 2 \text{ mol } (76.02 \, J \, K^{-1} \, mol^{-1}) + 1 \text{ mol } (205.03 \, J \, K^{-1} \, mol^{-1})$$

$$- 2 \text{ mol } (70.29 \, J \, K^{-1} \, mol^{-1})$$

$$= \textbf{+ 216.49 J K}^{-1}$$

As we saw repeatedly in Chapter 11, when ΔG^o equals zero:

$$T = \Delta H^o / \Delta S^o = 181{,}660 \, J / 216.49 \, J \, K^{-1} = 839 \, K = 566^oC$$

Because ΔH and ΔS are both greater than zero, the reaction will be spontaneous at higher temperatures: **T > 566°C**

12-43. The calculation of the mass of contained copper must take into account the number of grams of the copper compound in 1 metric ton of ore and the fraction of copper in a mass of the compound. There are 1,000,000 grams in one metric ton, so 1% of this amount is 10,000 grams. For neatness' sake, a table is best for keeping track of the contributions of the different compounds to the mass of copper in the ore.

ORE	grams in 1 metric ton	molar mass $(g\ mol^{-1})$	mass fraction Cu	mass of Cu (g)
$CuFeS_2$	11,000 g	183.525	0.346	3,809
CuS	4200 g	95.612	0.665	2,791
Cu_5FeS_4	5100 g	501.84	0.633	3,229
			TOTAL:	9,829 g = **9.8 kg**

12-45. Bronze has the advantage of decreased brittleness. It also has increased harness, which is usually (but not always) deisriable in materials.

12-47. Froth floatation uses a froth (finely divided bubbles) to separate copper ores from other materials that float to the bottom of the system.

12-49.(a)   #1: $Fe_2O_3\ (s) + 3\ CO\ (g) \rightarrow 2\ Fe\ (s) + 3\ CO_2\ (g)$

           #2: $Fe_3O_4\ (s) + 4\ CO\ (g) \rightarrow 3\ Fe\ (s) + 4\ CO_2\ (g)$

           #3: $FeCO_3\ (s) + CO\ (g) \rightarrow Fe\ (s) + 2\ CO_2\ (g)$

(b) For all three reactions, the term for $\Delta G_f^o(Fe)$ is omitted, because it is zero.

#1: $\Delta G_r^o = 3\ \Delta G_f^o(CO_2) - \Delta G_f^o(Fe_2O_3) - 3\ \Delta G_f^o(CO)$

           $= 3\ mol\ (-394.36\frac{kJ}{mol}) - 1\ mol\ (-742.2\frac{kJ}{mol}) - 3\ mol\ (-137.15\frac{kJ}{mol})$

           $= -29.43\ kJ$ for the reaction, $\Delta G^o =$ **-14.7 kJ (mol Fe)$^{-1}$**

#2: $\Delta G_r^o = 4\ \Delta G_f^o(CO_2) - \Delta G_f^o(Fe_3O_4) - 4\ \Delta G_f^o(CO)$

           $= 4\ mol\ (-394.36\ \frac{kJ}{mol}) - 1\ mol\ (-1015.5\ \frac{kJ}{mol}) - 4\ mol\ (-137.15\ \frac{kJ}{mol})$

           $= -13.34\ kJ$ for the reaction, $\Delta G^o =$ **-4.4 kJ (mol Fe)$^{-1}$**

#3: $\Delta G_r^o = 2\ \Delta G_f^o(CO_2) - \Delta G_f^o(FeCO_3) - \Delta G_f^o(CO)$

           $= 2\ mol\ (-394.36\ \frac{kJ}{mol}) - 1\ mol\ (-666.72\frac{kJ}{mol}) - 1\ mol\ (-137.15\ \frac{kJ}{mol})$

           $= +15.1\ kJ$ for the reaction, $\Delta G^o =$ **+15.1 kJ (mol Fe)$^{-1}$**

Per mole of iron produced, **Fe_2O_3** is thermodynamically easiest to reduce with CO as reductant.

12-51. Cu_5FeS_4 *(s)* + 7 $FeCl_3$ *(aq)* → 12 $CuCl$ *(s)* + $FeCl_2$ *(aq)* + 4 S *(s)*

12-53. Appendix D indicates that ΔG_f^o(CoO) is –214.22 kJ·mol^{-1}, With $n = 2$, so $\Delta G_f^o / n$ is – **107kJ mol^{-1}**. The entry for CoO should go between Fe and Ni, and we expect Co to be produced by smelting.

12-55. Limestone and dolomite are very basic minerals that react to absorb the amphoteric aluminum and silicon oxides present in the ore as aluminate and silicate salts.

12-57. The answers are found in the text's section on "Iron and Steel."

Si: Production: Si *(s)* + O_2 *(g)* → SiO_2 *(s)*

            Trapping: CaO *(s)* + Al_2O_3 *(s)* + SiO_2 *(s)* → $CaO \bullet Al_2O_3 \bullet (SiO_2)_2$ *(l)*

P: Production: P_4 *(s)* + 5 O_2 *(g)* → 2 P_2O_5 *(s)*

            Trapping: 3 CaO *(s)* + P_2O_5 *(s)* → $(CaO)_3 \bullet P_2O_5$ *(s)*

Both trapping reactions consist in part of a Lewis acid-base reaction between a basic oxide (CaO) and an acidic oxide (SiO_2 and P_2O_5). The actual species present are complex derivatives of silicate, SiO_4^{4-} and phosphate, PO_4^{3-}.

12-59. <u>Anode:</u>   2 Cl^- → Cl_2 *(g)* + 2 e^-

<u>Cathode:</u> $Na^+ + e^-$ → Na *(l)*

12-61. In the Hall-Heroult process 3 mol e$^-$ is required per mole of Al, or mol Al = 1/3 mol e$^-$. But in this case the charge flows through a *series* of cells, so each Coulomb of charge is "used" 100 times. The total number of electrons moved through *each* cell is

$$Q = 55,000 \text{ C s}^{-1} \times 86,400 \text{ s} = 4.75 \times 10^9 \text{ C}$$

$$\text{mol e}^- = (4.75 \times 10^9 \text{ C}) \div (96,485 \text{ C mol}^{-1}) = 49,250 \text{ mol e}^-$$

$$\text{mol Al} = 0.3333 \times 49,250 \text{ mol e}^- = 1.64 \times 10^4 \text{ mol Al}$$

In 100 cells, the mol of Al produced will be: $100 \times 1.64 \times 10^4 = 1.64 \times 10^4$ mol Al

$$\text{g Al} = (1.64 \times 10^6 \text{ mol Al}) \times (26.982 \text{ g mol}^{-1} \text{ Al}) = \mathbf{4.42 \times 10^7 \text{ g Al}}$$

12-63. Molten cryolite is used because it melts considerably lower than alumina. It also provides a wide spectrum of ions that facilitate the flow of electrical charge between the anodes and the cathodes.

12-65.   2 Mg *(s)* + $TiCl_4$ *(l)* → Ti *(s)* + 2 $MgCl_2$ *(s)*
For n Ti in 100 kg (100,000 g): mol Ti = 100,000 g/47.867 g mol^{-1} = 2,089.1 mol Ti
Two moles of Mg are required for every mole of Ti produced, so we need 4,178 mol Mg. This weighs 4,178 mol × 24.305 g mol^{-1} = 102,000 g = **102 kg**.

12-67. Dolomite is an inexpensive mineral that provides a means to remove the magnesium ions from seawater (in exchange for the calcium in the dolomite). The magnesium in the dolomite also becomes part of the magnesium for the further steps.

12-69. The cathode reaction is Zn^{2+} *(aq)* + 2 e⁻ → Zn *(s)*.

$$n\ Zn = 7.32\ g/65.39\ g\ mol^{-1} = 0.112\ mol\ Zn$$

$$n\ e^- = 2 \times n\ Zn = 0.224\ mol\ e^-$$

$$Q = 0.224\ mol\ e^- \times 96{,}485\ C\ mol^{-1} = 21{,}600\ C$$

$$I = Q\ /\ t$$

$$t = Q\ /\ I = 21{,}600\ C\ /\ 8.50\ C\ s^{-1} = 2540\ s = \textbf{42.4 min}$$

12-71. All of these reactions are balanced in the absence of water. The species that is oxidized is in bold. The species that is reduced is in italics. In the case of disproportionation reactions a bold italic species indicates the substance that disproportionates. Note that "species oxidized" means that one of the elements in the compound is oxidized, not necessarily every element.

(a) $\quad\quad\quad\quad 4\ N_2O_5\ (l) + \textbf{Au}\ (s) \rightarrow (NO_2)Au(NO_3)_4\ (s) + 3\ NO_2\ (g)$

(b) $\quad\quad\quad 3\ \textbf{TiCl}_4\ (l) + 2\ NO_2Cl\ (g) \rightarrow 2\ Cl_2\ (g) + 2\ TiOCl_2\ (s) + (NO)_2TiCl_6\ (s)$

(c) $\quad\quad\quad\quad\quad 8\ \textbf{NHBr}_2\ (s) \rightarrow 2\ NH_4Br\ (s) + 3\ N_2\ (g) + 7\ Br_2\ (g)$

(d) $\quad\quad\quad 3\ \textbf{NaN}_3\ (s) + 8\ NO_2Cl\ (g) \rightarrow 7\ N_2O\ (g) + 3\ NaNO_3\ (s) + 4\ Cl_2\ (g)$

12-73. $Cl_2O\ (g) + N_2O_5\ (g) \rightarrow 2\ NO_2Cl\ (g) + O_2\ (g)$

In this case, the species oxidized is the oxygen in the N_2O_5 and the species reduced is the chlorine in the Cl_2O.

$Cl_2O\ (aq) + 2\ HNO_3\ (aq) \rightarrow 2\ NO^+\ (aq) + 2\ Cl^-\ (aq) + 2\ O_2\ (g) + H_2O\ (l)$

In this case, the species oxidized is the oxygen in the HNO_3 and the species reduced are the chlorine in the Cl_2O and the nitrogen in HNO_3.

12-75. The presence of a large amount of hydroxide means this redox reaction is occurring in *basic, aqueous* solution:

$$2\ Al\ (s) + 2\ OH^-\ (aq) + 6\ H_2O\ (l) \rightarrow 2\ Al(OH)_4^-\ (aq) + 3\ H_2\ (g)$$

12-77. We can have this reaction proceed as the oxidation of NH_2OH to N_2 and the reduction of NO to N_2O:

oxidation: $\quad\quad\quad 2\ OH^-\ (aq) + 2\ NH_2OH\ (aq) \rightarrow N_2\ (g) + 4\ H_2O\ (g) + 2\ e^-$

reduction: $\quad\quad\quad 1\ e^- + H_2O\ (g) + 2\ NO\ (g) \rightarrow N_2O\ (g) + 2\ OH^-\ (aq)$

12-77. (continued)

We can also have the reaction proceed by the oxidation of NH_2OH to N_2O and the reduction of NO to N_2:

oxidation: $4\ OH^-\ (aq) + 2\ NH_2OH\ (aq)\ \rightarrow N_2O\ (g) + 5\ H_2O\ (g) + 4\ e^-$

reduction: $4\ e^- + 2\ H_2O\ (g) + 2\ NO\ (g)\ \rightarrow N_2\ (g) + 4\ OH^-\ (aq)$

12-79. The reaction to yield hydrogen and chlorate from aqueous sodium chloride is not favored thermodynamically. We can infer that the reverse reaction (of chlorate and hydrogen gas to give chloride ion and water) is spontaneous:

$$ClO_3^-\ (aq) + 3\ H_2\ (g)\ \rightarrow Cl^-\ (aq) + 3\ H_2O\ (l)$$

12-81. 1.83 g of Zn corresponds to $1.83\ /\ 65.39\ = 0.0280$ mol of Zn, and given 2 e^- passed for every Zn atom, the number of moles of electrons that passed through the cathode is 0.0560 mol e^-. This equals 5400 C of charge. But the cathode only sampled 0.25% of the electricity used. The total charge was 5400 C $/\ 0.025 = \mathbf{2.16 \times 10^6\ C}$.

12-83. The reaction in the cell is:

$$Zn\ (s) + Ni^{2+}\ (aq)\ \rightarrow Ni\ (s) + Zn^{2+}\ (aq)$$

(a) $\quad$ mol Zn = 32.68 g / 65.39 g mol^{-1} $\ = 0.500$ mol

$\quad\quad\quad$ mol Ni^{2+} = 0.575 L $\times$ 1.00 mol L^{-1} $\ = 0.575$ mol

The **Zn** is the limiting reactant.

(b) Oxidation of 0.500 mol of Zn will require 1.00 mol of electrons, or 96,485 C. The time for this amount of charge to flow, given a current of 0.0715 C s^{-1}, is 96,485 C / 0.0715 C s^{-1} = 1.35×10^6 s = 375 hr = **15.6 days**.

(c) The nickel electrode will gain 0.500 mol of Ni, with a mass of **29.3 g**.

(d) The Ni^{2+} solution will lose 0.500 mol of Ni^{2+}, leaving 0.075 mol of Ni^{2+} in 0.575 L of solution, for a final $[Ni^{2+}]$ = 0.075 mol / 0.575 L = **0.13 M**.

12-85. Anode: $\quad\quad\quad\quad\quad\quad\quad\quad\quad$ $2\ Cl^-\ \rightarrow Cl_2\ (g) + 2\ e^-$

$\quad\quad\quad$ Cathode: $\quad\quad\quad\quad\quad\quad\quad\quad$ $Na^+ + e^-\ \rightarrow Na\ (l)$

12-87. Assigning "oxidized" and "reduced" to particular elements in a compound can be impossible, and anyway it doesn't matter. The *substance* is what gets oxidized, not any element within it. So in this case the substance $CuFeS_2$ is oxidized, no matter how we divide up the event amont the elements in the compound.

12-89. There are three stages in the production process to compare and contrast: extraction / initial separation, preparation of electrolysis materials, and the final electrolysis process. The extraction of magnesium involves a solid base, for example the ore, dolomite (itself a source of Mg), and seawater, which contains free Mg^{2+} *(aq)*. The initial separation is in the form of *solid* magnesium hydroxide.

Base is also important in the extraction and separation of aluminum, and is used to *dissolve* the aluminum as $Al(OH)_4^-$. Further processing is needed to prepare both elements for electrolysis. For magnesium, this involves conversion to anhydrous $MgCl_2$ *(s)* by treatment with HCl and evaporation. For aluminum, the oxide is required for hydrolysis, and it is directly formed from the intermediate $Al(OH)_4^-$ by crystallization followed by drying. The final electrolysis to make Mg can be done on a $MgCl_2$ melt; a by-product is Cl_2, which is recycled. Al, in contrast, requires molten cryolite as a medium for the electrolysis. The other product of aluminum reduction is CO_2 from graphite.

12-91. (a) The reaction is balanced most easily by inspection:

Balancing Fe: $\qquad\qquad 2\,FeCr_2O_4 \rightarrow Fe_2O_3$

Balancing Cr : $\qquad\qquad 2\,FeCr_2O_4 \rightarrow Fe_2O_3 + 4\,Na_2CrO_4$

" $\quad$ Na : $\quad 4\,Na_2CO_3 + 2\,FeCrO_4 \rightarrow Fe_2O_3 + 4\,Na_2CrO_4$

" $\quad$ C: $\quad 4\,Na_2CO_3 + 2\,FeCrO_4 \rightarrow Fe_2O_3 + 4\,Na_2CrO_4 + 4\,CO_2$

" $\quad$ O: $7/2\,O_2 + 4\,Na_2CO_3 + 2\,FeCrO_4 \rightarrow Fe_2O_3 + 4\,Na_2CrO_4 + 4\,CO_2$

Converting to integers and inserting the appropriate phase descriptions yields:

$$7\,O_2\,(g) + 8\,Na_2CO_3\,(l) + 4\,FeCrO_4 \rightarrow 2\,Fe_2O_3\,(s) + 8\,Na_2CrO_4\,(l) + 8\,CO_2$$

(b)

	Data required*	Value	
$+8 \times \Delta H_f^o(Na_2CrO_4\ (l))$	8 mol $\times$ (-1342 kJ mol^{-1}) =	- 10,736	kJ
$+8 \times \Delta H_f^o(CO_2\ (g))$	8 mol $\times$ (-393.51 kJ mol^{-1}) =	- 3148.08	kJ
$+2\ \Delta H_f^o(Fe_2O_3\ (s))$	4 mol $\times$ (-824.2 kJ mol^{-1}) =	- 1648.6	kJ
$-8\ \Delta H_f^o(Na_2CO_3\ (l))$	-8 mol $\times$ (-1130.68 kJ mol^{-1}) =	+ 9045.44	kJ
$-4\ \Delta H_f^o(FeCr_2O_4\ (s))$	-4 mol $\times$ (-1445 kJ mol^{-1}) =	+ 5780	kJ
	ΔH^o =	**-707**	kJ

* We cannot get data easily for the enthalpy change associated with melting Na_2CO_3 or Na_2CrO_4. However, both of these substances are present in equal chemical amounts, one is a reactant and the other is a product and, finally, both are 2:1 sodium salts of an oxyanion. They should have similar molar enthalpies of fusion, and neglecting that number should have an equal and opposite effect: on the reaction.

(c) The sodium chromate should be leached from the mixture by water. Then it should be isolated by evaporation of the water, and perhaps used directly in some kind of electrolysis cell or reduced by an appropriate element - for example, C, Si, or Mg.

12-93. (a) From Appendix D, we obtain $\Delta H^o = +208.19$ kJ and $\Delta S^o = +213.63$ J K^{-1} for the reaction as written. We can estimate the temperature at which K equals 1 by dividing ΔH^o by ΔS^o, which gives $T = 975$ K.

(b) Dissolving the Mg in antimony produces the product in a solution. This will *increase* the positive entropy of the reaction, since almost all dissolution reactions have a positive entropy change. Though the enthalpy change may be a problem, and is very hard to predict, we expect that the positive entropy change would lower the temperature for the reaction.

CHAPTER 13 - ELECTROCHEMISTRY AND CELL VOLTAGE

13-1. $150 \text{ watts} \times \dfrac{1 \text{ kilowatt}}{1000 \text{ watts}} \times \dfrac{\$0.10}{1 \text{ kilowatt-hour}} \times 10 \text{ hours} = \textbf{\$0.15}$

13-3. One volt is one J C^{-1}. One amp is one C sec^{-1}. The rate of energy production $(1.34 \text{ J C}^{-1}) \times (0.800 \text{ C s}^{-1}) = 1.072 \text{ J s}^{-1}$. The time to produce 1 J of energy is:

$\dfrac{1.00 \text{ J}}{1.072 \text{ J s}^{-1}} = \textbf{0.933 s}.$

13-5. (a) $\qquad\qquad\qquad \text{Pb } (s) + \text{Cu}^{2+} (aq) \rightarrow \text{Pb}^{2+} (aq) + \text{Cu } (s)$

$\Delta G^\circ = -n\mathcal{F}\Delta\mathcal{E}^\circ \qquad\quad (\text{Note: } n = 2 \text{ mol e}^- / \text{mol reaction})$

$\qquad = -(2) (96{,}485 \text{ C (mol e}^-)^{-1}) (0.471 \text{ J C}^{-1}) = -90{,}880 \text{ J} = \textbf{-90.9 kJ}$

(b) $\quad \text{MnO}_4^- (aq) + 8 \text{ H}^+ (aq) + 5 \text{ Fe}^{2+} (aq) \rightarrow 4 \text{ H}_2\text{O} (l) + \text{Mn}^{2+} (aq) + 5 \text{ Fe}^{3+} (aq)$

$\Delta G^\circ = -n\mathcal{F}\Delta\mathcal{E}^\circ \qquad\quad (\text{Note: } n = 5 \text{ mol e}^- / \text{mol reaction})$

$\qquad = -(5) (96{,}485 \text{ C (mol e}^-)^{-1}) (0.736 \text{ J C}^{-1})$

$\qquad = -355{,}000 \text{ J} = \textbf{-355 kJ}$

(c) $\qquad\qquad\qquad 2 \text{ Ag}^+ (aq) + \text{Co}^{2+} (aq) \rightarrow \text{Co } (s) + 2 \text{ Ag } (s)$

$\Delta G^\circ = -n\mathcal{F}\Delta\mathcal{E}^\circ \qquad\quad (\text{Note: } n = 2 \text{ mol e}^- / \text{mol reaction})$

$\Delta G^\circ = -(2) (96{,}485 \text{ C (mol e}^-)^{-1}) (-1.08 \text{ J C}^{-1}) = +208400 \text{ J} = \textbf{+208 kJ}$

13-7. The reaction is: $\text{Ni } (s) + 2 \text{ Ag}^+ (aq) \rightarrow \text{Ni}^{2+} (aq) + 2 \text{ Ag } (s)$

Note: $n = 2 \text{ mol e}^- / \text{mol reaction}$ in the reaction as written.

$\Delta G^\circ = -(2) (96{,}485 \text{ C (mol e}^-)^{-1}) (+1.057 \text{ J C}^{-1}) = -203969 \text{ kJ}$

$w_{elec,max} = \Delta G \; (= \Delta G^\circ \text{ in this case with all substances in standard states})$

$w_{elec,max} = -\dfrac{203{,}969 \text{ J}}{2 \text{ mol Ag}} \times \dfrac{1 \text{ mol Ag}}{107.87 \text{ g Ag}} \times 1.00 \text{ g Ag} = \textbf{-945.4 J}$

This is the work absorbed by the system. The maxiumum work *done by the system* is **+945 J g^{-1}**.

13-9. (a) $\qquad$ Cathode: $\qquad\qquad \text{Br}_2 (l) + 2e^- \rightarrow 2 \text{ Br}^- (aq)$

$\qquad\qquad$ Anode: $\qquad\qquad\qquad \text{Co } (s) \rightarrow \text{Co}^{2+} (aq) + 2e^-$

$\qquad\qquad$ Overall: $\qquad \text{Br}_2 (l) + \text{Co } (s) \rightarrow 2 \text{ Br}^- (aq) + \text{Co}^{2+} (aq)$

(b) $\Delta\mathcal{E}^\circ = \mathcal{E}^\circ (\text{Br}_2 \mid \text{Br}^-) - \mathcal{E}^\circ (\text{Co}^{2+} \mid \text{Co}) = 1.066 \text{ V} - (-0.28 \text{ V}) = \textbf{+1.35 V}$

13-11. (a) Indium plating out implies indium is the cathode and zinc is anode.

Cathode: $\quad\quad$ In^{3+} *(aq)* $+ 3e^- \rightarrow In$ *(s)*

Anode: $\quad\quad$ Zn *(s)* $\rightarrow Zn^{2+}$ *(aq)* $+ 2e^-$

(b) $\quad\quad \Delta \mathcal{E}^o = \mathcal{E}^o (In^{3+} | In) - \mathcal{E}^o (Zn^{2+} | Zn)$

$\mathcal{E}^o (In^{3+} | In) = \Delta \mathcal{E}^o + \mathcal{E}^o (Zn^{2+} | Zn) = (+0.425 \text{ V}) + (-0.762 \text{ V}) = $ **-0.337 V**

13-13. The possible reduction half-reaction reaction is: Mn^{2+} *(aq)* $+ 2e^- \rightarrow Mn$ *(s)*

The possible oxidation half-reaction is: Mn^{2+} *(aq)* $\rightarrow Mn^{3+}$ *(aq)* $+ e^-$

The cell potential would be: $\quad \Delta \mathcal{E}^o = \mathcal{E}^o (Mn^{2+} | Mn) - \mathcal{E}^o (Mn^{3+} | Mn^{2+})$

$= (-1.185 \text{ V}) - (+1.542 \text{ V}) = $ **-2.727 V**

The reaction is **non-spontaneous**, because $\Delta \mathcal{E}^o$ is negative. We do not expect Mn^{2+} to disproportionate in its standard state.

13-15. $\quad\quad\quad\quad\quad\quad$ $2 \, Fe^{3+}$ *(aq)* $+ Fe$ *(s)* $\rightarrow 3 \, Fe^{2+}$ *(aq)*

Reduction: $\quad\quad\quad$ Fe^{3+} *(aq)* $+ e^- \rightarrow Fe^{2+}$ *(aq)*

Oxidation: $\quad\quad\quad$ Fe *(s)* $\rightarrow Fe^{2+}$ *(aq)* $+ 2 \, e^-$

$\Delta \mathcal{E}^o = \mathcal{E}^o (Fe^{3+} | Fe^{2+}) - \mathcal{E}^o (Fe^{2+} | Fe) = (+0.771) - (-0.447) = $ **1.218 V**

13-17. Solid aluminum will readily lose electrons. It will be a good **reducing** agent.

13-19. Compare $\mathcal{E}^o (Br_2 | Br^-) = +1.066$ V and $\mathcal{E}^o (Cl_2 | Cl^-) = +1.358$ V. **Chlorine** is a stronger oxidant, and should be a better disinfectant.

13-21. The reduction half-reaction is:

MnO_4^- *(aq)* $+ 8H^+$ *(aq)* $+ 5e^- \rightarrow 4H_2O$ *(l)* $+ Mn^{2+}$ *(aq)* $\quad\quad \mathcal{E}^o = +1.507$ V

If a reductant appears in a reduction half reaction with $\mathcal{E}^o > +1.507$ V, then MnO_4^- *(aq)* will *not* oxidize it.

Here:

	reductant	$\mathcal{E}^o$ (ox | red)	MnO_4^- capable of oxidizing it in the standard state?
(a)	F^- *(aq)*	+3.05 V	**No**
(b)	Cl^- *(aq)*	+1.358 V	**Yes**
(c)	Cr^{3+} *(aq)*	+1.232 V	**Yes**
(d)	ClO_4^- *(aq)*	*	**No**

* ClO_4^- is not listed in appendix E anywhere as a reductant. It is "impossible" to oxidize.

13-23. (a) The whole cell reaction that is important here has a $P_4 \mid PH_3$ cathode and a $H_2PO_2^- \mid P_4$ anode. $\Delta\mathcal{E}^o = \mathcal{E}^o\,(\,P_4 \mid PH_3\,) - \mathcal{E}^o\,(\,H_2PO_2 \mid P_4\,) = -0.89 - (-2.05) = +1.16$ V

Yes, P_4 will disproportionate at pH = 14.

(b) **P_4 is the stronger reducing agent,** because $-(-2.05$ V$) > -(-0.89$ V$)$.

13-25. The activity series ranks reducing ability from lowest to highest. La^{3+} / La has a $\mathcal{E}^o$ of -2.304, just above that of Mg^{2+} / Mg. So La / La^{3+} will be just *below* Mg / Mg^{2+} in the activity series but above Al / Al^{3+}. $Eu^{3+} / Eu\ (-1.991$ V$)$ is below Al^{3+} / Al, so Eu / Eu^{3+} will be above Al / Al^{3+} in the activity series. Finally, Ga / Ga^{3+} will be above Fe / Fe^{2+} but below Cr / Cr^{2+} and Sm / Sm^{3+} will be be below Al / Al^{3+} but above Mn / Mn^{2+}.

13-27. For this reaction:

$$\Delta\mathcal{E}^o = \mathcal{E}^o\,(\,Pb^{2+} \mid Pb\,) - \mathcal{E}^o\,(\,Cr^{3+} \mid Cr^{2+}\,)$$

$$= (-0.126) - (-0.407) = +0.281 \text{ V}$$

$$Q = \frac{\left[Cr^{3+}\right]^2}{\left[Cr^{2+}\right]^2\left[Pb^{2+}\right]} = \frac{(0.0030)^2}{(0.20)^2(0.15)} = 0.00150$$

$n = 2$ for this reaction, so:

$$\Delta\mathcal{E} = \Delta\mathcal{E}^o - \frac{0.0257 \text{ V}}{2}\ln Q$$

$$= 0.281 - \frac{0.0257 \text{ V}}{2}\ln(0.00150) = +\mathbf{0.365 \text{ V}}$$

13-29. (a)

$$Cu\ (s) + Pb^{2+}\ (aq) \rightarrow Pb\ (s) + Cu^{2+}\ (aq)$$

$$\Delta\mathcal{E}^o = \mathcal{E}^o\,(\,Cu^{2+} \mid Cu\,) - \mathcal{E}^o\,(\,Pb^{2+} \mid Pb\,)$$

$$= 0.345 \text{ V} - (-0.126 \text{ V}) = +0.471 \text{ V}$$

$$Q = \frac{\left[Pb^{2+}\right]}{\left[Cu^{2+}\right]} = \frac{.0020}{.075} = 0.02667$$

$$\Delta\mathcal{E} = +0.471 \text{ V} - \frac{0.0257 \text{ V}}{2}\ln 0.02667 = +\mathbf{0.52 \text{ V}}$$

(b) $MnO_4^-\ (aq) + 8\,H^+(aq) + 5\,Fe^{2+}(aq) \rightarrow 5\,Fe^{3+}(aq) + 4\,H_2O(l) + Mn^{2+}\ (aq)$

$$\Delta\mathcal{E}^o = \mathcal{E}^o\,(\,MnO_4^- \mid Mn^{2+}\,) - \mathcal{E}^o\,(\,Fe^{3+} \mid Fe^{2+}\,)$$

$$= 1.507 \text{ V} - (0.771 \text{ V}) = 0.736 \text{ V}$$

$$Q = \frac{\left[Mn^{2+}\right]\left[Fe^{3+}\right]^5}{\left[MnO_4^-\right]\left[Fe^{2+}\right]^5\left[H^+\right]^8} = \frac{(0.95)(0.010)^5}{(0.050)(0.020)^5(0.10)^8} = 5.94 \times 10^7$$

Note: $n = 5$

$$\Delta\mathcal{E} = +0.736 \text{ V} - \frac{0.0257 \text{ V}}{5}\ln\left(5.94 \times 10^7\right) = +\mathbf{0.644 \text{ V}}$$

13-29. (continued)

(c) $\quad$ Co (s) + 2 Ag$^+$ (aq) $\rightarrow$ 2 Ag (s) + Co^{2+} (aq)

$$\Delta \mathcal{E}^o = \mathcal{E}^o \,(\, Ag^+ \mid Ag \,) - \mathcal{E}^o \,(\, Co^{2+} \mid Co \,) = 0.800 \text{ V} - (+0.28) = +0.52 \text{ V}$$

$$Q = [Co^{2+}] \,[Ag^+]^{-2} = (0.14)\,(0.10)^{-2} = 14; \quad (n = 2)$$

$$\Delta \mathcal{E} = +0.52 \text{ V} - \frac{0.0257 \text{ V}}{2} \ln(14) = +0.52 - 0.034 = \textbf{+0.486 V}$$

13-31. The standard reduction potential for Cr^{3+} (aq) + e^- $\rightarrow$ Cr^{2+} (aq) is -0.407 V. This will be modified by non-standard-state conditions. We use the Nernst equation for half-reactions the same way we use it for whole reactions.

$$\mathcal{E} = \mathcal{E}^o - \quad \text{and:} \quad Q_{hc} = \frac{\left|Cr^{2+}\right|}{\left|Cr^{3+}\right|} = \frac{0.0019}{0.15} = 0.0127$$

$$\mathcal{E} = -0.407 - \frac{0.0257 \text{V}}{1} \ln(0.0127) = \textbf{-0.29 V}$$

13-33. The standard reduction potential for a Cd^{2+} $\mid$ Cd half-reaction is -0.403 V. The half cell with the *more negative* reduction potential will be the anode in a Galvanic cell. This is the **Cd^{2+} $\mid$ Cd** standard half cell.

13-35. $\quad$ H$_2$ (g) + I$_2$ (s) $\rightarrow$ 2I$^-$ (aq) + 2H$^+$ (aq) $\qquad$ $n = 2$

$$\Delta \mathcal{E}^o = \mathcal{E}^o \,(\, I_2 \mid I^- \,) - \mathcal{E}^o \,(\, H^+ \mid H_2 \,) = +0.536 - 0 = +0.536 \text{ V}$$

$$\Delta \mathcal{E} = \Delta \mathcal{E}^o - \frac{0.0257}{2} \ln Q$$

$$\ln Q = 2 \frac{\Delta E - \Delta E^o}{0.0257} = -\frac{2(0.841 - 0.536)}{0.0257} = -23.73$$

$$Q = e^{-23.73} = 4.97 \times 10^{-11} =$$

$$[H^+] = \sqrt{\frac{Q \times P_{H_2}}{[I^-]^2}} = \sqrt{\frac{4.96 \times 10^{-11} \times 1.00}{(1.00)^2}} = 7.04 \times 10^{-6} \text{ M}; \quad \textbf{pH = 5.15}$$

13-37. (a) $\quad \Delta \mathcal{E}^o = \mathcal{E}^o \,(\, HClO \mid Cl_2 \,) - \mathcal{E}^o \,(\, Cr_2O_7^{2-} \mid Cr^{3+} \,) = (+1.611) - (1.232) = \textbf{+0.379}$

(b) $\quad \Delta \mathcal{E} = \Delta \mathcal{E}^o - \dfrac{0.0257 \text{ V}}{6} \ln Q$

$$\ln Q = -\frac{(\Delta E - \Delta E^o)6}{0.0257} = -\frac{(0.379 - 0.351)6}{0.0257} = +6.54$$

13-37. (continued)

$$Q = e^{+6.54} = 690.19 = \frac{\left[H^+\right]^8 \left[Cr_2O_7^{2-}\right] P_{Cl_2}^3}{\left[Cr^{3+}\right]^2 \left[HClO\right]^6}$$

$$\left[Cr^{3+}\right] = \sqrt{\frac{\left[H^+\right]^8 \left[Cr_2O_7^{2-}\right] P_{Cl_2}^3}{Q\left[HClO\right]^6}} = \sqrt{\frac{(1)^8 (0.80)(0.10)^3}{(690)(0.20)^6}} = \textbf{0.13 M}$$

13-39. The standard potential difference for the reaction is

$$\Delta \mathcal{E}^o = \mathcal{E}^o \left(Ag^+ \mid Ag \right) - \mathcal{E}^o \left(Cd^{2+} \mid Cd \right) +0.800 - (-0.403) = +1.203 \text{ V}$$

For these conditions, with $n = 2$:

$$Q = \frac{\left[Cd^{2+}\right]}{\left[Ag^+\right]^2} = \frac{0.75}{0.0010^2} = 750000$$

$$\Delta \mathcal{E} = +1.203 \text{ V} - \frac{0.0257}{2} \ln 750000 = \textbf{+1.029 V}$$

13-41. (a) Anode: $\quad\quad\quad\quad Ni\ (s) \rightarrow Ni^{2+}\ (aq) + 2e^-$

Cathode: $\quad\quad\quad Ag^+\ (aq) + e^- \rightarrow Ag\ (s)$

Overall: $\quad\quad\quad 2Ag^+\ (aq) + Ni\ (s) \rightarrow Ni^{2+}\ (aq) + 2Ag\ (s)$

(b) $\quad\quad \Delta \mathcal{E}^o = \mathcal{E}^o \left(Ag^+ \mid Ag \right) - \mathcal{E}^o \left(Ni^{2+} \mid Ni \right) +0.800 - (-0.257) = +1.057 \text{ V}$

$$\Delta \mathcal{E} = 1.057 - \frac{0.0257}{2} \ln \frac{(0.10)}{(0.10)^2} = 1.057 - 0.030 \text{ V} = \textbf{1.027 V}$$

(c) $\quad\quad\quad\quad \Delta \mathcal{E}^o = \frac{0.0257}{n} \ln K$

$$\ln K = \frac{n \Delta E^o}{0.0257} = \frac{2 \times 1.057}{0.0257} = 82.25 \quad K = e^{82.25} = \textbf{5.3} \times \textbf{10}^{\textbf{34}}$$

13-43. $\Delta \mathcal{E}^o = \mathcal{E}^o \left(HClO_2 \mid HClO \right) - \mathcal{E}^o \left(Cr_2O_7^{2-} \mid Cr^{3+} \right) = (+1.645) - (1.232) = \textbf{+0.413}$

We note that $n = 6$. Therefore:

$$\ln K = \frac{6 \times 0.413}{0.0257} = 96.4 \Rightarrow K = e^{+96.4} = 7.5 \times 10^{41}$$

This large equilibrium constant indicates the reaction will go essentially all the way to the right. The Cr^{3+} : $HClO_2$ molar ratio is 1 : 2, compared to a ratio in the reaction of 2 : 3. Hence, there is *excess HClO_2*. The Cr^{3+} will all be consumed and the solution after the reaction will be **orange**.

13-45.
$$\Delta \mathcal{E}^\circ = \mathcal{E}^\circ (\ \text{In}^+ \mid \text{In}\ ) - \mathcal{E}^\circ (\ \text{In}^{3+} \mid \text{In}\ ) = (-0.21\ \text{V}) - (-0.40\ \text{V}) = +0.19\ \text{V}$$

$$\ln K = \frac{n\Delta E^\circ}{0.0257} = \frac{(2)(0.19)}{0.0257} = 14.79 \implies K = 10^{+14.79} = \mathbf{2.6 \times 10^6}$$

13-47. The whole cell reaction is:

$$2\ \text{H}^+ (1\ \text{M}) + \text{H}_2\ (1\ \text{atm}) \rightarrow 2\ \text{H}^+ (\text{unknown}) + \text{H}_2\ (1\ \text{atm})$$

Note that the standard cell potential for this reaction is 0.00 V.

$$\Delta \mathcal{E} = 0.00 - \frac{0.0257}{2} \ln \frac{\left[\text{H}^+ (\text{unknown}) \right]^2 (1.00)}{(1.00)^2 (1.00)}$$

$$= -0.01285 \ln \left[\text{H}^+ (\text{unknown}) \right]^2$$

$$= -(2)(0.01285) \ln \left[\text{H}^+ (\text{unknown}) \right]$$

We want to relate this to the pH instead of the natural logarithm of the H^+. We use the relationship that $\log_{10} x = 2.303 \ln x$. We then have:

$$\Delta \mathcal{E} = -0.0592 \log [\text{H}^+] = 0.0592\ \text{pH (unknown)}$$

$$\text{pH (unknown)} = \frac{\Delta E}{0.0592} = \frac{0.150}{0.0592} = \mathbf{2.53}$$

A buffer made with equal concentrations of a weak acid and its conjugate base has the characteristic that $pK_a \approx \text{pH}$. Here, $pK_a = 2.53$, so
$K_a = 10^{-2.53} = \mathbf{0.0029}$.

13-49. (a)
$$\Delta \mathcal{E}^\circ = \mathcal{E}^\circ (\ \text{Br}_2 \mid \text{Br}^-\ ) - \mathcal{E}^\circ (\ \text{H}^+ \mid \text{H}_2\ )$$
$$= 1.065 - 0 = +\mathbf{1.065\ V}$$

(b)
$$\Delta \mathcal{E} = \Delta \mathcal{E}^\circ - \frac{.0257\ \text{V}}{2} \ln \left(\frac{\left[\text{Br}^- \right]^2 \left[\text{H}^+ \right]^2}{P_{\text{H}_2}} \right)$$

$$1.710\ \text{V} = 1.065\ \text{V} - \frac{.0257\ \text{V}}{2} \ln \left(\frac{\left[\text{Br}^- \right]^2 (1)^2}{(1)} \right)$$

$$0.645\ \text{V} = -\frac{0.0257\ \text{V}}{2} \ln \left[\text{Br}^- \right]^2$$

$$\ln [\text{Br}^-] = \frac{0.645}{-0.0257} \frac{0.645\ \text{V}}{-0.0592\ \text{V}} = -25.10$$

$$[\text{Br}^-] = e^{-25.10} = \mathbf{1.3 \times 10^{-11}\ M}.$$

13-51.
$$\Delta \mathcal{E}^o = \mathcal{E}^o(PbO_2 \mid PbSO_4) - \mathcal{E}^o(PbSO_4 \mid Pb)$$
$$= 1.691\ V - (-0.359\ V) = \mathbf{+2.050\ V}$$

Six such cells in series would generate an electrical potential of
$6(2.050\ V) = \mathbf{12.300\ V}$.

13-53. The reaction yields $2e^-$ per Pb reacted.
$$10\ kg\ of\ Pb = 10{,}000\ g / 207.2\ g\ mol^{-1} = 48.26\ mol\ Pb$$
$$mol\ of\ e^- = 48.26\ mol \times 2 = 96.5\ mol\ e^-$$
$$C\ of\ e^- = 96.5\ mol \times 96{,}485\ C\ mol^{-1} = \mathbf{9.3 \times 10^6\ C}.$$

13-55. In problem No. 51 we discovered that the voltage for the six cells in series was six times that for one cell. But it is easier to evaluate the maximum work we can obtain by treating the system as 10,000 g of Pb. This reacts at a single cell potential of +2.041 V. Then, for the charge (determined in No. 53) passed in this system:
$$w_{elec,max} = \Delta G = -\ charge \times potential$$
$$= -9.3 \times 10^6\ C \times 2.041\ V\ J\ C^{-1}$$
$$= 1.90 \times 10^7\ J = \mathbf{-19{,}000\ kJ}$$

13-57. Not entirely, because the cell reaction also consumes Pb and PbO_2. If these are not replenished, the battery will remain rundown. On the other hand, if excess Pb and PbO_2 are present, replenishment with more H_2SO_4 will restore the battery's action (until something is depleted again).

13-59. Alkaline cells have no dissolved species, and therefore they have no species whose changing concentration, as explained by the Nernst equation, causes a change in voltage. Of course, even alkaline cells run out, and the voltage drops off at the end of the battery's life.

13-61. The substances formed on reaction involve complex salt mixtures. These do not have the same ability to react with an electrode as do the starting materials.

13-63.
$$\Delta \mathcal{E}^o = \mathcal{E}^o(O_2 \mid H_2O) - \mathcal{E}^o(H_2 \mid H^+) = 1.229\ V$$
$$w_{max} = \left(1.229 \frac{J}{C}\right)\left(\frac{96485\ C}{mol\ e^-}\right)\left(\frac{2\ mole^-}{1\ mol H_2O}\right)\left(\frac{1\ mol\ H_2O}{18.015\ g\ H_2O}\right)$$
$$= 13{,}200\ J\ g^{-1} \qquad 60\%\ of\ this\ is\ \mathbf{7{,}900\ J\ g^{-1}}$$

Chapter 13

13-65. Since the reaction is done in basic aqueous conditions, we must use the water / hydrogen half cell for those conditions.

$$\Delta \mathcal{E}^o = \mathcal{E}^o \,(\, H_2O \mid H_2 \,) - \mathcal{E}^o \,(\, Fe^{2+} \mid Fe \,) = (-0.828) - (-0.447) = \textbf{-0.381 V}$$

This reaction is **non-spontaneous**. Because hydroxide is a product, the reaction will become more favorable as [OH⁻] decreases--as one moves to pH's lower than 14.

13-67. Sodium would make an excellent sacrificial anode, except for the fact that it is far too reactive *with water* to be stable when exposed to a water-containing environment.

13-69. (a) For this we proceed from the ΔG^o_f data to get ΔG^o_r and then use this to get $\Delta \mathcal{E}^o$

$$\Delta G^o_r = 2 \times \Delta G^o_f \,(H_2O,\, l) + \Delta G^o_f \,(CH_4,\, g) - \Delta G^o_f \,(CO_2,\, g) - \Delta G^o_f \,(H_2,\, g)$$

$$= 2 \times (-237.18) + (-50.75) - (-394.36) - 2 \times 0 = -130.69 \text{ kJ} = 130690 \text{ J}$$

For the cell potential: $\Delta \mathcal{E}^o = \Delta G^o_r \,/\, n\, F = \textbf{0.169 V}$ (note, $n = 8$)

(b) Under these conditions, $\Delta \mathcal{E} = \Delta \mathcal{E}^o - \dfrac{0.0592}{8} \log_{10} \dfrac{P_{CH_4}}{P_{CO_2} P^4_{H_2}}$. Substition in this gives

$$\Delta \mathcal{E} = 0.169 - 0.011 = \textbf{0.158 V}$$

13-71. Corrosion is being used to clean water by the introduction of filters that contain iron and other similar metals. Contaminated water passes over these and the high activity iron reduces the contaminants, often forming innocuous iron salts.

13-73.

$$Q = \left(\frac{2.1 \times 10^{13} \text{ g Al}}{26.98 \text{ g mol}^{-1}} \right) \left(\frac{3 \text{ mol e}^-}{1 \text{ mol Al}} \right) \left(\frac{96,485 \text{ C}}{1 \text{ mol e}^-} \right) = 2.25 \times 10^{17} \text{ C.}$$

$$\text{Energy} = 2.25 \times 10^{17} \text{ C} \times 5 \frac{J}{C} \times \frac{1 \text{ kW - hr}}{3.6 \times 10^6 \text{ J}} = 3.13 \times 10^{11} \text{ kW-hr}$$

$$\text{Cost} = 3.13 \times 10^{11} \text{kW-hr} \times (\$0.10 \text{ (kW-hr)}^{-1}) = \$3.13 \times 10^{10} = \textbf{\$31 billion.}$$

13-75. (a) Appendix E gives these two reduction half-reactions:

$$MnO_2 \,(s) + 4H^+ \,(aq) + 2e^- \rightarrow Mn^{2+} \,(aq) + 2H_2O \,(l) \qquad \mathcal{E}^o = +1.224 \text{ V}$$
$$2H^+ \,(aq) + O_2 \,(g) + 2e^- \rightarrow H_2O_2 \,(aq) \qquad \mathcal{E}^o = +0.695 \text{ V}$$

Since the reaction is carried out under standard state conditions, the cathode is the reaction at more positive $\mathcal{E}^o$--the $MnO_2 \mid Mn^{2+}$ reaction.

Cathode: $MnO_2 \,(s) + 4H^+ \,(aq) + 2e^- \rightarrow Mn^{2+} \,(aq) + 2H_2O \,(l)$

Anode: $H_2O_2 \,(aq) \rightarrow 2H^+ \,(aq) + O_2 + (g) + 2e^-$

Whole cell: $MnO_2(s) + 2H^+(aq) + H_2O_2(aq) \rightarrow Mn^{2+}(aq) + 2H_2O(l) + O_2(g)$

(b) $\Delta \mathcal{E}^o = (+1.224 \text{ V}) - (0.695 \text{ V}) = \textbf{+0.529 V}$

13-77. We need to drive the non-spontaneous reaction:

Ni^{2+} *(aq)* + 2I⁻ *(aq)* → Ni *(s)* + I_2 *(s)*

For which: $\Delta \mathcal{E}^o = \mathcal{E}^o$ (Ni^{2+} | Ni) - $\mathcal{E}^o$ (I_2 | I⁻)= (-0.257- (0.536V) = **-0.793 V**

A potential of +0.793 V must be applied to drive this reaction.

Anode: 2I⁻ *(aq)* → I_2 *(s)* + 2e⁻

Cathode: Ni^{2+} *(aq)* + 2e⁻ → Ni *(s)*

13-79. The standard reduction potential for reduction to PH_3 *(g)* is more negative than the standard reduction potential for reduction to HS⁻ *(g)*. Therefore, it is harder to form PH_3, so PH_3 must be a stronger reducing agent.

13-81.

Cathode: Ag^+ *(aq)* + e⁻ → Ag *(s)*

Anode: Zn *(s)* → Zn^{2+} *(aq)* + 2e⁻

$$\Delta \mathcal{E}^o = \mathcal{E}^o (Ag^+ | Ag) - \mathcal{E}^o (Zn^{2+} | Zn)$$
$$= (+0.800 \text{ V}) - (-0.762 \text{ V}) = +1.56 \text{ V}$$

If she makes a cell that uses the solutions that she has without altering them, then the voltage would be as shown on the next page.

$$\Delta \mathcal{E} = \Delta \mathcal{E}^o - \frac{0.0257 \text{ V}}{2} \ln \frac{\left| Zn^{2+} \right|}{\left[Ag^+ \right]^2}$$

$$= 1.56 \text{ V} - (0.01285 \text{ V}) \ln \left(\frac{.10}{(.01)^2} \right) = 1.56 \text{ V} - (0.01285 \text{ V} \times 6.91) = 1.47 \text{ V}.$$

This is *not* the voltage she needs, so one of the solutions must be diluted to give the correct voltage. Being a careful student of the Nernst equation, she knows the logarithmic term is the source of possible variations. So she writes:

$$1.50 \text{ V} = 1.56 \text{ V} - \frac{0.0592 \text{ V}}{2} \log_{10} Q$$

Rearranging: $+\dfrac{0.060 \text{ V}}{0.0296 \text{ V}} = \log_{10} Q$

$$10^{+2.027} = Q = 106 \text{ This is the reaction quotient she needs.}$$

The Q for the solutions available is: $Q = \dfrac{\left[Zn^{2+} \right]}{\left[Ag^+ \right]^2} = 1000$

So Q must be decreased. She decides to lower the concentration of $[Zn^{2+}]$ so $Q = 106$. In this case, keeping $[Ag^+] = 0.0100$ M:

$$[Zn^{2+}] = 106 [Ag^+]^2 = 106 (0.010)^2 = 0.0106$$

13-81. (continued)

So she prepares a solution with $[Zn^{2+}] = 0.0106$ from her stock solution and places it in the apparatus sketched below.

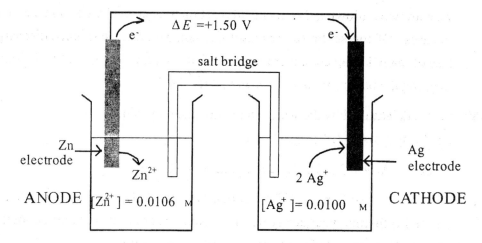

13-83(a) The spontaneous reaction will tend to *decrease* the concentration in the cell with higher $[Pb^{2+}]$ (0.10 M) and *increase* the concentration in the cell with lower $[Pb^{2+}]$ (0.010 M). So the whole cell reaction will be:

$$Pb^{2+} (0.10 \text{ M}) + Pb \ (s) \rightarrow Pb \ (s) + Pb^{2+} (0.010 \text{ M})$$

The product Pb^{2+} is in the cell with the lower $[Pb^{2+}]$. Thus, for a spontaneous reaction, we want to write $Q = \dfrac{\left[Pb^{2+}\right]_{lower}}{\left[Pb^{2+}\right]_{higher}} = \dfrac{(0.010)}{(0.100)} = 0.10$

$$\Delta \mathcal{E} = 0.0 - (0.0257 \text{ V} / 2) \ln 10^{-1} = \textbf{+0.030 V}$$

(b) After the addition of the sulfate, a significant amount of the Pb^{2+} in the solution will precipitate as lead sulfate. There will be a small concentration of Pb^{2+}, controlled by the common ion effect. The spontaneous reaction will now produce Pb^{2+}.

$$[Pb^{2+}]_{lower} = \frac{K_{sp}\left(PbSO_4\right)}{\left[SO_4^{2-}\right]} = \frac{1.1 \times 10^{-8}}{0.10 \text{ M}} = 1.1 \times 10^{-7} \text{ M}$$

Then: $\Delta \mathcal{E} = 0.0 - 0.01285 \text{ V} \ln \dfrac{\left(1.1 \times 10^{-7}\right)}{(0.010)} = \textbf{+ 0.146 V}$

13-85 The spontaneous reaction in the Leclanché cell is:

$$Zn \ (s) + 2 \ MnO_2 \ (s) + 2 \ NH_4^+(aq) \rightarrow Zn^{2+}(aq) + Mn_2O_3 \ (s) + 2 \ NH_3 \ (aq) + H_2O \ (l)$$

Electrons are released from the zinc anode -- the battery's *negative* terminal and taken up at the MnO_2 | graphite cathode -- the *positive* terminal.

13-87. The batteries assembled here produce electricity through a chemical reaction, as is true of all batteries. The batteries assembled here already contain the chemical substances in the form needed for this reaction to occur. The "electricity" is available by the discharge of the battery by allowing this chemical reaction to occur. There is no need to charge the batteries. Of course, for safety reasons certain batteries are shipped discharged, as is the case of many laptop computer batteries. But even then, we do not add a substance; charging the battery involves a chemical reaction.

13-89. ΔG^o is determined by the water and hydrazine only, since
$$\Delta G^o_f\ (O_2\ (g)) = \Delta G^o_f\ (N_2\ (g)) = 0.$$

Thus: $\Delta G^o = 2\ \Delta G^o_f (H_2O\ (l)) - \Delta G^o_f (N_2H_4\ (aq))$

$= 2$ mol $(-237.18$ kJ mol$^{-1}) - 1$ mol$(+128.1$ kJ mol$^{-1}) = -602.4$ kJ

The cell is thermodynamically feasible. And what the seller claims about the products is correct. N_2 is part of air and liquid water is, well, water.

The cell potential in this case, since four electrons are transferred, is:

$$\Delta \mathcal{E}^o = -\frac{\Delta G^o}{nF} = -\frac{-602,400J}{(4\ \text{mol e}^-)(96,485C\ (\text{mol e}^-)^{-1})} = +1.56\ \frac{J}{C} = \textbf{+1.56 V}$$

13-91. The potentials are:

$Sn^{2+} + 2e^- \rightarrow Sn$ $\quad \mathcal{E}^o = -0.138$ V $\qquad Fe^{2+} + 2e^- \rightarrow Fe \qquad \mathcal{E}^o = -0.447$ V

$Zn^{2+} + 2e^- \rightarrow Zn$ $\quad \mathcal{E}^o = -0.762$ V

Sn is harder to oxidize than Fe. Thus, given access to Fe *(s)*, oxidants will more readily attack Fe. But in an intact tin can the oxidant can't reach the Fe, sot the Sn protects by coating. With a galvanized object, in contrast, the zinc acts as a sacrificial anode as well as a coating. A exposed section of iron will be protected by the oxidation of Zn, even if the Zn is some distance away.

13-93. In all these cases the reaction involves a conversion of AgX into Ag + X$^-$. We can compare this to the solubility reaction, involving AgX into Ag$^+$ + X$^-$. The electrochemical reaction can be viewed as "adding" the reaction Ag$^+$ + e$^-$ → Ag to the solubility reaction. This means that, since the thermodynamic prefernce for Ag$^+$ + e$^-$ → Ag will be the same in al cases, the trend in the stanard reduction potentials of AgX will be directly related to the solubility of AgX. We see that for AgCl, AgBr, and AgI the reduction becomes harder and harder, which matches will with the decreasing K_{sp} in the same series. However, the radioactive halide astatide (At$^-$) forms a salt that is easier to reduce, and therefore we expect AgAt to have the highest K_{sp}.

13-95. The K_a reaction for HF is HF *(aq)* $\rightarrow$ H$^+$ *(aq)* + F$^-$ *(aq)* . We can get this reaction by combining:

CATHODE:	F$_2$ *(g)* + 2 e^- $\rightarrow$ 2 F$^-$ *(aq)*	$\mathcal{E}^o$ = 2.866 V
ANODE:	F$_2$ *(g)* + 2 H$^+$ + 2 e^- $\rightarrow$ 2 HF *(aq)*	$\mathcal{E}^o$ = 3.053 V
OVERALL	2 HF $\rightarrow$ 2 F$^-$ *(aq)* + 2 H$^+$ *(aq)*	$\Delta\mathcal{E}^o$ = -0.187 V

We then get: $K = e^{\left(nF\Delta E^o / RT\right)} = e^{\left(\frac{2\times^-0.187\times96485}{8.3145\times298}\right)} = e^{-14.56} = 4.73 \times 10^{-7}$

Of course, this is for *twice* the K_a reaction, so $K_a = \sqrt{K} = \sqrt{4.73\times10^{-7}} = 6.8\times10^{-4}$

CHAPTER 14 - CHEMICAL KINETICS

14-1. The first mile marker was passed at 13 minutes and 16 seconds before noon; the second marker was passed at 2 minutes and 24 seconds *after* noon. So the time elapsed was 15 minutes and 40 seconds, or 15.667 minutes.

The distance traveled was 30.0 - 11.5 = 18.5 miles.

The rate of travel in miles per minute was: $\dfrac{18.5 \text{ miles}}{15.667 \text{ minutes}} = \textbf{1.18 mile min}^{-1}$.

This is equal to $(1.181 \text{ mile min}^{-1}) \times (60 \text{ min hr}^{-1}) = \textbf{70.9 mile hr}^{-1}$.

14-3.

TIME ELAPSED	MILES TRAVELED	RATE OF TRAVEL
11 hours	550 miles	**50** miles per hour
12 hours	550 miles	**46** miles per hour
13 hours	550 miles	**42** miles per hour
18.5 hours	550 + 300= 850 miles	**46** miles per hour

14-5. A line drawn as a tangent to the curve at t = 200 seconds has the following parameters. These are conveniently determined by extending the tangent line segment to the ends of the graph. Such a segment intersects the left ($t = 0$ s) of the graph at a concentration of 0.0205 M and the right of the graph ($t = 300$ s) at a concentration of 0.0325 M. We then get:

$$\text{concentration change} = 0.0325 - 0.0205 = 0.0120 \text{ M}$$
$$\text{time change} = 300 \text{ seconds}$$

$$\text{rate} = \frac{\text{concentration change}}{\text{time change}} = \frac{0.0120 \text{ M}}{300 \text{ s}} = \textbf{4.0} \times \textbf{10}^{-5} \textbf{ mol L}^{-1} \textbf{ s}^{-1}$$

A similar exercise at $t = 50$ seconds gives a value of 1.9×10^{-4} mol L^{-1} s^{-1}, which is the value given in the Appendix to the first printing of the book.

14-7. The related expressions for the rate reflect the stoichiometry of the reaction.

$$\text{rate} = -\frac{\Delta[N_2]}{\Delta t} = -\frac{1}{3}\frac{\Delta[H_2]}{\Delta t} = -\frac{1}{2}\frac{\Delta[NH_3]}{\Delta t}$$

14-9. (a) As determined in the last problem, the relationship between the expressions of the rate are:

$$-\frac{1}{3}\frac{\Delta[H_2]}{\Delta t} = +\frac{1}{2}\frac{\Delta[NH_3]}{\Delta t}$$

$$\frac{\Delta[H_2]}{\Delta t} = -\frac{3}{2}\left(-0.150\frac{\text{mol}}{\text{L} \cdot \text{s}}\right) = \textbf{0.225 mol L}^{-1} \textbf{ s}^{-1}$$

(b) $-\dfrac{\Delta[N_2]}{\Delta t} = +\dfrac{1}{2}\dfrac{\Delta[NH_3]}{\Delta t} \Rightarrow \dfrac{\Delta[N_2]}{\Delta t} = -\dfrac{1}{3}\left(0.150\dfrac{\text{mol}}{\text{L} \cdot \text{s}}\right) = \textbf{0.075 mol L}^{-1} \textbf{ s}^{-1}$

14-11. (a) **Second** order -- indicated by the lone exponent "2" on $[NO_2]$

(b) **First** order -- indicated by the implied exponent "1" on $[I_2]$

14-13. The overall rate expression will have: Rate = k [species A, B, C, ...]^3

The units of rate are mol L^{-1} sec^{-1}. The units of concentration are mol L^{-1}.

Therefore, we can rearrange for

$$k = \frac{\text{rate}}{[\text{species A, B, C..}]^3} = \frac{\text{mol L}^{-1}\text{s}^{-1}}{\left(\text{mol L}^{-1}\right)^3} = \frac{\text{L}^2}{\text{mol}^2\ \text{s}} = \textbf{L}^2\ \textbf{mol}^{\textbf{-2}}\ \textbf{s}^{\textbf{-1}}$$

14-15. (a) As in Example 15-2, we can determine the order of a reaction with respect to a reactant by examining the dependence of the ratio of the rate on the ratio of the concentrations of that reactant. Using the first two data points:

$$\frac{\text{rate}_2}{\text{rate}_1} = \left(\frac{[\text{NOBr}]_2}{[\text{NOBr}]_1}\right)^n$$

$$\frac{6.2 \times 10^{-5}}{1.35 \times 10^{-4}} = \left(\frac{0.0088}{0.0130}\right)^n$$

$$0.4593 = (0.6769)^n$$

Clearly, n is not 1, and we must try out different values for n. If $n = 2$, then $(0.6769)^n = (0.6769)^2 = 0.4582$. This equals the experimental rate ratio, so we conclude the reaction is second order in NOBr. We can verify this by checking the rate ratios vs. concentration ratios for the second and third data points:

$$\frac{\text{rate}_3}{\text{rate}_2} = \left(\frac{[\text{NOBr}]_3}{[\text{NOBr}]_2}\right)^2$$

$$\frac{1.92 \times 10^{-5}}{6.2 \times 10^{-5}} = \left(\frac{0.0049}{0.0088}\right)^2$$

$$0.3097 = 0.3100 \quad \text{This verifies our conclusion that } n = 2.$$

The rate expression is then: **rate = k [NOBr]^2**

(b) The rate constant can be determined from the experimental rate and the concentration of NOBr:
$$k = \frac{\text{rate}}{[\text{NOBr}]^2}$$

For the first datum:
$$k = \frac{1.35 \times 10^{-4}\ \text{mol L}^{-1}\text{s}^{-1}}{\left(0.0130\ \text{mol L}^{-1}\right)^2} = 0.799\ \text{L mol}^{-1}\ \text{s}^{-1}$$

For the second: $k = 0.801$ L mol^{-1} s^{-1};

For the third: $k = 0.801$ L mol^{-1} s^{-1}

The average of these three is **0.800 L mol^{-1} s^{-1}.**

14-17. (a) Increasing [I⁻] by a factor of ten will increase the rate of the reaction by a factor of ten, since the reaction is first order in [I⁻].

(b) Decreasing [H₂O₂] by a factor of four will decrease the rate by a factor of four, since the reaction is first order in [H₂O₂].

14-19. (a) The experimental data imply the following about the rate expression:

Cutting $[H_2]$ in half $\Rightarrow$ cuts rate in half	The reaction is first order in $[H_2]$
Raising $[NO]$ by 10 $\Rightarrow$ raises rate by 100	The reaction is second order in $[NO]$

$$\text{Rate} = k\,[H_2]\,[NO]^2$$

The units of k for this third-order reaction are $L^2\,mol^{-2}\,s^{-1}$

(b) Raising $[H_2]$ by factor of 2 $\Rightarrow$ Raise rate by factor of 2

Raising $[NO]$ by 3 $\Rightarrow$ Raise rate by factor of $3^2 = 9$

TOGETHER: Raise rate by factor of $9 = 18$

14-21. (a) The first two data points differ in the concentration of *both* reactants, so we can't use them (yet). The second and third data differ in the concentration of CH₃I only, so they can be used to determine the order of the reaction in CH₃I:

$$\frac{\text{rate}_3}{\text{rate}_2} = \left(\frac{[CH_3I]_3}{[CH_3I]_2}\right)^n$$

$$\frac{6.0 \times 10^{-6}}{3.0 \times 10^{-6}} = \left(\frac{4.00 \times 10^{-4}}{2.00 \times 10^{-4}}\right)^n$$

$$2.0 = (2.00)^n, \quad \text{so } n = 1$$

We must now compare the first two data to obtain the order of the reaction order with respect to C₅H₅N. To do this analytically, we must include the both reactants in the ratio, but of course we now know the reaction order in CH₃I:

$$\frac{\text{rate}_2}{\text{rate}_1} = \left(\frac{[CH_3I]_2}{[CH_3I]_1}\right)^1 \times \left(\frac{[C_5H_5N]_2}{[C_5H_5N]_1}\right)^n$$

14-21. (continued)

$$\frac{3.0 \times 10^{-6}}{7.5 \times 10^{-7}} = \left(\frac{2.00 \times 10^{-4}}{1.00 \times 10^{-4}}\right)^1 \times \left(\frac{2.00 \times 10^{-4}}{1.00 \times 10^{-4}}\right)^n$$

$$4.0 = (2.00) \times (2.00)^n$$

$$2.0 = (2.00)^n, \quad \text{so } n = 1$$

The overall rate expression is: rate = k **[C5H5N] [CH3I]**.

(b) We can solve for k by putting the rate and concentration of any data point in the

equation: $$k = \frac{\text{rate}}{[C_5H_5N][CH_3I]}$$

For the first data point: $$k = \frac{7.5 \times 10^{-7} \text{ mol L}^{-1} \text{ s}^{-1}}{\left(1.00 \times 10^{-4} \text{ mol L}^{-1}\right)\left(1.00 \times 10^{-4} \text{ mol L}^{-1}\right)}$$

$$= \textbf{75 L mol}^{-1} \textbf{ s}^{-1}$$

(c) rate = $(75 \text{ L mol}^{-1} \text{ s}^{-1})(5.0 \times 10^{-5} \text{ mol L}^{-1})(2.0 \times 10^{-5} \text{ mol L}^{-1})$

$$= \textbf{7.5} \times \textbf{10}^{-8} \textbf{ mol L}^{-1} \textbf{ s}^{-1}$$

14-23. (a) Comparing data in the first two lines, we get the order of the reaction in SCN⁻:

$$\frac{\text{rate}_2}{\text{rate}_1} = \left(\frac{[SCN^-]_2}{[SCN^-]_1}\right)^n$$

$$\frac{3.5 \times 10^{-9}}{1.9 \times 10^{-9}} = \left(\frac{0.084}{0.045}\right)^n$$

$$1.84 = (1.87)^n \quad \text{so } n = 1 \text{ for [SCN}^-]$$

Comparing the data in the 2nd and 3rd lines we get the order of the reaction in

Cr(H2O)6³⁺: $$\frac{\text{rate}_3}{\text{rate}_2} = \left(\frac{[Cr(H_2O)_6^{3+}]_3}{[Cr(H_2O)_6^{3+}]_2}\right)^n$$

$$\frac{8.9 \times 10^{-9}}{3.5 \times 10^{-9}} = \left(\frac{0.053}{0.021}\right)^n$$

$$2.54 = (2.53)^n \quad \text{so } n = 1 \text{ for } [Cr(H_2O)_6^{3+}]$$

Together: rate = $k \left[Cr(H_2O)_6^{3+}\right][SCN^-]$

14-23. (continued)

(b) For the first data point: $k = \dfrac{rate}{\left[Cr(H_2O)_6^{3+}\right]\left[SCN^-\right]}$

$$= \frac{1.9 \times 10^{-9} \text{ mol L}^{-1}\text{ s}^{-1}}{\left(0.021 \text{ mol L}^{-1}\right)\left(0.045 \text{ mol L}^{-1}\right)}$$

$$= \mathbf{2.0 \times 10^{-6} \text{ L mol}^{-1}\text{ s}^{-1}.}$$

14-25. (a) There is a general relationship between the rate constant k and the half-life $t_{1/2}$ for a first order reaction:

$$k = \frac{\ln 2}{t_{1/2}} = \frac{0.6931}{t_{1/2}} = \frac{0.6931}{4.5 \times 10^4 \text{ s}} = 1.54 \times 10^{-5} \text{ s}^{-1} = \mathbf{1.5 \times 10^{-5} \text{ s}^{-1}}$$

(b) Note that 16.2 hours = 58320 seconds. We seek the fraction x of an original amount that remains.

This means: $\quad x\,[SO_2Cl_2]_0 = [SO_2Cl_2]_0\, e^{-kt} \quad$ Note the common $[SO_2Cl_2]_0$

Then: $\qquad\qquad x = e^{-\left(1.54\times10^{-5}\text{ s}^{-1}\right)\left(58320s\right)} = e^{-0.8983} = \mathbf{0.41}$

14-27. We can rearrange the basic equation for the time dependence of concentration to solve for k. We then use the experimental numbers given in the problem to get the value for k. Note that, as is usually the case, it is easier to rearrange the equation before inserting any numbers. Starting at $t = 0$, we have:

$$[C_2H_5Cl] = [C_2H_5Cl]_0\, e^{-kt}$$

$$\frac{[C_2H_5Cl]}{[C_2H_5Cl]_0} = e^{-kt}$$

$$\ln\left(\frac{[C_2H_5Cl]}{[C_2H_5Cl]_0}\right) = -kt$$

$$-\frac{1}{340\text{ s}}\ln\left(\frac{0.0016 \text{ mol L}^{-1}}{0.0098 \text{ mol L}^{-1}}\right) = k = \mathbf{0.0053 \text{ s}^{-1}}$$

To get the half life, we use the relation between half-life and k for a first order system:

$$t_{1/2} = \frac{\ln 2}{k} = \frac{0.6931}{0.0053 \text{ s}^{-1}} = \mathbf{130 \text{ s}}$$

14-29. (a) This is **true**, because rate is proportional to the concentration of A, which is decreasing with time.

(b) This is **false**. The rate constant merely indicates the "velocity" of the reaction, not the time it takes to complete the reaction.

14-29. (continued)

(c) This is **true**, because the reaction is first order in A.

(d) This is **false**. The half life of a first-order reaction is $t_{1/2} = {}^{0.6931}/_k$.

(e) This is **false**. See the answer to (d) for the reason.

14-31. (a) For a second order reaction:

$$\frac{1}{[I]} = \frac{1}{[I]_o} + 2kt = \frac{1}{1.00 \times 10^{-4}\,\text{mol L}^{-1}} + 2(8.2 \times 10^9\,\text{L mol}^{-1}\,\text{s}^{-1})(2.0 \times 10^{-6}\,\text{s})$$

$$= (1.00 \times 10^4\,\text{L mol}^{-1}) + (3.28 \times 10^4\,\text{L mol}^{-1}) = 42{,}800\,\text{L mol}^{-1}$$

$$[I] = \frac{1}{42{,}800\,\text{L mol}^{-1}} = \mathbf{2.3 \times 10^{-5}\,mol\ L^{-1}}$$

14-33. The units of the rate constant -- s^{-1} -- indicate that this is a first-order reaction.

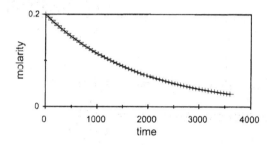

Note that the concentration decays exponentially, and reaches half its original value after 1300 seconds, the half-life for this reaction.

14-35.(a) bimolecular   rate $= k\,[\text{HCO}][\text{O}_2]$

    (b) termolecular   rate $= k\,[\text{CH}_3][\text{O}_2][\text{N}_2]$

    (c) unimolecular   rate $= k\,[\text{HO}_2\text{NO}_2]$

14-37.(a) The first step is unimolecular, the others are all bimolecular.

(b) If we simply add all four reactions, we get:

$\text{H}_2\text{O}_2 + \textbf{O} + \textbf{CF}_2\textbf{Cl}_2 + \textbf{ClO} + \text{O}_3 + \textbf{Cl} + \textbf{ClF}_2\textbf{Cl}$

$$\rightarrow \text{H}_2\text{O} + \textbf{O} + \textbf{ClO} + \textbf{CF}_2\textbf{Cl} + \textbf{Cl} + 2\,\text{O}_2 + \textbf{CF}_2\textbf{Cl}_2$$

Compounds in bold are common to both sides and can be eliminated in getting the overall equation: $\text{H}_2\text{O}_2 + \text{O}_3 \rightarrow \text{H}_2\text{O} + 2\,\text{O}_2$

(c) An intermediate species is one that is *formed and then destroyed* during a reaction. In this reaction O, ClO, CF$_2$Cl, and Cl are intermediates. Note that CF$_2$Cl$_2$ is *destroyed and then re-formed* in the reaction. It is **not** an intermediate.

14-39. For a pair of elementary steps such as this, we have

$$K = \frac{\text{rate constant forward}}{\text{rate constant backward}}$$

$$\textit{rate constant backward} = \frac{\text{rate constant forward}}{K} = \frac{1.3 \times 10^{10} \text{ L mol}^{-1} \text{ s}^{-1}}{5.0 \times 10^{10}}$$

$$= 0.26 \text{ L mol}^{-1} \text{ s}^{-1}$$

14-41. (a) Overall: $A + B + E \rightarrow D + F$

From the rate determining step: $\quad \text{rate} = k_2[C][E]$

From the equilibrium: $\quad K = \dfrac{k_1}{k_{-1}} = \dfrac{[C][D]}{[A][B]}$

Rearranging: $\quad [C] = \dfrac{k_1}{k_{-1}} \dfrac{[A][B]}{[D]}$

Combining: $\quad \text{rate} = \left(\dfrac{k_1 k_2}{k_{-1}}\right) \dfrac{[A][B][E]}{[D]}$

(b) Overall: $A + D \rightarrow B + F$

From the rate determining step: $\quad \text{rate} = k_3[E]$

From the second equilibrium: $\quad K_2 = \dfrac{k_2}{k_{-2}} = \dfrac{[E]}{[C][D]}$

Rearranging: $\quad [E] = \dfrac{k_2}{k_{-2}}[C][D]$

From the first equilibrium: $\quad K_1 = \dfrac{k_1}{k_{-1}} = \dfrac{[B][C]}{[A]}$

Rearranging: $\quad [C] = \dfrac{k_1}{k_{-1}} \dfrac{[A]}{[B]}$

Therefore: $\quad [E] = \dfrac{k_2}{k_{-2}} \dfrac{k_1}{k_{-1}} \dfrac{[A][D]}{[B]}$

And: $\quad \text{rate} = \dfrac{k_1 k_2 k_3}{k_{-1} k_{-2}} \dfrac{[A][D]}{[B]}$

14-43. Only mechanism **(b)** is consistent with the rate expression. In it, the intermediates H_2Cl_2 and $CH_3CHClCH_2$, which both participate in the elementary rate-determining step, are generated in fast equilibria dependent on $[HCl]^2$ and $[HCl][CH_3CHCH_2]$. When these expressions are combined the dependence will be $[HCl]^3[CH_3CHCH_2]$.
Mechanism (a) is not consistent with the rate expression, because the last step is fast and is after the slow step, so it makes no difference to the reaction rate. Mechanism (c) also has two pre-equilibria, but only two molecules of HCl are involved.

Chapter 14

14-45. Only mechanism (a) is consistent with the rate expression. Only the first step in it will matter to the rate expression, and this step is a unimolecular reaction involving just NO_2Cl. Both mechanisms (b) and (c) predict a second-order dependence on NO_2Cl.

14-47. Determining the activation energy and A can be done by choosing any two experimental points to get E_a and A.

$$k_1 = 5.49 \times 10^6 \qquad T_1 = 5000\ K \qquad \frac{1}{T_1} = 2.0 \times 10^{-4}\ K^{-1}$$

$$k_3 = 5.09 \times 10^9 \qquad T_2 = 15{,}000\ K \qquad \frac{1}{T_2} = 6.67 \times 10^{-5}\ K^{-1}$$

These are then used in the Arrhenius equation, cast as in Example 15-9 to make the determination of E_a and A convenient:

$$\ln\left(\frac{k_1}{k_3}\right) = -\frac{E_a}{R}\left(\frac{1}{T_1} - \frac{1}{T_3}\right)$$

$$\ln\left(\frac{5.49 \times 10^6}{5.09 \times 10^9}\right) = -\frac{E_a}{8.3145\ J\ mol^{-1}\ K^{-1}}\left(2.0 \times 10^{-4}\ K^{-1} - 6.67 \times 10^{-5}\ K^{-1}\right)$$

$$-6.832 = -E_a \times (1.6032 \times 10^{-5}\ mol\ J^{-1})$$

$$E_a = \frac{6.832}{1.6032 \times 10^{-5}\ mol\ J^{-1}} = \mathbf{4.25 \times 10^5\ J\ mol^{-1}}$$

(b) Using point 3 only:

$$\ln A = \ln k_3 + \frac{E_a}{RT_3} = \ln(5.09 \times 10^9) + \frac{4.25 \times 10^5\ J\ mol^{-1}}{(8.3145\ J\ mol^{-1}\ K^{-1})(15000K)}$$

$$\ln A = +22.35 + 3.416 = 25.766$$

$$A = e^{25.766} = \mathbf{1.54 \times 10^{11}\ L\ mol^{-1}\ s^{-1}}$$

14-47. (continued)

One may also get both $\ln A$ and E_a from a graph of $\ln k$ vs. $+1/T$:

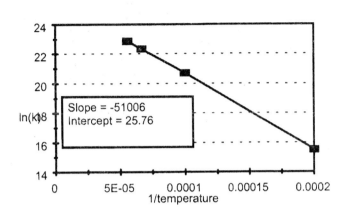

Slope = -51006
Intercept = 25.76

The slope of the graph equals $-\dfrac{E_a}{R}$, and

has units of Kelvins and then:

$E_a = -slope \times R = 4.24 \times 10^5$ J mol^{-1}

We also have intercept $= \ln A$,

so $\quad A = e^{intercept} = e^{25.76}$

$\qquad\qquad = 1.54 \times 10^{11}$ L mol^{-1} s^{-1}

14-49. We have two expressions at the two temperatures:

$$k_1 = A\,e^{-E_a/RT_1}$$
$$k_2 = A\,e^{-E_a/RT_2}$$

If we divide the second by the second, we obtain:

$$\frac{k_2}{k_1} = \frac{Ae^{-E_a/RT_2}}{Ae^{-E_a/RT_1}} = e^{-E_a/RT_2}e^{+E_a/RT_1} = e^{-\left[\frac{E_a}{R}\left(\frac{1}{T_1}-\frac{1}{T_2}\right)\right]}$$

$$k_2 = k_1\, e^{-\left[\frac{E_a}{R}\left(\frac{1}{T_1}-\frac{1}{T_2}\right)\right]}$$

$$= (1.94 \times 10^{-4}\text{ L mol}^{-1}\text{s}^{-1})\,e^{-\left[\frac{161,000\text{ J mol}^{-1}}{8.3145\text{ J mol}^{-1}\text{ K}^{-1}}\left(\frac{1}{313.15\text{K}}-\frac{1}{303.15\text{K}}\right)\right]}$$

$$= (1.94 \times 10^{-4}\text{ L mol}^{-1}\text{s}^{-1})\,e^{+2.04} = \mathbf{1.49 \times 10^{-3}\text{ L mol}^{-1}\text{ s}^{-1}}$$

(b) We know that the reaction is second order because of the units of k. The half life for

a second-order reaction with both reactants present in equal amounts is: $t_{1/2} = \dfrac{1}{2k[X]_o}$

This will apply both at T_1 and T_2

14-49. (continued)

Dividing the expression for k at T_2 by the one for k at T_1 we get:

$$\frac{t_{1/2}(2)}{t_{1/2}(1)} = \left(\frac{1}{2k_2[X]_o}\right)\left(\frac{2k_1[X]_o}{1}\right) = \frac{k_1}{k_2}$$

$$t_{1/2}(2) = t_{1/2}(1)\frac{k_1}{k_2} = 1.00 \times 10^4 \, s \frac{\left(1.94 \times 10^{-4} \, L \, mol^{-1} \, K^{-1}\right)}{\left(1.49 \times 10^{-3} \, L \, mol^{-1} \, K^{-1}\right)} = 1300 \, s$$

14-51. (a) $\quad A = k \, e^{+E_a/RT} = \left(0.41 \, s^{-1}\right) e^{\left[\left(\frac{161,000 J \, mol^{-1}}{8.3145 \, J \, mol^{-1} K^{-1}}\right)\left(\frac{1}{600 \, K}\right)\right]} = 4.3 \times 10^{13} \, s^{-1}$

(b) $\quad k_2 = A \, e^{-\left[\frac{E_a}{R}\left(\frac{1}{T_2}\right)\right]} = 4.3 \times 10^{13} \, s^{-1} e^{-\left[\frac{161,000 J \, mol^{-1}}{8.3145 J \, mol^{-1} \, K^{-1}}\left(\frac{1}{1000K}\right)\right]}$

$\quad = 1.7 \times 10^5 \, s^{-1}$

14-53. The height of the transition state complex on an energy plot is 3.5 kJ mol^{-1} above the energy of "OH *(g)* + HCl *(g)*" and *these* substances are 66.8 kJ higher in energy than "H$_2$O *(g)* + Cl *(g)*". In that case, the transition state complex must be must be 66.8 + 3.5 = **70.3 kJ mol^{-1}** for the reaction   H$_2$O *(g)* + Cl *(g)* → OH *(g)* + HCl *(g)*".

14-55. The substance CF$_2$Cl$_2$ is consumed in the second step and regenerated in the fourth. It is therefore an essential part of the reaction but is neither consumed nor produced over the whole mechanism. This means it plays a **catalytic** role.

14-57. Catalysis involves, in its best manifestations, the use of very small amounts of materials to cause a reaction to proceed. This is instead of, and superior to, having to introduce stiochiometric amounts of substances to effect a chemical reaction, perhaps through the creation of a totally separate intermediate chemical substance. Note that as catalysis avoids this extra stiochiometric reagent, and perhaps also the extra cost of isolating intermediates, there may also be a considerable savings in cost. This is why it is possible to also justify "green" chemistry as good business practice.

14-59. We assume that the train slows at a constant rate. At the beginning of any given time interval, the average speed in miles per hour will be: 60 - x, where x is the number of minutes from the start of the hill. After y minutes later, the speed will be 60 - x - y.

14-59. (continued)

The average speed in an interval will be very close to:
$$\text{average speed} = \frac{(60-x)+(60-x-y)}{2} = 60 - x - \frac{y}{2}$$

We can then determine the average speed for any interval starting x minutes from the start of the hill and ending y minutes later.

Start time (x)	Interval length (y)	Average speed (mph)
0	1	$60 - 0 - \dfrac{1}{2} = \mathbf{59.5}$
0	3	$60 - 0 - \dfrac{3}{2} = \mathbf{58.5}$
3	3	$60 - 3 - \dfrac{3}{2} = \mathbf{55.5}$
0	20	$60 - 0 - \dfrac{20}{2} = \mathbf{50.0}$

14-61. $\text{rate} = k\,[\text{Hb}][\text{O}_2] = (4 \times 10^7 \text{ L mol}^{-1} \text{ s}^{-1})(2 \times 10^{-9} \text{ mol L}^{-1})(5 \times 10^{-5} \text{ mol L}^{-1})$
$$= \mathbf{4 \times 10^{-6} \text{ mol L}^{-1} \text{ s}^{-1}}$$

14-63. (a) The data for the plot are as follows:

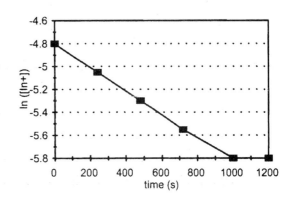

TIME	ln [In$^+$]
0	-4.80
240	-5.05
480	-5.29
720	-5.55
1000	-5.80
1200	-5.80
10000	-5.80

The last point is omitted from the graph.

The data has two areas - a steady drop in ln[In$^+$] until about 1000 seconds, then a constant value because the reaction has reached equilibrium. The nature of dynamic equilibrium means that as soon as some In *(s)* and In^{3+} *(aq)* have formed, they will begin to react give In$^+$ back. We must look at the very beginning of the data for the "pure" kinetics of the reaction of In$^+$ alone.

14-63. (continued)

The In^+ reaction is first order in $[In^+]$, so it will fit the equation with the point when $t = 240$ s:

$$\ln[In^+] = \ln[In^+]_0 - kt$$

Rearranging:

$$k = -\frac{1}{t}\ln\left(\frac{[In^+]_t}{[In^+]_0}\right)$$

$$k = -\frac{1}{240\,s}\ln\left(\frac{0.00641\,M}{0.00823\,M}\right) = \mathbf{1.04 \times 10^{-3}\,s^{-1}}$$

(b) The half life for a first-order process is $t_{1/2} = \dfrac{0.6931}{k} = \dfrac{0.6931}{0.00104\,s^{-1}} = \mathbf{666\,s}$

(c) At equilibrium, we have 0.00303 M In^+ and have generated a concentration of In^{3+} equal to one-third the concentration of consumed In^+. Therefore,

$$[In^{3+}] = {}^1\!/_3(0.00823\,M - 0.00303\,M) = 0.00173\,M$$

These concentrations are used to get the equilibrium constant:

$$K = \frac{\left|In^{3+}\right|}{\left[In^+\right]^3} = \frac{0.00173}{(0.00303)^3} = \mathbf{62{,}200}$$

14-65. Raising the pH will decrease the value for $[H^+]$. Therefore, the term $\dfrac{1}{[H^+]}$ will **increase**, and the rate will increase also. A pH change from 3.00 to 4.00 will be a change of 0.10 the initial concentration, so the effect will be $1 / 0.10 = 10$. The rate will increase tenfold.

14-67. The information in the problem gives us $k_f = 3.7 \times 10^9$ L mol^{-1} s^{-1} for the reaction of hydroxide and cyanide ions. The reaction in this case is best written:

$$OH^-\ (aq) + HCN\ (aq) \underset{k_f}{\overset{k_r}{\rightleftharpoons}} CN^-\ (aq) + H_2O\ (l)$$

This reaction is the opposite of the reaction for hydrolysis of cyanide anion -- the "K_b" reaction.

$$K_b\,(CN^-) = \frac{K_w}{K_a(HCN)} = \frac{1.0 \times 10^{-14}}{4.93 \times 10^{-10}} = 2.03 \times 10^{-5}$$

Given the general relation $K = {}^{rate\ forward}\!/_{rate\ reverse}$ we can solve for k_f:

$$k_f = K_b\,k_r = (2.03 \times 10^{-5})(3.7 \times 10^9\ L\ mol^{-1}\ s^{-1})$$

$$= \mathbf{7.5 \times 10^4\,s^{-1}}$$

14-69. In this case, we can use the values for the initial $[I_2]$ and the time for the color to fade to calculate the $\dfrac{\Delta [I_2]}{\Delta t}$ for each set of initial conditions. Because I_2 is a reactant, we know that rate $= -\dfrac{\Delta [I_2]}{\Delta t}$. For example, the first experiment has a $[I_2] = 2.0 \times 10^{-4}$ and $\Delta [I_2] = -2.0 \times 10^{-4}$ M. We also have $t = 100$ s. Therefore, the rate of the reaction in this first experiment is $\dfrac{2.0 \times 10^{-4} \text{ mol L}^{-1}}{100 \text{ s}} = 2.0 \times 10^{-6} \text{ mol L}^{-1} \text{ s}^{-1}$.

So we can tabulate:

Experiment	$[I_2]$	$[CH_3COCH_3]$	rate
1	2.0×10^{-4}	0.80	$2.0 \times 10^{-6} \text{ mol L}^{-1} \text{ s}^{-1}$
2	1.6×10^{-4}	0.80	$1.6 \times 10^{-6} \text{ mol L}^{-1} \text{ s}^{-1}$
3	1.2×10^{-4}	0.40	$0.60 \times 10^{-7} \text{ mol L}^{-1} \text{ s}^{-1}$

We can compare the first two experiments and see that the reaction is first order in the $[I_2]$. If the reaction was independent of the $[CH_3COCH_3]$, then the rate for the last reaction would be $1.2 \times 10^{-6} \text{ mol L}^{-1} \text{ s}^{-1}$. But is is half that amount, and we realize that the reaction has slowed by half as the $[CH_3COCH_3]$ has been cut in half. So the reaction is first order in $[CH_3COCH_3]$ also. We now have our rate law:

$$\text{rate} = k\, [I_2]\, [CH_3COCH_3]$$

We solve for k using any of the experiments to get $k = 0.0125 \text{ L mol}^{-1} \text{ s}^{-1}$.

14-71.

Initiation step:	$CH_3CHO \rightarrow CH_3 + CHO$
Propagation steps:	$CH_3 + \mathbf{CH_3CHO} \rightarrow CH_4 + CH_2CHO$
(also form most of product)	$CH_2CHO \rightarrow CO + CH_3$
Termination steps:	$CH_3 + CH_3 \rightarrow CH_3CH_3$
(may or may not form product)	$CHO + CH_3 \rightarrow CH_4 + CO$

Most of the reactants undergo transformation to products in the propagation steps, as indicated in bold. The propagation steps regenerate the methyl radical intermediate, as they should in a true chain reaction.

The termination steps include one that merely terminates the chain, in this case making a small amount of ethane by product. The other actually forms product, but this is, it should be emphasized, *not* an important product forming step. It is most important as a chain-breaking step.

14-73. (a) The cooking of food is kinetically complex, but we will treat as if it is a process with a mechanism that does not change. The same relationship will hold for the rate constants. Then:

$$\ln\left(\frac{k_{373}}{k_{385}}\right) = -\frac{E_a}{R}\left(\frac{1}{373} - \frac{1}{385}\right)$$

$$\ln\left(\frac{1}{2}\right) = -\frac{E_a}{8.3145 \text{ J mol}^{-1}\text{ K}^{-1}}\left(2.6810\times10^{-3}\text{K}^{-1} - 2.5974\times10^{-3}\text{K}^{-1}\right)$$

$$-0.6931 = -E_a \times (1.005 \times 10^{-5} \text{ mol J}^{-1})$$

$$E_a = \frac{0.6931}{1.005\times10^{-5} \text{ mol J}^{-1}} = \textbf{69,000 J mol}^{-1}$$

(b) This activation energy is all that is needed for the calculation of the relative rate at 94.4°C (367.55 K):

$$\ln\left(\frac{k_{373}}{k_{367.55}}\right) = -\frac{69000 \text{ J mol}^{-1}}{8.3145\text{J mol}^{-1}\text{ K}^{-1}}\left(\frac{1}{373\text{ K}} - \frac{1}{367.55\text{ K}}\right) = -0.3300$$

$$\left(\frac{k_{373}}{k_{367.55}}\right) = e^{+0.3399} = 1.$$

Since the average rate constant at 373 K is 1.391 times the rate at 267.55 K, the time required for cooking at 267.55 K is 1.391 times the time at 373 K:

time at 267.55 K = 10 min × 1.391 = **14 min**

14-75. We are witnessing the combustion of the hydrogen by the oxygen in the air, catalyzed by the planiuim powder, which glows white-hot as a reault of the heat of combustion.

14-77. The thermodynamic data from Appendix D (remember that I2 is in the *gas* phase, so the enthalpy of formation for I2 *(g)* is not zero) give: ΔH^o = -9.48 kJ and ΔS^o = 21.81 J K^{-1}. Then at 1000 K:

$$\Delta G^o_{1000} = -9,480 \text{ J} - 1000 \text{ K} (21.81 \text{ J K}^{-1})$$
$$= -31,290 \text{ J}$$

If we make the approximation that the change in enthalpy ΔH^o is equal to the change in energy ΔE^o (reasonable for a gas phase reaction with equal moles of gaseous products and reactants) then ΔE^o = - 9.48 kJ.

This means that the energy of the products lies 9,480 J *below* the energy of the reactants. We know that the transition state energy is 165,000 J *above* the energy of the reactants. Therefore, the transition state energy is

165,000 + 9,480 = 177,480 J = 177 kJ above the energy of the products, for an activation energy for the reverse reaction of **177 kJ mol^{-1}**

14-77. (continued)

We cannot use this E_a to determine k_r directly because we do not know the pre-exponential factor A. But the free energy change allows us to calculate the equilibrium constant for the reaction:

$$K = e^{-\Delta G^\circ / RT} = e^{-(-31,290\,\text{J mol}^{-1})/(8.3145\,\text{J mol}^{-1}\,\text{K}^{-1})(1000\,\text{K})}$$
$$= 43.1$$

We can then use the relationship of forward and reverse rate constants and the equilibrium constant to get k_r.

$$K = 43.1 = \frac{k_f}{k_r}$$

Rearranging: $\quad k_r = \dfrac{240\,\text{L mol}^{-1}\,\text{s}^{-1}}{43.1} = \textbf{5.56 L mol}^{-1}\,\textbf{K}^{-1}$

14-79. 1. If this occurred in one elementary step, this reaction would involve the simultaneous reaction of 27 species, an astoundingly unlikely event.

2. The presence of "2" octane molecules is an artifact of our convention that the reaction stoichiometry have integer values for all coefficients. Of course, if we considered what happens to one octane molecule alone, we would need 25/2 O_2 molecules to be in the elementary step, and we can't have half an O_2!

3. Octane, and other hydrocarbons, have structures that have surface hydrogens around the carbons; the carbon atoms are generally not accessible for direct reaction. It is unlikely the reaction would proceed with the oxygen molecules being able to directly pull CO_2 out from inside all the hydrogens.

14-81. (a) $5\,H^+ + 3\,NH_2CH_2COOH + 2\,MnO_4^- \rightarrow 2\,MnO_2 + 3\,CH_2O + 3\,CO_2 + 3\,NH_4^+ + H_2O$

(b) Such odd behavior indicates that something -- a catalyst since it is not among the continuously depleted reactants -- is forming in the course of the reaction that causes it to speed up, but then for some reason this accelerating species loses its effectiveness. There are three possible explanations. First, the catalyst may be an intermediate, whose rapid formation causes it to reach higher and higher concentrations until the rate of the initial reaction slows down as the reactants are depleted. Second, the catalyst may be an intermediate or a product, but the step it catalyzes is dependent on relatively high concentrations of the reactants. As the reactants are depleted there is a change in the mechanism and the rate is no longer dependent on the catalyst or intermediate. Third, the reaction is consuming hydrogen ions and producing ammonium. There may be a buffering effect that occurs in the initial system that is depleted as the reaction proceeds.

14-83. Catalysis implies that the reaction runs faster than without catalysis, but also that there is a different mechanism. In this case, we see that the catalyzed pathway has a higher activation energy E_a, but that is OK. The preexponential factor A must be much larger in the catalyzed reaction.

14-85. (a) We have to make a few preliminary notes. First, the rapid reaction of hydrogen sulfite with mercury (II) ion means to make the $HgSO_3$ concentration is depending on which of the reagents is limiting. We note that the initial concentration has ten times more Hg^{2+} than hydrogen sulfite, so the initial concentration of $HgSO_3$ is 4.00×10^{-5} M. Since this will *not* be replenished in any way during the reaction, we can answer (a) as a simple first order graph of the $[HgSO_3]$ starting from this concentration and the first order rate constant we are given for its elementary reaction:

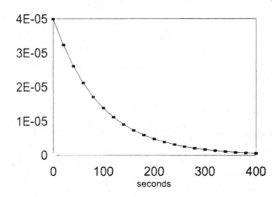

(b) We have the Hg (*aq*) being produced by the first reaction. The concentration of the Hg^{2+} ion will, we assume, be the same as it was initially, since it was in a much larger concentration than the hydrogen sulfite. We have $[Hg^{2+}] = 3.6 \times 10^{-4}$ M at the "start". At 4 seconds into the reaction, the concentration of Hg (*aq*) will be 7.6×10^{-6} M. These will react at a rate of:

rate $= k_2[Hg][Hg^{2+}] = (5.9 \times 10^8)(7.6 \times 10^{-6}$ M$)(3.6 \times 10^{-4}$ M$) = 1.6$ mol L^{-1} s^{-1}

We notice immediately that the rate is many orders of magnitude higher than the concentrations of the two reactant species. In other words, at this rate constant, *all of the Hg (aq) is converted to Hg$_2^{2+}$ as soon as it is formed.*

The concentration curve of the $[Hg_2^{2+}]$ will then be simple to calculate from the previous curve:

14-85. (continued)

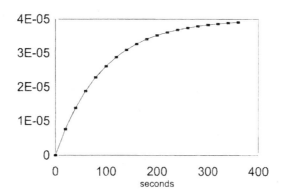

14-87. (a) rate = $(6.4 \times 10^6 \text{ L}^2 \text{ mol}^{-2} \text{ s}^{-1})(6.0 \times 10^{-4} \text{ mol L}^{-1})^2(6.0 \times 10^{-4} \text{ mol L}^{-1})$

= 0.00138 mol L^{-1} s^{-1}

(b) We use the relationship between the change in concentration of each reactant and the rate to determine the change in concentration. We neglect the change in rate that may occur over this time period.

$$\text{rate} = -\frac{\Delta[O_2]}{\Delta t}$$

$\Delta[O_2] = -\text{rate} \times \Delta t = (0.00138 \text{ mol L}^{-1} \text{ s}^{-1})(0.01 \text{ s}) = -1.38 \times 10^{-5} \text{ M}$

$[O_2] = 6.0 \times 10^{-4} \text{ M} - 1.4 \times 10^{-5} \text{ M} = 5.86 \times 10^{-4} \text{ M} = \mathbf{5.9 \times 10^{-4} \text{ M}}$

$$\text{rate} = -\frac{1}{4}\frac{\Delta[NO]}{\Delta t}$$

$\Delta[NO] = -4 \text{ rate} \times \Delta t = 4(0.00138 \text{ mol L}^{-1} \text{ s}^{-1})(0.01 \text{ s}) = -5.52 \times 10^{-5} \text{ M}$

$[O_2] = 6.0 \times 10^{-4} \text{ M} - 5.52 \times 10^{-5} \text{ M} = 5.45 \times 10^{-4} \text{ M} = \mathbf{5.4 \times 10^{-4} \text{ M}})$

We take the values from (b) and calculate, as in (a), to a new rate of **1.11 × 10^{-3} mol L^{-1} s^{-1}.**

(d) The new value will be **0.92 × 10^{-3} mol L^{-1} s^{-1}.**

(e) We can iterate the process we have done for the first two intervals. We find that the concentration of NO reaches 3.0×10^{-4} after **0.11 seconds.**

14-89. The question does not ask for the half life. It only asks for the temperature when the half-life has a value one half of the value at 1300 K. If we think about the relationship of half life and the first order rate constant, $t_{1/2} = \dfrac{0.6931}{k}$ we realize that this is the same as asking "when will k be twice the value at 1300 K". Let's say that 1300 K is T_1 and the other temperature is T_2. We want $k_2 = 2 \times k_1$.

Chapter 14

14-89. (continued)

We can take the ratio of the two rate constants and work to the temperature T_2 as follows:

$$\frac{k_2}{k_1} = \frac{Ae^{(-E_a/RT_2)}}{Ae^{(-E_a/RT_1)}}$$

$$\frac{2k_1}{k_1} = \frac{e^{(-E_a/RT_2)}}{e^{(-E_a/RT_1)}}$$

$$2 = e^{(E_a/R)(1/T_1 - 1/T_2)}$$

$$\ln 2 = \left(\frac{E_a}{R}\right)\left(\frac{1}{T_1} - \frac{1}{T_2}\right)$$

$$\left(\frac{R}{E_a}\right)\ln 2 = \frac{1}{T_1} - \frac{1}{T_2}$$

$$\frac{1}{T_2} = \frac{1}{T_1} - 0.6931\left(\frac{R}{E_a}\right) = \frac{1}{1300\text{K}} - 0.6931\frac{8.3145 \text{ J mol}^{-1}\text{ K}^{-1}}{263,000 \text{ J}}$$

$$\frac{1}{T_2} = 0.000769231 \text{ K}^{-1} - 0.00002191 \text{ K}^{-1} = 0.000747318 \text{ K}^{-1}$$

$$T_2 = 1338 \text{ K}$$

CHAPTER 15 - NUCLEAR CHEMISTRY

15-1. (a) The unknown particle must have mass number $2 \times 12 - 1 = 23$ and atomic number

$2 \times 6 - 0 = 12$: $2\,{}^{12}_{6}C \rightarrow {}^{23}_{12}Mg + {}^{1}_{0}n$

(b) The unknown particle must have mass number $12 + 4 - 1 = 15$ and atomic number

$6 + 2 - 1 = 7$: ${}^{15}_{7}N + {}^{1}_{1}H \rightarrow {}^{12}_{6}C + {}^{4}_{2}He$

(c) The unknown particle must have mass number $2 \times 3 - 2 \times 1 = 4$ and atomic number

$2 \times 2 - 2 \times 1 = 2$ $2\,{}^{3}_{2}He \rightarrow {}^{4}_{2}He + {}^{1}_{1}H$

15-3. We must add the mass of the separated components and compare it to the mass of the whole atom (in u). The difference in the masses is the mass defect.

(a) ${}^{40}_{20}Ca$

20 electrons:	$20 \times 0.00054858 =$	0.01097160
20 protons:	$20 \times 1.00727647 =$	20.14552934
20 neutrons:	$20 \times 1.00866490 =$	20.17329832
	Sum of Parts:	40.32979926
	Observed Mass:	39.9625912
	Difference:	0.36720806

The difference represents the mass that "disappeared" (in fact, it was released as energy) when the atom was formed. It is equal to the binding energy of the nucleus. The relationship between mass and energy is given by Einstein's relation, from which we get a conversion factor: $\dfrac{931.494 \text{ MeV}}{1 \text{ u}}$.

The relationship between the energy units MeV and kJ is $\dfrac{1.602177 \times 10^{-13} \text{ J}}{1 \text{ MeV}}$.

For the ${}^{40}_{20}Ca$ nucleus, the binding energy for the whole nucleus is:

BE (nucleus) $= 0.36721000 \text{ u} \times 931.494 \text{ MeV u}^{-1} = 342.0539 \text{ MeV} = \textbf{342.1 MeV}$

BE (mole) $= 342.0539 \text{ MeV} \times \dfrac{1.602177 \times 10^{-13} \text{ J}}{1 \text{ MeV}} \times 6.0221 \times 10^{23} \text{ atoms mol}^{-1}$

$= \textbf{3.300} \times \textbf{10}^{\textbf{10}} \textbf{ kJ mol}^{\textbf{-1}}$

There are 40 nucleons in this nucleus, so

BE (nucleon) $= {}^{342.0539 \text{ MeV}}/{40 \text{ nucleons}} = \textbf{8.551 MeV nucleon}^{\textbf{-1}}$.

15-3. (continued)

(b) $^{87}_{37}\text{Rb}$

37 electrons:	37 × 0.00054858 =	0.02029746
37 protons:	37 × 1.00727647 =	37.26922939
50 neutrons:	50 × 1.00866490 =	50.43324500
	Sum of Parts:	87.72277185
	Observed Mass:	86.90918400
	Mass Defect:	0.81358785

BE(nucleus) **757.9 MeV**

BE(nucleon) **8.711 MeV nucleon^{-1}**

BE(mol) **7.312 × 10^{10} kJ mol^{-1}**

(c) $^{238}_{92}\text{U}$ Sum of Parts: 239.98498000 u Observed Mass: 238.05078600 u

Mass Defect: 1.9341940 u BE(nucleon): **7.570 MeV nucleon^{-1}**

BE(nucleus): **1801.7 MeV** BE(mol): **1.738 × 10^{11} kJ mol^{-1}**

15-5. The difference in the mass is the mass of a $^{4}_{2}\text{He}$ atom plus the mass change reflected in the change in energy. A $\Delta E = -9.23$ MeV is equal to a mass change of -0.0099088132 u, so the total mass difference is: $\Delta m = -0.0099088132$ u -4.00260324 u $=$ **4.01251 u.**

15-7. The mass of the $^{8}_{4}\text{Be}$ atom is 8.0053052 u, while that of two $^{4}_{2}\text{He}$ atoms is 8.0052065 u. The two helium atoms have less mass than a single $^{8}_{4}\text{Be}$ atom, by a difference of **0.0000987 u.** Therefore, the **two helium atoms are more stable.**

15-9. (a) $\quad ^{37}_{17}\text{Cl} \rightarrow ^{37}_{18}\text{Ar}^{+} + ^{0}_{-1}e^{-} + \tilde{\nu}$

(b) $\quad ^{22}_{11}\text{Na} \rightarrow ^{22}_{10}\text{Ne}^{-} + ^{0}_{1}e^{+} + \nu$

(c) $\quad ^{224}_{88}\text{Ra} \rightarrow ^{220}_{86}\text{Rn} + ^{4}_{2}\text{He}$

(d) $\quad ^{82}_{38}\text{Sr}^{+} + ^{0}_{-1}e^{-} \rightarrow ^{82}_{37}\text{Rb} + \nu$

Note the following rules for balancing nuclear chemical reactions:

i) The number of nucleons (the top number in the notation for a nuclide) must balance through the reactions.

ii) The number of nuclear charges + charges on positrons or electrons (the bottom number of the notation for a particle) must balance.

iii) The creation of an electron is accompanied by the formation of an *anti*-neutrino.

iv) The creation of a positron is accompanied by the formation of a neutrino.

v) The overall charge will balance (this is important in β decay and electron capture processes)

15-11. (a) Because ^{15}C has too many neutrons, it is most likely to produce a beta particle and become ^{15}N.

(b) Because ^{237}Np is such a heavy atom, it is most likely to produce an alpha particle and become ^{233}Pa.

(c) Because ^{11}C is deficient in neutrons, it is most likely to produce a positron, to become ^{11}B.

15-13. Neutron activation: $\qquad\qquad\qquad$ $^{30}_{14}Si + {}^{1}_{0}n \rightarrow {}^{31}_{14}Si$

Decay to ^{31}P: $\qquad\qquad\qquad\qquad$ $^{31}_{14}Si \rightarrow {}^{31}_{15}P^{+} + {}^{0}_{-1}e^{-} + \tilde{\nu}$

15-15. (a) The kinetic energy of the products will come from the release of nuclear energy, which is indicated by the difference in mass of the products and reactants. As with other thermodynamic calculations, we take the

Σ (products) - Σ (reactants), where the quantities summed are the masses.

IMPORTANT NOTE: A beta particle formed in this reaction, but its mass is included in the mass of the product $^{40}_{20}Ca$ atom, so we do not list the mass of the electron separately:

PRODUCT MASSES		
(all masses in u)	$^{40}_{20}Ca$ atom:	+39.962591
	anti-neutrino:	+0.000000
-REACTANT MASSES	$^{40}_{19}K$ atom:	-39.9639987
	Δ(mass)	-0.0014075 u

This decrease in mass results in a release of energy to the surroundings equal to

$- (-0.0014075 \text{ u}) \times \left(\dfrac{931.494 \text{ MeV}}{1 \text{ u}} \right) = \textbf{1.311 MeV}$. This energy appears initially as the

kinetic energy of the emitted electron, and this energy is therefore the maximum kinetic energy the electron can have. Lower values for the kinetic energy of the emitted electrons mean some of the energy has been transferred to other places including the energy of the neutrino and the "recoil" energy of the product atom motion.

(b) The average kinetic energy is an experimental number, reflecting the actual beta particles detected. These do not have the maximum kinetic energy because some of the energy in the release of nuclear energy is diverted in varying amounts to the neutrinos produced and to the "recoil" energy of the product nucleus.

15-17. <u>PRODUCT MASSES</u> $^{19}_{9}F$ atom: +18.9984032

         (all masses in u)   electron on fluoride: + 0.0005487990

                                      positron: + 0.0005487990

                                      neutrino: +0.0

   -<u>REACTANT MASSES</u> $^{19}_{10}Ne$ atom: $-$ 19.0018798

                                    Δ(mass)   -0.002379002 u

This corresponds to an energy release of **2.2164 MeV**, which is the maximum kinetic energy for the postitron produced.

15-19. <u>PRODUCT MASSES</u> $^{231}_{91}Pa$ atom: +231.035880

                                    neutrino: +0.0

   -<u>REACTANT MASSES</u> $^{231}_{92}U$ atom: $-$ 231.0363

                                    Δ(mass)   -0.00042 u

This corresponds to an energy release of **-0.38 MeV**.

15-21. A film badge detects the radiation exposure of the user over time. It is more immediately sensititve to radiation than the body, and therefore a darkening of the developed film warns of overexposure to radiation long before any health problems are evident.

15-23. In these and subsequent problems we will need to conveniently convert units of time. For long time spans, we take the length of the year as 365.25 days, to allow for leap years. Then we have:

TIME CONVERSIONS

1 <u>year</u> = 365.25 days = 8766 hrs   = 5.2596×10^5 min = 3.15576×10^7 s

1 <u>day</u> = 24 hrs   = 1440 min = 86,400 s

1 <u>hour</u> = 60 minutes = 3600 s

1 <u>min</u> = 60 s

The general relation for activity is:

$$A = kN$$

where N is the number of particles and k is the first-order rate constant. It is more convenient to express k in the form of the half-life $t_{1/2}$: $k = \dfrac{0.6931}{t_{1/2}}$.

15-23. (continued)

Here, $$k = \frac{0.6931}{t_{1/2}} = \left(\frac{0.6931}{103 \text{ yr}}\right)\left(\frac{1 \text{ yr}}{5.260 \times 10^5 \text{ min}}\right) = 1.279 \times 10^{-8} \text{ min}^{-1}$$

And: $$N = \left(0.0010 \text{ g } ^{210}\text{Po}\right)\left(\frac{1 \text{ mol } ^{210}\text{Po}}{210 \text{ g } ^{210}\text{Po}}\right)\left(\frac{6.0221 \times 10^{23} \text{ atoms} ^{210}\text{Po}}{1 \text{ mol } ^{210}\text{Po}}\right)$$

$$= 2.868 \times 10^{18} \text{ atoms } ^{210}\text{Po}$$

Then: $$A = (1.279 \times 10^{-8} \text{ min}^{-1})(2.868 \times 10^{18} \text{ atoms})$$

$$= 3.7 \times 10^{10} \text{ atoms min}^{-1} = \mathbf{3.7 \times 10^{10} \text{ min}^{-1}}.$$

15-25. (a) In this problem, we want to solve for N. Note that 1 Bq (one becquerel) is equal to 1 disintegration per second):

$$N = A \frac{t_{1/2}}{0.6931} = \left(2.5 \times 10^4 \text{ s}^{-1}\right)\left(\frac{29 \text{ s}}{0.6931}\right) = \mathbf{1.0 \times 10^6 \text{ atoms } ^{19}\text{O}.}$$

(b) There will be a significant decay in two minutes. We must use the first order kinetic equation:

$$N = N_0\, e^{-(0.6931/t_{1/2})t} = 1.0 \times 10^6\, e^{-(0.6931/29 \text{ s})120 \text{ s}}$$

$$= 1.0 \times 10^6\, e^{-2.868} = \mathbf{5.9 \times 10^4 \text{ atoms } ^{19}\text{O}.}$$

15-27. The number of atoms of ^{137}Cs is:

$$N = \left(0.00100 \text{ g } ^{137}\text{Cs}\right)\left(\frac{1 \text{ mol } ^{137}\text{Cs}}{137 \text{ g } ^{137}\text{Cs}}\right)\left(\frac{6.0221 \times 10^{23} \text{ atoms} ^{137}\text{Cs}}{1 \text{ mol } ^{137}\text{Cs}}\right)$$

$$= 4.396 \times 10^{18} \text{ atoms } ^{137}\text{Cs}$$

The initial activity is: $(3.19 \times 10^9 \text{ s}^{-1})(3.156 \times 10^7 \text{ s yr}^{-1}) = 1.01 \times 10^{17} \text{ yr}^{-1}$

$$t_{1/2} = \frac{0.6931N}{A} = \frac{(0.6931)\left(4.396 \times 10^{18}\right)}{1.01 \times 10^{17} \text{ yr}^{-1}} = \mathbf{30.3 \text{ yr}}$$

15-29. We must apply the general formula for radioactive decay, for the decline of 1.00 mol of neutrons to 0.90 mol neutrons:

$$0.90 \text{ mol} = 1.00 \text{ mol } e^{-(0.6931/t_{1/2})t}$$

$$0.90 = e^{-(0.6931/t_{1/2})t}$$

$$\ln(0.90) = -(0.6931 / t_{1/2})\, t$$

$$-0.1054 = -(0.6931 / t_{1/2})\, t$$

$$t_{1/2} \times (^{0.1054}/_{0.6931}) = t = 0.152\, t_{1/2}$$

Based on old $t_{1/2}$: $t = (0.152)(1100 \text{ s}) = \mathbf{167 \text{ s}}$

Based on new $t_{1/2}$: $t = (0.152)(876 \text{ s}) = \mathbf{133 \text{ s}}$

Chapter 15

15-31. The number of atoms of ^{219}At is:

$$N = \left(0.044 \text{ g } ^{219}\text{At}\right)\left(\frac{1 \text{ mol } ^{219}\text{At}}{219.01 \text{ g } ^{219}\text{At}}\right)\left(\frac{6.0221 \times 10^{23} \text{atoms}^{219}\text{At}}{1 \text{ mol } ^{219}\text{At}}\right)$$

$$= 1.209 \times 10^{20} \text{ atoms } ^{219}\text{At}$$

$$A = \left(\frac{0.6931}{54 \text{ s}}\right)\left(1.209 \times 10^{20} \text{ atoms}\right) = 1.553 \times 10^{18} \text{ s}^{-1} = \mathbf{1.6 \times 10^{18} \text{ Bq}}$$

15-33. (a) If there are 0.333 mol of ^{206}Pb for every 1.000 mol ^{238}U, then the number of mol of ^{238}U at the start was 1.333 mol ^{238}U. Thus, the fraction remaining from the original amount of ^{238}U is (1.333 - 1.000)/1.333 = 0.750. The fraction that has decayed is 1.000 - 0.750 = **0.250 (25.0%)**.

(b) Rearranging the equation for a first-order decay we obtain (see the textbook immediately after Example 15-7):

$$\frac{N}{N_{initial}} = e^{-kt}$$

$$-\ln\left(\frac{N}{N_{initial}}\right) = kt = \left(\frac{0.6931}{t_{1/2}}\right)t$$

$$t = -\ln\left(\frac{N}{N_{initial}}\right)\left(\frac{t_{1/2}}{0.6931}\right)$$

Here, $\dfrac{N}{N_{initial}} = 0.750$, so: $t = -\ln(0.750)\left(\dfrac{4.468 \times 10^9 \text{ yr}}{0.6931}\right) = \mathbf{1.85 \times 10^9 \text{ yr}}$

15-35. As developed in Example 15-6:

$$\ln\left(\frac{A}{A_o}\right) = -\left(\frac{0.6931}{t_{1/2}}\right)t = -\left(\frac{0.6931}{5730 \text{ yr}}\right)t$$

$$\ln\left(\frac{0.153 \text{ Bq g}^{-1}}{0.255 \text{ Bq g}^{-1}}\right) = -(1.210 \times 10^{-4} \text{ yr}^{-1})\,t$$

$$0.5108 = (1.210 \times 10^{-4} \text{ yr}^{-1})\,t$$

$$t = \mathbf{4220 \text{ yr}}$$

15-37.
$$\ln\left(\frac{0.0005 \text{ Bq g}^{-1}}{0.255 \text{ Bq g}^{-1}}\right) = -(1.210 \times 10^{-4} \text{ yr}^{-1})\,t$$

$$6.234 = (1.210 \times 10^{-4} \text{ yr}^{-1})\,t$$

$$t = \mathbf{5.1 \times 10^4 \text{ yr}}$$

15-39. The upper atmosphere is absorbing cosmic rays and producing ^{14}C at a constant rate. This ^{14}C quickly equilibrates with the rest of the atmospheric carbon (primarily CO_2) and makes its way at a constant rate into the photosynthetic chain. Since all living organisms on the earth are taking in carbon from this pool, either directly or indirectly, this pool is being kept a constant level in all carbon-consumnig organisms. Once the organism dies, of course, carbon uptake stops and the ^{14}C pool begins its inexorable decay.

15-41.
$$^{11}_{6}C \rightarrow ^{11}_{5}B + ^{0}_{+1}e^- + \nu$$
$$^{15}_{8}O \rightarrow ^{15}_{7}N + ^{0}_{+1}e^+ + \nu$$

15-43. The dosage called the *rem* is approximately equal (for positron emission) to the rad, itself defined as that dosage that deposits 0.01 J of nuclear energy in 1 kg of tissue. The energy release is determined by the energy released by the radioactive nuclide over the time of the exposure:

Energy release = (# of disintegrations) × (ΔE per disintegration)
= activity × ΔE

In this problem we are concerned about the *ratio* of the dosage, assuming that all of the material decays in the body.

Here:
$$\frac{\text{dosage from } ^{15}O}{\text{dosage from } ^{11}C} = \frac{N(^{15}O) \times \Delta E(^{15}O)}{N(^{11}C) \times \Delta E(^{11}C)}$$

Note that the chemical amounts (N) are equal.

Then:
$$\frac{\text{dosage from } ^{15}O}{\text{dosage from } ^{11}C} = \frac{\Delta E(^{15}O)}{\Delta E(^{11}C)} = \frac{1.72 \text{ MeV}}{0.99 \text{ MeV}} = \textbf{1.74}$$

15-45. (a) The number of atoms of ^{131}I ingested is 2.3×10^{16} atoms ^{131}I
The half life in seconds is: $8.041 \text{ days} \times 86{,}400 \text{ s day}^{-1} = 6.947 \times 10^5$ s
The activity is then: $A = N(0.6931 / t_{1/2}) = 2.3 \times 10^{16}(0.6931 / 6.947 \times 10^5 \text{ s})$
$= 2.3 \times 10^{10} \text{ s}^{-1} = \textbf{2.3} \times \textbf{10}^{\textbf{10}} \textbf{ Bq}$

(b) The average energy deposited per disintegration is:
$$E = 0.40 \text{ MeV} \times \frac{1.6022 \times 10^{-13} \text{ J}}{1 \text{ MeV}} = 6.409 \times 10^{-14} \text{ J}.$$

The rate of dosage at the first moment of ingestion is:
$6.409 \times 10^{-14} \text{ J} \times 2.3 \times 10^{10} \text{ s}^{-1} = 0.001474 \text{ J s}^{-1}$
Per kilogram of body mass this equals: $0.001474 \text{ J s}^{-1} / 60 \text{ kg} = 2.5 \times 10^{-5} \text{ J kg}^{-1} \text{ s}^{-1}$
Given 1 rad $= 10^{-2}$ J kg^{-1} and 1 millirad = 0.01 rad, the dosage is
$2.5 \times 10^{-5} \text{ J kg}^{-1} = \textbf{2.5 mrad}$

Chapter 15

15-45. (continued)

(c) The rate of dosage is 2.45 mrad (0.00245 rad) per second. How long will it take to give a dose (500 rad) that exceeds the LD_{50}? If we assume a constant activity, then the time to deposit 500 rad of energy will be

$$500 \text{ rad}/0.00245 \text{ rad s}^{-1} = 2.0 \times 10^5 \text{ s} = 57 \text{ hours.}$$

This dose will **surely become lethal** in within two to three weeks. Some of the nuclide will decay -- one-half in 8.04 days -- but not until a massive dose has been delivered.

15-47. (a) First decay: $\quad {}^{90}_{38}\text{Sr} \rightarrow {}^{90}_{39}\text{Y}^+ + {}^{0}_{-1}e^- + \tilde{\nu}$

Second decay: $\quad {}^{90}_{39}\text{Y}^+ \rightarrow {}^{90}_{40}\text{Zr}^{2+} + {}^{0}_{-1}e^- + \tilde{\nu}$

Overall $\quad {}^{90}_{38}\text{Sr} \rightarrow {}^{90}_{40}\text{Zr}^{2+} + 2\,{}^{0}_{-1}e^- + \tilde{\nu}$

(b) Note that we can include the mass of the emitted electrons in the mass of the zirconium atom. The anti-neutrino is massless. Then the energy released is:

$$\Delta E = -(\text{mass }{}^{90}_{40}\text{Zr} - \text{mass }{}^{90}_{38}\text{Sr}) = (89.9043 \text{ u} - 89.9073) = 0.0030 \text{ u}$$

$$= 0.0030 \text{ u} \times 931.494 \text{ u MeV}^{-1}$$

$$= \textbf{2.8 MeV}$$

(c) A 1.00 g mass of ${}^{90}_{38}\text{Sr}$ corresponds to 6.698×10^{21} atoms. The half life in seconds is $28.1 \text{ yr} \times 3.156 \times 10^7 \text{ s yr}^{-1} = 8.8677 \times 10^8 \text{ s.}$

Then the activity is: $\quad A = {}^{90}_{39}\text{Y} + {}^{0}_{-1}\beta + \tilde{\nu} N$

$$= \frac{0.6931}{8.8677 \times 10^8 \text{ s}} 6.698 \times 10^{21}$$

$$= 5.24 \times 10^{12} \text{ s}^{-1} = \textbf{5.24} \times \textbf{10}^{\textbf{12}} \textbf{ Bq}$$

(d) $\quad A = A_0\, e^{-\left(\frac{0.6931}{t_{1/2}}\right)t}$

$$= (5.24 \times 10^{12} \text{ Bq})\, e^{-\left(\frac{0.6931}{28.1 \text{ yr}}\right)100 \text{ yr}} = \textbf{4.45} \times \textbf{10}^{\textbf{11}} \textbf{ Bq.}$$

15-49. The isotope with the longest half life, ${}^{238}\text{U}$, has the largest isotopic mass. So all U will become enriched in this isotope, and the atomic mass of natural U will **increase**.

15-51. We perform a products - reactants analysis again. Note that, although in a nuclear process it *is* important to indicate the presence of a neutron as a reactant, it is *not* important to the overall thermodynamics.

For this problem, we can write the reaction as:

$${}^{235}_{92}\text{U} \rightarrow {}^{94}_{36}\text{Kr} + {}^{139}_{56}\text{Ba} + 2\,{}^{1}_{0}\text{n}$$

15-51. (continued)

PRODUCT PARTICLES	$1 \; {}^{94}_{36}Kr$	+	93.919
	$1 \; {}^{139}_{56}Ba$	+	138.909
	$2 \; {}^{1}_{0}n$	+	2.0173
REACTANT PARTICLE	$-1 \; {}^{235}_{92}U$	-	235.0439
	$\Delta m(total)$	=	-0.1986 u

This lost mass appears as energy released. It corresponds to

$(0.1986 \; u)(931.494 MeV \; u^{-1})(1.602177 \times 10^{-13} \; J \; MeV^{-1}) = 2.964 \times 10^{-11} \; J$

This is the energy released *per disintegration*. The number of atoms of ${}^{235}_{92}U$ in

1 g of sample -- which all disintegrate under the conditions of this problem -- are:

$$\left(\frac{1 \; mol \; {}^{235}_{92}U}{235.05 \; g \; {}^{235}_{92}U} \right) \left(\frac{6.0221 \times 10^{23} \; atoms \; {}^{235}_{92}U}{1 \; mol \; {}^{235}_{92}U} \right) = 2.562 \times 10^{21} \; atoms \; g^{-1}$$

The energy released when this number of ${}^{235}_{92}U$ atoms react is:

$(2.562 \times 10^{21} \; atoms \; g^{-1})(2.964 \times 10^{-11} \; J \; atom^{-1})$

$= 7.59 \times 10^{10} \; J \; g^{-1} = \mathbf{7.59 \times 10^{7} \; kJ \; g^{-1}}$

15-53. Moderators are light elements in reactors that slow down neutrons passing through them. Neutrons emitted in fission are generally moving too quickly to cause other fission events, and this decreases the efficiency of the process. Moderators boost the number of neutrons that can react with other nuclei and induce further fission. Another important role of moderators is to keep the reaction from going too fast. This is important, because controlling the energy and the number of neutrons with moderators allows for the control of the overall rate of fission. This allows what is essentially a chain reaction that would increase in rate catastrophically to be kept to a safe and steady rate. It also allows a fission reactor to be "turned off" when needed.

15-55. Early in a star's life the fusion reaction of hydrogen dominates, for hydrogen is essentially the only element present. As the star ages helium begins to build up, and fusion reactions with helium become important. At higher concentrations of helium and higher temperatures, fusion of helium nuclei to even heavier elements begins. The nuclei of all of the elements in the universe except hydrogen were formed in this manner. This includes the carbon, nitrogen, oxygen, and other elements in the human body. As one researcher in the origin of life used to tell his classes, "You are recycled stardust!"

15-57. As noted in the text, iodide is preferentially taken up in the thyroid gland, so that the radiation given off by decay of iodine-131 is concentrated in one small organ. Most of the iodine ingested will eventually decay within that gland. Tritium, on the other hand, is diluted throughout the body and decays much more slowly. Most of the tritium ingested will probably be excreted long before it decays.

15-59.(a) From Appendix D:

$$\Delta H^o = 2 \times \Delta H_f^o (H_2O, \ (g)) - \Delta H_f^o (N_2H_4, \ (l))$$

$$= (2 \ mol \times -241.82 \ kJ \ mol^{-1}) - (1 \ mol \times 50.63 \ kJ \ mol^{-1}) = \qquad \textbf{-534.27 kJ}$$

(b) As we saw in Chapter 11:

$$\Delta E^o = \Delta H^o - RT\Delta n_g$$

Where Δn_g = change in No. of moles of gas...in this case, +2.

So: $\Delta E^o = -534,270 \ J - (8.3145 \ J \ mol^{-1} \ K^{-1})(298 \ K)(2 \ mol) = \textbf{-539,230 J}$

15-61. Decay by positron emission leaves behind a negatively charged atom, and the electron that is typically ejected, though it comes from the electrons of the atom, not from the nucleus, must be accounted for among the products.

$$^{64}_{29}Cu \rightarrow \ ^{64}_{28}Ni + \ ^{0}_{+1}e^+ + \ ^{0}_{-1}e^-$$

PRODUCTS

$$1 \ ^{0}_{+1}e^+ \ + \ 0.00054858$$

$$1 \ ^{0}_{-1}e^- \ + \ 0.00054858$$

$$1 \ ^{64}_{28}Ni \ + \ 63.92796$$

REACTANTS

$$1 \ ^{64}_{29}Cu \ - \ 63.92976$$

Net mass change: $\qquad -0.000703 \ u$

This corresponds to $(0.000703 \ u) \times (931.494 \ MeV \ u^{-1}) = \textbf{-0.655 MeV}$

15-63. For positron emission the reaction will be: $\qquad ^{231}_{92}U \rightarrow \ ^{231}_{91}Pa^- + \ ^{0}_{+1}e^+$

For electron capture the reaction will be: $\qquad ^{231}_{92}U^+ + \ ^{0}_{-1}e^- \rightarrow \ ^{231}_{91}Pa$

(b) As we saw above, the mass change in positron emission involves the mass of the emitted positron and an ejected "normal" electron:

$$\Delta m = mass(^{231}_{91}Pa) + mass(^{0}_{+1}e^+) + mass(^{0}_{-1}e^-) - mass(^{231}_{92}U)$$

$$= 231.035880 + 2 \ (0.00054858) - 231.0363 = + 0.00068 \ u$$

There is an *increase* in the mass of the system, indicating **a non-spontaneous process.**

15-63. (continued)

For the electron capture the mass balance is

$$\Delta m = \text{mass}(^{231}_{91}\text{Pa}) - \text{mass}(^{0}_{-1}e^{-}) - \text{mass}(^{231}_{92}\text{U})$$

$$= 231.035880 - 0.00054858 - 231.0363 = -\mathbf{0.00097\ u}$$

There is a net decrease in the mass of the system, indicating **spontaneity**.

The critical difference about positron emission is that the mass change includes the production of a positron from the nucleus and an electron from the atom's electron cloud. These work *against* spontaneity. Electron capture, on the other hand, effects the same change from U to Pa, but it also "uses" the mass in one of the uranium electrons.

15-65. The numbers in the problem are more useful if we note that

$$0.255\ \text{Bq g}^{-1} = 0.255\ \text{s}^{-1}\ \text{g}^{-1}$$

and

$$t_{1/2} = 5730\ \text{yr} \times 3.1558 \times 10^{7}\ \text{s yr}^{-1} = 1.808 \times 10^{11}\ \text{s}$$

The equation relating activity and number of ^{14}C nuclei is the basis for determining N:

$$A = kN$$

$$N = \frac{A}{k} = \frac{At_{1/2}}{0.6931} = \frac{\left(0.255\ \text{s}^{-1}\ \text{g}^{-1}\right)\left(1.808 \times 10^{11}\ \text{s}\right)}{0.6931}$$

$$= \mathbf{6.653 \times 10^{10}\ atoms\ ^{14}C\ g^{-1}}$$

(b) The total number of atoms of all types of carbon is, per gram:

$$N_{\text{total}} = \left(1\ \text{g C}\right)\left(\frac{1\ \text{mol C}}{12.011\ \text{g C}}\right)\left(\frac{6.0221 \times 10^{23}\ \text{atoms C}}{1\ \text{mol C}}\right)$$

$$= 5.014 \times 10^{22}\ \text{atoms C g}^{-1}$$

The ratio is: $\dfrac{N}{N_{\text{total}}} = \dfrac{6.653 \times 10^{10}\ \text{atoms}\ ^{14}\text{C g}^{-1}}{5.014 \times 10^{22}\ \text{atoms C g}^{-1}} = \mathbf{1.33 \times 10^{-12}}$

Put another way, there are $\dfrac{5.014 \times 10^{22}\ \text{atoms C g}^{-1}}{6.665 \times 10^{10}\ \text{atoms}\ ^{14}\text{C g}^{-1}} = 7.5 \times 10^{11}\ \text{atoms}$

(or 750 billion atoms) of non-radioactive carbon for every atom of ^{14}C.

15-67 In this problem, since we are everywhere concerned with *ratios* of atoms, we may work in moles. The amount of helium tells us how many ^{238}U nuclei *used* to be in the rock, but which have decayed. There are 8 mol of He for every mol of decayed ^{238}U, and the number of moles of He is determined by recalling that one mole of He gas occupies 22.4 L at STP:

$$\text{decayed } ^{238}\text{U} = \left(9.0 \times 10^{-5}\,\text{cm}^3\,\text{He}\right)\left(\frac{1\,\text{L He}}{1000\,\text{cm}^3\,\text{He}}\right)\left(\frac{1\,\text{mol He}}{22.4\,\text{L He}}\right)\left(\frac{1\,\text{mol }^{238}\text{U}}{8\,\text{mol He}}\right)$$

$$= 5.022 \times 10^{-10}\,\text{mol }^{238}\text{U}$$

The number of moles of ^{238}U remaining is determined directly:

$$\text{remaining } ^{238}\text{U} = \left(2.0 \times 10^{-7}\,\text{g }^{238}\text{U}\right)\left(\frac{1\,\text{mol }^{238}\text{U}}{238.01\,\text{g }^{238}\text{U}}\right) = 8.403 \times 10^{-10}\,\text{mol}$$

$$N = 8.403 \times 10^{-10}\,\text{mol}$$

The amount of ^{238}U originally in the rock, per gram, is:

$$N_O = \text{remaining }^{238}\text{U} + \text{decayed }^{238}\text{U} = 13.425 \times 10^{-10}\,\text{mol}$$

For the decay of any nuclide, the kinetics indicate:

$$-\frac{0.6931\,t}{t_{1/2}} = \ln\left(\frac{N}{N_o}\right) = \ln\left(\frac{8.401 \times 10^{-10}\,\text{mol }^{238}\text{U}}{13.425 \times 10^{-10}\,\text{mol }^{238}\text{U}}\right) = -0.4688$$

$$t = \frac{\left(4.47 \times 10^9\,\text{yr}\right)(0.4688)}{(0.6931)} = \mathbf{3.0 \times 10^9\ yr\ (3\ billion\ years)}$$

15-69 (a) As we saw earlier, under the assumption that 1 rem = 1 rad and that *all* of the nuclide decays in the body:

$$\frac{\text{dosage from }^{60}\text{Co}}{\text{dosage from }^{131}\text{I}} = \frac{\Delta E(^{60}\text{Co})}{\Delta E(^{131}\text{I})} = \frac{0.32\,\text{MeV}}{0.60\,\text{MeV}} = \mathbf{0.53}$$

(b) If we must consider only a limited time of the decay, then the radiation dosage will also be proportional to the rate constant for the decay. Note also that the rate constant is *inversely* proportional to the half life, and that the half-life for ^{60}Co = 5.27 yr = 1924.6 days. Then:

$$\frac{\text{dosage from }^{60}\text{Co}}{\text{dosage from }^{131}\text{I}} = \frac{k(^{60}\text{Co}) \times \Delta E(^{60}\text{Co})}{k(^{131}\text{I}) \times \Delta E(^{131}\text{I})} = \frac{t_{1/2}(^{131}\text{I}) \times \Delta E(^{60}\text{Co})}{t_{1/2}(^{60}\text{Co}) \times \Delta E(^{131}\text{I})}$$

$$= \frac{(8.041\,\text{days})(0.32\,\text{MeV})}{(1924.6\,\text{days})(0.60\,\text{MeV})} = \mathbf{0.023}$$

15-71.
$$^{10}_{5}\text{B} + ^{1}_{0}\text{n} \rightarrow ^{4}_{2}\text{He} + ^{7}_{3}\text{Li}$$

15-73. Note that we are not told the mass number of the actinium, and indeed that doesn't matter if all we want is the atomic number of the unknown nucleus. In working a problem like this, it is sometimes best to use a backwards approach: "What element would result if actinium combined with an electron or with a helium nucleus?" Then we can determine the atomic number and mass number of the unknown element directly:

$$\beta \text{ decay (backwards):} \qquad _{89}Ac + _{-1}e \rightarrow _{88}Ra$$

$$\text{(forwards):} \qquad _{88}Ra \rightarrow _{89}Ac + _{-1}e$$

$$\alpha \text{ decay (backwards):} \qquad _{89}Ac + _{2}He \rightarrow _{91}Pa$$

$$\text{(forwards):} \qquad _{91}Pa \rightarrow _{89}Ac + _{2}He$$

Since radium was already known, and no radium isotopes decay by β emission, the logical source of actinium was indeed element 91, suitably named "proto-actinium" = protactinium.

15-75.(a) The energy output of the sun is much greater than the amount absorbed by the earth, specifically:

$$\text{total output} = \frac{3.4 \times 10^{17} \text{ J s}^{-1}}{4.5 \times 10^{-10}} = 7.56 \times 10^{26} \text{ J s}^{-1}$$

This energy is coming from the conversion of some mass of the sun into energy, specifically, by Einstein's relation:

$$\Delta m = \frac{\Delta E}{c^2} = -\frac{7.56 \times 10^{26} \text{ J s}^{-1}}{\left(2.998 \times 10^8 \text{ m s}^{-1}\right)^2} = -8.4 \times 10^9 \text{ J} \frac{\text{s}}{\text{m}^2}$$

Converting J:
$$= -8.4 \times 10^9 \frac{\text{kg m}^2}{\text{s}^2} \frac{\text{s}}{\text{m}^2} = \mathbf{-8.4 \times 10^9 \text{ kg s}^{-1}}$$

The decrease in the mass of the sun is due to the reaction "hydrogen burning" first suggested by Bethe and van Weizsäcker. The individual steps, involving helium-3 intermediates, are discussed in the text. The *overall* nuclear reaction is (massless γ rays and neutrinos are also formed): $4 \, _1^1H \rightarrow _2^4He + 2 \, _{+1}^0\beta$

The corresponding mass change is :

$$\Delta m = \text{mass } _2^4He + 2 \text{ mass } _{+1}^0\beta - 4 \times \text{mass } _1^1H$$

$$= 4.0026034 \text{ u} + (2 \times 0.0054858 \text{ u}) - (4 \times 1.00782504 \text{ u}) = -0.0275996 \text{ u}$$

In kg: $\Delta m = -0.0275996 \text{ u} \times 1.6606 \times 10^{-24} \text{ g u}^{-1}$

$$= -4.583 \times 10^{-26} \text{ g} = \mathbf{-4.583 \times 10^{-29} \text{ kg}}$$

15-75. (continued)

How many times is this reaction proceeding per second? We relate the mass change per reaction and the sun's mass change per second to find out:

$$\text{Reactions per second} = \frac{-8.4 \times 10^9 \text{ kg s}^{-1}}{-4.580 \times 10^{-29} \text{ kg}} = 1.833 \times 10^{38} \text{ s}^{-1}$$

Four hydrogen atoms, which together have a mass of 6.69×10^{-27} kg, are involved in each reaction. The mass of the reacting hydrogen is, per second:

$$\text{mass } {}^1_1\text{H} = 1.835 \times 10^{38} \text{ s}^{-1} \times 6.69 \times 10^{-27} \text{ kg}$$

$$= \mathbf{1.22 \times 10^{12} \text{ kg } {}^1_1\text{H s}^{-1}}.$$

Since there are one thousand kg in a metric ton, this is equivalent to 1.22×10^9 metric tons per second.

CHAPTER 16 - QUANTUM MECHANICS AND THE HYDROGEN ATOM

16-1. The frequency of the waves is once every 3.2 s, or $\nu = \dfrac{1}{3.2}$ sec^{-1} = 0.312 s^{-1}. The wavelength--the crest-to-crest distance--is λ = 2.1 m. So the velocity is:

$v = \nu\lambda = (0.312 \text{ s}^{-1})\ (2.1 \text{ m}) = $ **0.66 m s^{-1}**.

16-3. For light: $c = \nu\lambda$ $\Rightarrow$ $\lambda = \dfrac{c}{\nu} = \dfrac{2.998\times10^8 \text{ ms}^{-1}}{9.86\times10^7 \text{ s}^{-1}} = $ **3.04 m**

16-5. (a) $\nu = \dfrac{c}{\lambda} = \dfrac{2.998\times10^8 \text{ ms}^{-1}}{6.00\times10^2 \text{ m}} = $ **5.00 $\times$ 10^5 s^{-1}**

(b) time $= \dfrac{\text{distance}}{\text{velocity}} = \dfrac{8.0\times10^{10} \text{ m}}{2.998\times10^8 \text{ m s}^{-1}} = 267$ s = **4.4 min**

16-7. $\lambda = \dfrac{\text{speed}}{\text{frequency}} = \dfrac{343.3 \text{ m s}^{-1}}{261.6 \text{ s}^{-1}} = $ **1.313 m**

time $= \dfrac{\text{distance}}{\text{velocity}} = \dfrac{30.0 \text{ m}}{343.3 \text{ m s}^{-1}} = $ **0.0873 s**

16-9. **Blue** light has a shorter wavelength than green light, and therefore more energy per photon. It will give the electrons more energy than green light.

16-11. A wavelength of 671 nm corresponds to **red** light.

16-13. The Planck relation is: $E = h\nu = \dfrac{hc}{\lambda}$

Rearranging: $\lambda = \dfrac{hc}{E} = \dfrac{\left(6.6261\times10^{-34} \text{ J s}\right)\left(2.998\times10^8 \text{ m s}^{-1}\right)}{3.6\times10^{-19} \text{ J}}$

$= 5.52 \times 10^{-7}$ m = **550 nm**

This corresponds to **green** light.

16-15. When handling wavelengths in nm, it is useful (and perfectly accurate) to merely tack "$\times$ 10^{-9} m" onto the number instead of "nm." You can even put the number into your calculator that way. Thus, we can write:

"589.3 nm = 589.3 $\times$ 10^{-9} m."

16-15. (continued)

(a) $E = \dfrac{hc}{\lambda} = \dfrac{\left(6.6261\times10^{-34}\text{J s}\right)\left(2.998\times10^{8}\text{m s}^{-1}\right)}{589.3\times10^{-9}\text{ m}} = \mathbf{3.371 \times 10^{-19}\ J}$

(b) $\left(3.371\times10^{-19}\dfrac{\text{J}}{\text{photon}}\right)\left(6.0221\times10^{23}\dfrac{\text{photon}}{\text{mol}}\right) = 203{,}000\text{ J mol}^{-1}$

$= \mathbf{203.0\ kJ\ mol^{-1}}$

(c) We know that one mole of sodium atoms emits 203,000 J of energy. To produce 1000 J of energy, we will need:

$$\dfrac{1000\text{ J}}{203{,}000\text{ J mol}^{-1}} = 4.296 \times 10^{-3}\text{ mol.}$$

If this energy is emitted in one second, then the rate of sodium atom emission is **4.296 × 10^{-3} mol s^{-1}**.

16-17. The minimum photon energy required to eject electrons from cesium is the energy of the work function, 3.43×10^{-19} J. This corresponds to:

$$\lambda = \dfrac{hc}{E} = \dfrac{\left(6.6261\times10^{-34}\text{J - s}\right)\left(2.998\times10^{8}\text{m s}^{-1}\right)}{3.43\times10^{-19}\text{J}}$$

$$= 5.79 \times 10^{-7}\text{ m} = 579\text{ nm}$$

Light with wavelength below this wavelength--all blue, green, and part of the yellow regions--will eject electrons from Cs. For Se, the maximum wavelength is 209 nm. This lies below the visible region. No visible light will eject electrons from a Se surface.

16-19. (a) $E_{\text{kinetic,max}} = E_{\text{light}} - \Phi$

$$E_{\text{light}} = \dfrac{hc}{\lambda} = \dfrac{\left(6.6261\times10^{-34}\text{J s}\right)\left(2.998\times10^{8}\text{m s}^{-1}\right)}{2.50\times10^{-7}\text{ m}}$$

$$= 7.95 \times 10^{-19}\text{ J}$$

$$\Phi = 7.21 \times 10^{-19}\text{ J}$$

$$E_{\text{kinetic,max}} = 7.95 \times 10^{-19}\text{ J} - 7.21 \times 10^{-19}\text{ J}$$

$$= 0.74 \times 10^{-19}\text{ J} = \mathbf{7.4 \times 10^{-20}\ J}$$

16-19. (continued)

(b)

$$E_{kinetic,} = \frac{1}{2}m_e v^2 \qquad \text{Note } 1 \text{ J} = \text{kg m}^2 \text{ s}^{-2}$$

$$v = \sqrt{\frac{2E}{m_e}} = \sqrt{\frac{2\left(7.4\times10^{-20}\text{kg m}^2\text{s}^{-2}\right)}{9.1093\times10^{-31}\text{kg}}}$$

$$= \sqrt{1.625\times10^{-11}\frac{\text{m}^2}{\text{s}^2}} = \mathbf{4.0 \times 10^5 \text{ m s}^{-1}}$$

16-21. For clarity, the answers to (a) and (b) are given in a common table. The energy in Rydbergs is calculated directly from Z and n while the energy in J is calculated from : E_n (J) = E_n (Ry) $\times$ 2.18 $\times$ 10^{-18} J Ry^{-1}.

The radius in units of a_o is determined from n^2/Z (with $Z = 1$ for hydrogen) and in units of meters by substitution of $a_o = 0.529 \times 10^{-10}$ m. The largest atom is the one with $n = 8$.

n	$E_n = -\dfrac{Z^2}{n^2}$ Ry	E_n (J)	$r_o = \dfrac{n^2}{Z}a_o$	r_o (m)
3	- 0.111 Ry	-2.42 $\times$ 10^{-19} J	9 a_o	4.76 $\times$ 10^{-10} m
6	- 0.0278 Ry	-6.06 $\times$ 10^{-20} J	36 a_o	19.0 $\times$ 10^{-10} m
8	-0.0156 Ry	-3.41 $\times$ 10^{-20} J	64 a_o	33.9 $\times$ 10^{-10} m

16-23. For the energy change, we use the equation derived in Section 17-3 of the textbook.

	$n_{initial}$	n_{final}	$\Delta E = \left(\dfrac{1}{n_{final}^2} - \dfrac{1}{n_{initial}^2}\right)$ Ry	ΔE (J)
(a)	7	4	+0.0421 Ry	9.18 $\times$ 10^{-20} J
(b)	4	3	+0.0486 Ry	1.06 $\times$ 10^{-19} J
(c)	3	4	-0.0486 Ry	-1.06 $\times$ 10^{-19} J

16-25. (a) The photon is emitted in this case. It has wavelength

$$\lambda = \frac{hc}{\Delta E} = \frac{\left(6.62621\times10^{-34}\text{J s}\right)\left(2.998\times10^8\text{m s}^{-1}\right)}{9.18\times10^{-20}\text{J}}$$

$$= \mathbf{2.16 \times 10^{-6} \text{ m}} = 2,160 \text{ nm}$$

16-25. (continued)

(b) Again, a photon is emitted. Its wavelength, determined as in (a), is

1.87×10^{-6} m (= 1870 nm).

(c) Here, the photon must be absorbed. It has the same wavelength as (b),
1.87×10^{-6} m.

16-27. The B^{4+} ion has one electron, so we may use the Bohr model with $Z = 5$.

$$E_3 = -\frac{Z^2}{n^2}\text{Ry} = -\frac{25}{9}2.18\times10^{-18}\text{J} = \textbf{-6.06} \times \textbf{10}^{\textbf{-18}}\textbf{ J}$$

$$r_3 = \frac{n^2}{Z}a_o = \frac{9}{5}5.29\times10^{-11}\text{m} = 9.52 \times 10^{-11}\text{ m} = \textbf{0.0952 nm}$$

$$E_3 \text{ (mole)} = (6.06 \times 10^{-18}\text{ J})(6.0221 \times 10^{23}\text{ mol}^{-1}) = \textbf{3,650 kJ mol}^{-1}$$

The energy of an atom in the $n = 2$ state will be:

$$E_2 = -\frac{25}{4}2.18\times10^{-18}\text{J} = -13.62 \times 10^{-18}\text{ J}$$

For the $3 \rightarrow 2$ transition,

$$\Delta E = E_2 - E_3 = -13.62 \times 10^{-18} - (-6.06 \times 10^{-18}) = 7.56 \times 10^{-18}\text{ J}$$

For the frequency of the emitted light:

$$\nu = \frac{\Delta E}{h} = \frac{7.56\times10^{-18}\text{ J}}{6.6261\times10^{-34}\text{ J s}} = \textbf{1.14} \times \textbf{10}^{\textbf{16}}\textbf{ s}^{-1}$$

For wavelength: $$\lambda = \frac{c}{\nu} = \frac{2.998\times10^8\text{ m s}^{-1}}{1.14\times10^{16}\text{ s}^{-1}} = \textbf{2.63} \times \textbf{10}^{\textbf{-8}}\textbf{ m = 26.3 nm}$$

16-29. Since $\Delta E = -Z^2\left(\frac{1}{n_{final}^2} - \frac{1}{n_{initial}^2}\right)\text{Ry}$ and in this problem n_{final} and $n_{initial}$ is the same in

both cases: $$\frac{\Delta E_H}{\Delta E_{Li}} = \frac{Z_H^2}{Z_{Li}^2} = \frac{hc/\lambda_H}{hc/\lambda_{Li}} = \frac{\lambda_{Li}}{\lambda_H}$$

Therefore: $$\lambda_{Li} = \lambda_H\frac{Z_H^2}{Z_{Li}^2} = (656.1\text{nm})\left(\frac{1^2}{3^2}\right) = \textbf{72.90 nm}.$$

This lies in the **ultraviolet** region of the spectrum.

16-31. (a) The equation relating the number of the harmonic, the wavelength, and length is

$n\dfrac{\lambda}{2}=L$. Rearranging, we get: $\lambda=\dfrac{2L}{n}$.

For $n = 1$ $\lambda = 2L = $ **100 cm** and for $n = 3$ $\lambda = \dfrac{2L}{n} = $ **33 cm**

(b) The third harmonic has $n - 1 = 3 - 1 = $ **2 nodes**.

16-33. (a) $\lambda = \dfrac{h}{mv} = \dfrac{6.6261\times10^{-34}\,\text{Js}}{\left(9.109\times10^{-31}\text{kg}\right)\left(1.00\times10^{3}\text{m s}^{-1}\right)}$

$= 7.27 \times 10^{-7}\,\text{J}\dfrac{\text{s}^2}{\text{kg-m}} = $ **7.27 × 10⁻⁷ m**

Note: $J = \dfrac{\text{kg m}^2}{\text{s}^2}$, so $(\text{J})\left(\dfrac{\text{s}^2}{\text{kg m}}\right)=\left(\dfrac{\text{kg m}^2}{\text{s}^2}\right)\left(\dfrac{\text{s}^2}{\text{kg m}}\right) = \text{m}$

(b) $\lambda = \dfrac{6.6261\times10^{-34}\,\text{J s}}{\left(1.6726\times10^{-27}\text{kg}\right)\left(1.00\times10^{3}\text{m s}^{-1}\right)} = $ **3.96 × 10⁻¹⁰ m**

(c) Note: 170.0 g = 0.1700 kg; 95 km hr⁻¹ = 26.4 m s⁻¹.

$0.095\text{m}\left(\dfrac{1\text{hr}}{60\text{min}}\right)\left(\dfrac{1\text{min}}{60\text{sc}}\right)= 2.64\times10^{-5}\dfrac{\text{m}}{\text{s}}$
26.4m;

$\lambda = \dfrac{6.6261\times10^{-34}\,\text{J s}}{\left(0.1700\text{kg}\right)\left(26.4\text{m s}^{-1}\right)} = $ **1.48 × 10⁻³⁴ m**

16-35. (a) 4*p*   (b) 2*s*   (c) 6*f*

16-37.

	(a)	(b)	(c)
nodes = $n - 1$	3	1	5
angular nodes = ℓ	1	0	3
radial nodes: $n - \ell - 1$	2	1	2

16-39. An electron in the $3p_y$ orbital is most likely to be found <u>along the y-axis</u>. The probability of finding the electron <u>in the x-z plane</u> equals zero. The wave-function also has one radial node, which is a <u>spherical surface</u>. The probability of finding the electron there is also zero.

16-43. A quantum number defines some quantized physical quantity. For the Cl^{16+} ion, which is hydrogen-like, there are three electronic quantum numbers.

16-45. Answers are written in bold

n	ℓ	label	number of orbitals
2	1	**2p**	**3**
3	**2**	3d	5
4	**3**	4f	7

Allowed: (b) only.

(a): ℓ must be less than or equal to $n-1$.   (c) $|m_\ell|$ must be less than or equal to ℓ.

(d) ℓ must be greater than or equal to zero.

16-47.
$$\lambda = \frac{speed}{\nu} = \frac{343\,\text{ms}^{-1}}{440\,\text{s}^{-1}} = \textbf{0.780 m}$$

$$time = \frac{distance}{speed} = \frac{10.0\,\text{m}}{343\,\text{m s}^{-1}} = \textbf{0.0292 s}$$

16-49. The observed colors appear from a combination of wavelengths, which extend from the red region through the yellow, then the green, and finally the blue. A green appearance would require emission a narrow band of wavelengths in the center of the visible region of the electromagnetic spectrum. This is inconsistent with the continuous nature of black-body radiation. Emission of green light does occur from a black body, but always in combination with light of longer wavelengths, and sometimes with light of shorter wavelengths, too.

16-51.   x-ray: E (photon) $= \dfrac{hc}{\lambda} = \dfrac{\left(6.6261\times10^{-34}\,\text{J s}\right)\left(2.998\times10^8\,\text{m s}^{-1}\right)}{0.20\times10^{-9}\,\text{m}}$

$$= \textbf{9.9} \times \textbf{10}^{\textbf{-16}}\ \textbf{J photon}^{\textbf{-1}}$$

E (mole) $= 9.9 \times 10^{-16}\ \text{J} \times 6.0221 \times 10^{23}\ \text{mol}^{-1}$

$$= \textbf{6.0} \times \textbf{10}^{\textbf{8}}\ \textbf{J mol}^{\textbf{-1}}$$

AM radio: E (photon) $= \dfrac{h\nu}{\lambda} = \dfrac{\left(6.6261\times10^{-34}\,\text{J s}\right)\left(2.998\times10^8\,\text{m s}^{-1}\right)}{200\,\text{m}}$

$$= 9.9 \times 10^{-28}\ \text{J}$$

E (mol) $= 9.9 \times 10^{-28}\ \text{J} \times 6.0221 \times 10^{23}\ \text{photon mol}^{-1}$

$$= \textbf{6.0} \times \textbf{10}^{\textbf{-4}}\ \textbf{J mol}^{\textbf{-1}}$$

The x-ray radiation carries about 600 million joules per mole. The radio radiation only 0.6 thousandths. The x-ray photon can easily crack bonds. The radio photon is much weaker.

16-53. The Lyman series involves transitions to $n_{final} = 1$. With $Z = 1$, this corresponds to transitions to an energy level of $E_H = -1.0$ Ry.

Such an energy level *is* encountered with He$^+$ -- for $Z = 2$, and $n = 2$,

$E_{He^+} = -1$ Ry also.

Therefore, a He$^+$ atom in the $n = 2$ state could absorb light emitted in a Lyman transition of hydrogen. One specific example of a match in energy of a He$^+$ energy level and a H atom energy level is:

$$H_{n=2} \rightarrow H_{n=1} \qquad \Delta E = \left(\frac{1}{4} - 1\right) Ry = -\frac{3}{4} Ry$$

$$He^+_{n=2} \rightarrow He^+_{n=4} \qquad \Delta E = \left(1 - \frac{2^2}{4^2}\right) Ry = +\frac{3}{4} Ry$$

16-55. Each photon must provide half the energy needed for ionization. For He$^+$, $n = 2$ (other quantum numbers do not matter in this hydrogenic orbital) the energy for ionization is:

$$\Delta E_{ionization} = E_2 = +\frac{(2)^2}{(2)^2} \, 2.18 \times 10^{-18} \text{ J} = 2.18 \times 10^{-18} \text{ J}$$

Each photon must have an energy of 1.09×10^{-18} J. The wavelength will be:

$$\lambda = \frac{hc}{\Delta E} = 1.82 \times 10^{-7} \text{ m} = \textbf{182 nm.}$$

16-57. The Bohr quantization of angular momentum is: angular momentum $= mvr = n\dfrac{h}{2\pi}$

Rearranging:

$$n = \frac{2\pi m v r}{h} = \left(\frac{2\pi \left(1.5 \times 10^{11} \text{ m}\right)\left(6.0 \times 10^{24} \text{ kg}\right)\left(3.0 \times 10^4 \text{ ms}^{-1}\right)}{\left(6.626 \times 10^{-34} \text{ J-s}\right)}\right)$$

$$= \textbf{2.6} \times \textbf{10}^{\textbf{74}}$$

If n increases by 1 in the angular momentum mvr--currently equal to 2.7×10^{40} J-s--will increase by $\dfrac{h}{2\pi}$, or 2.1×10^{-34} J-s, less than one part in 10^{76}. No one would notice.

16–59. This is an energy conservation question: the energy contained in the microwave photons becomes the thermal energy of the water.

We begin by calculating the energy change of the water (we are going to assume that there is no work involved). Recall that a temperature change of $+85°C$ is the same as a change of $+85$ K.

16-59. (continued)

$$\Delta E_{water} = q + w \approx q_{water} = c_s m \Delta t$$
$$= 4.2 \text{ J K}^{-1} \text{ g}^{-1} \times 100.0 \text{ g} \times 85 \text{ K}$$
$$= 35,700 \text{ J}$$

The energy of one photon in this case is:

$$\Delta E_{one\ photon} = h\nu = \left(6.6261 \times 10^{-34} \text{ J s}\right)\left(2.45 \times 10^9 \text{ s}^{-1}\right) = 1.62 \times 10^{-24} \text{ J photon}^{-1}$$

To provide 35700 J we will need: $35,700 \text{ J} \times \dfrac{1 \text{ photon}}{1.62 \times 10^{-24} \text{ J}} = 2.2 \times 10^{28}$ photons

This is almost 40,000 moles of photons!

16-61. The problem outlined here is the problem of stabilizing the electron in a position away from the nucleus, in some kind of motion about the nucleus. The Bohr atom did this by establishing a rule: the electron could only move between specific states and, therefore, the continuous decay outlined in the objection could not occur. It also posited that none of these states could be at the nucleus ($n = 0$), and therefore the electron would never be able to "drop down" into the nucleus. Quantum mechanics deals with this issue through the wave properties of all matter, including the electron. The lowest energy a particle can occupy has a wavelength, and this wavelength must have an integer relationship to the space occupied by the wave. Even the lowest energy particle wave must have a spatial distribution.

16-63. They *look* the same--with nodal planes bisecting the *xy* planes. But the orbital on oxygen will be almost 50 times smaller.

CHAPTER 17 - MANY ELECTRON ATOMS AND CHEMICAL BONDING

17-1. (a) $1s^2\, 2s^2\, 2p^6\, 3s^2\, 3p^2 = [\text{Ne}]\, 3s^2\, 3p^2$

(b) $1s^2\, 2s^2\, 2p^6\, 3s^2\, 3p^4 = [\text{Ne}]\, 3s^2\, 3p^4$

(c) $1s^2\, 2s^2\, 2p^6\, 3s^2\, 3p^6\, 3d^7\, 4s^2 = [\text{Ar}]\, 3d^7\, 4s^2$

17-3.

Be^+	$1s^2\, 2s^1$	paramagnetic
C^-	$1s^2\, 2s^2\, 2p^3$	paramagnetic
Ne^{2+}	$1s^2\, 2s^2\, 2p^4$	paramagnetic
Mg^+	$1s^2\, 2s^2\, 2p^6\, 3s^1 = [\text{Ne}]\, 3s^1$	paramagnetic
P^{2+}	$1s^2\, 2s^2\, 2p^6\, 3s^2\, 3p^1 = [\text{Ne}]\, 3s^2\, 3p^1$	paramagnetic
Cl^-	$1s^2\, 2s^2\, 2p^6\, 3s^2\, 3p^6 = [\text{Ne}]\, 3s^2\, 3p^6$	not paramagnetic
As^+	$1s^2\, 2s^2\, 2p^6\, 3s^2\, 3p^6\, 3d^{10}\, 4s^2\, 4p^2$	
	$= [\text{Ar}]\, 3d^{10}\, 4s^2\, 4p^2$	paramagnetic
I^-	$1s^2\, 2s^2\, 2p^6\, 3s^2\, 3p^6\, 3d^{10}\, 4s^2\, 4p^6\, 4d^{10}\, 5s^2\, 5p^5$	
	$= [\text{Kr}]\, 4d^{10}\, 5s^2\, 5p^5$	not paramagnetic

17-5. (a) In   (b) P^{3-}

(c) V^{2+} (note that when ions are formed from transition element atoms, the

s electrons are removed first!)

17-7. Elements are currently being discovered in the seventh row of the periodic table, and they

have atomic numbers that are 32 greater than the atomic numbers of the corresponding

elements in the sixth row of the periodic table. This reflects the filling of $7s$, $7p$, $6d$, and

$5f$ subshells. Assuming that the $5g$ subshell does not get filled, the atomic number of the

next halogen will be 32 greater than the atomic number for At, so we expect: $85 + 32 =$

117.

17-9. A closed shell is the main feature of the electronic configuration of the noble gases.

Without the spin quantum number, closed shells would occur for the following

configurations: $1s^1$: $Z = 1$ $1s^1\, 2s^1\, 2p^3$: $Z = 5$ $1s^1\, 2s^1\, 2p^3\, 3s^1\, 3p^3$ a$Z = 9$

17-11. There will be six subshells in the $n = 6$ level:

2s electrons, 6 p electrons, 10 d electrons, 14 f electrons, 18 g electrons, 22 h electrons

This is a total of 72 electrons!

17-13. We expect this element to start a new inner transition series—below Ac. This would have 32 more protons than Ac, or 121 protons. The atomic number would be 121. (Of course, if the 5g shell is filled before the 6f, then we would need 18 more. The atomic number would now be 139.)

17-15. (a) Sr - it is just to the right of Rb in a row.

(b) Rn - it is two elements to the right of Po in a row.

(c) Xe - is a noble gas while Cs is an alkali metal.

(d) Sr - it is higher in a group than Ba.

17-17. (a) Cs - Xe is a noble gas with a closed electron shell that will have a negative electron affinity, while Cs has an opening in the s subshell.

(b) F - it is at the top right of the periodic table. This is the position of elements with the highest electron affinities.

(c) K - it can attain a filled s subshell with an extra electron, whereas Ca has a filled subshell already.

(d) At- it is to the right within a row. Adding one electron allows it to attain a closed shell. Polonium will still have an open shell with an extra electron.

17-19.

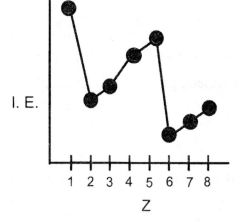

17-21. This energy is per mole of cesium atoms. We need the value per individual atom.

$$\Delta E = \frac{375,500 \text{ J mol}^{-1}}{6.0221 \times 10^{23} \text{ atom mol}^{-1}} = 6.2354 \times 10^{-19} \text{ J atom}^{-1}$$

As we saw in chapter 17:
$$\lambda = \frac{hc}{\Delta E} = \frac{(6.6261 \times 10^{-34} \text{ J s})(2.998 \times 10^8 \text{ m s}^{-1})}{6.2354 \times 10^{-19} \text{ J atom}^{-1}}$$

$$= 3.185 \times 10^{-7} \text{ m} = \textbf{318.5 nm}$$

17-23. We must recall that Be has a closed subshell configuration ($2s^2$) while B has its highest energy electron in an open shell ($2p^1$). While the trend towards higher ionization potential is present here, the Be atom gets an additional stabilization because of the closed subshell. The lone $2p^1$ electron is not well shielded in its isolation.

17-25. We have discussed how the first ionization energies of these two elements can be compared on the basis of the electron configuration of the neutral atom. The same can be done with the second ionization energy, but we note that for Be the second ionization occurs from a $2s^1$ configuration while for B it is from a $2s^2$ configuration. The removal of an electron from a filled subshell is disfavored because the electrons in that subshell do not shield each other well. In the case of Be, the electron removed in the second ionization comes from a half-filled shell, relatively well shielded by the remaining two electrons in the $1s$ level.

17-27. The radii increase from top to bottom because more electrons must be accommodated within the atom. Because electron shells of smaller principal quantum number n are already filled, shells of higher quantum number must be used. These have a larger radius that shells of smaller quantum number. Pauli repulsion -- whereby electrons may not occupy the same region of space -- also causes electrons added later to stay away from the core of the atom's electron cloud.

17-29. (a) Rb: it is lower than Li in the same group.

(b) K: a cation of an element is always smaller than a neutral atom.

(c) Kr: with an identical electron configuration, the species with the smaller nuclear charge is larger.

(d) K: both of these atoms have valence electrons in the *3s* subshell, so the atom with the smaller nuclear charge is larger.

(e) O^{2-}: with an identical electron configuration, the less positively charged species is larger.

17-31. (a) S^{2-}: it is lower in the group *and* has two negative charges

(b) Ti^{2+}: both of these atoms have valence electrons in the *3d* subshell, so the atom with the smaller nuclear charge is larger

(c) Mn^{2+}: for a given element, the size of ions always decreases with increasing positive charge.

(d) Sr^{2+}: it is lower in the same group

17-33. For Mn we have [Ar] $3d^5 4s^2$. For Eu we have [Xe] $4f^7 6s^2$. We see that both have a half-filled subshell as their outermost shell. This half-filled shell will involve dispersal of the electrons in all available *d* or *f* orbitals, creating an atom with a well-separated (and relatively large) cloud of electrons.

17-35. If we keep one element constant and let it bond to the members of a group, then the bond will be longer for the lower members of the group, as atoms increase in size down a column. Bonds will weaken as they get longer, so the bonds from the element to the lower members of a group will be weaker. Exceptions occur rarely, and only with certain exceptional elements, for example F, which forms a weaker bond to O than Cl does (see Chapter 11).

17-37. We expect bonds to arsenic to lie in between the lengths of comparable bonds to phosphorus and antimony. This would place the As-H bond at about 1.5-1.6 Å. The weakest bond will be the longest bond: SbH_3.

17-39. Additional shortening beyond that expected on the basis of covalent interactions indcates that some extra attraction--an ionic one--is present. The closeness of the bond length to the simple sum of the covalent radii suggests that there is little ionic character to this bond, and that the bond will have a low polarity.

17-41. (a) CI_4, a molecule with mostly covalent C-I bonds and weak intermolecular attractions should have the higher vapor pressure.

(b) OF_2 contains bonds between two elements of similar electronegativity. It should have more covalent bonding and higher vapor pressure.

(c) SiH_4 should have the higher vapor pressure; NaH, though formed of two elements from the same group (Group I) is a "saline hydride," indicating its significant salt-like (ionic) character.

17-43 (a) K (g) + Cl (g) → K^+ (g) + Cl^- (g)

I.E. (K) - E.A.(Cl) = +419 kJ mol^{-1} - +349 kJ mol^{-1} = **+70 kJ mol^{-1}**

K (g) + Cl (g) → K^- (g) + Cl^+ (g)

I.E. (Cl) - E.A.(K) = +1251 kJ mol^{-1} - +48 kJ mol^{-1} = **+1203 kJ mol^{-1}**

(a) Na (g) + Cl (g) → Na^+ (g) + Cl^- (g)

I.E. (Na) - E.A.(Cl) = +496 kJ mol^{-1} - +349 kJ mol^{-1} = **+147 kJ mol^{-1}**

Na (g) + Cl (g) → Na^- (g) + Cl^+ (g)

I.E. (Cl) - E.A.(Na) = +1251 kJ mol^{-1} - +53 kJ mol^{-1} = **+1198 kJ mol^{-1}**

17-45.

MOLECULE	E.N.(A)	E.N. (X)	ΔE.N.	% IONIC (Calc)	% IONIC (Observed))
HF	2.20	3.98	1.78	40	41
HCl	2.20	3.16	0.96	19	18
HBr	2.20	2.96	0.76	14	12
HI	2.20	2.66	0.46	8	6
CsF	0.79	3.98	3.19	87	70

17-45. (continued)

The quality of the agreement is good. This strongly suggests that electronegativity is a "real" parameter that accurately depicts some fundamental characteristics of the elements.

17-47. (a) The problem does not distinguish the direction of the polarity, so bonds are listed in order of decreasing absolute value for Δ E.N. :

N-**P** (ΔE.N= 0.85) > N-**C** (ΔE.N.= 0.49) > **N**-O (ΔE.N.= 0.40) > N-N (ΔE.N.= 0)

(b) For the heteronuclear species, a positive charge will be located on the atom with the lower electronegativity. These are indicated in **bold** in part (a).

17-49. All of these element should reach the top oxidation state available to their group: V^{5+}; P^{5+}; I^{7+}; Sr^{2+}.

17-51. The $6s$ electrons in Pb are in a shell that is closer to the nucleus, in part, than the many d and f electrons. Therefore, they can be "inert" with respect to bonding. The $2s$ electrons in C, on the other hand, are truly on the periphery of that atom's electron cloud.

17-53. With a high oxidation state on the metal, we expect these oxides to have significant covalent character to their bonding, forming molecular solids, not extended ionic solids.

17-55. Formic acid can be considered in the group of "$XO(OH)_m$" acids in Table 17-5. It should lie towards the bottom of the range, because carbon is not very electronegative. A K_a in the neighborhood of 1×10^{-4} to 1×10^{-3} is expected. This compares well with the experimental value of 1.77×10^{-4}. At pH 14, well above the pK_a of formic acid (3.75) $HCOO^-$ should predominate.

17-57. We predict that higher (No. lone O atoms) : (No. B atoms) ratios will correlate with greater acid strength. The number of lone O atoms can be determined by assuming that all H's are bound to an O. Then the number of lone O's becomes: No. O atoms - No. H atoms = No. lone O atoms.

We expect acidity to increase (the ratio is given in parentheses)

$H_3B_3O_3$ (0 : 3) < $H_5B_3O_7$ (2 : 3) < $H_6B_4O_9$ (3 : 4) < $H_2B_4O_7$ (5 : 4).

17-59.

SN = 4 for S; Pyramidal
derived from tetrahedron
H_2SO_3

SN = 4 for S; Tetrahedral
H_2SO_4

Acids are of the "$XO(OH)_m$" type, like H_2SO_3, are weaker than acids of the "$XO_2(OH)_m$" type, like sulfuric acid.

Two reasonable explanation are that the extra oxygen atom in sulfuric acid will serve to delocalize the negative charge that results when a hydrogen ion is removed. Also, the higher oxidation state on the sulfur in sulfuric acid will stablize the negative charge better.

17-61. The ground state of a sodium atom is [Ne] $3s^1$. The $6s^1$ configuration is an excited electronic state.

17-63. The ground state of a transition metal ion has $3d$ electrons not $4s$. The Cr^{4+} ion has two valence electrons. Thus, the configuration should be $[Ar]3d^2$.

17-65. (a) Li has a small first ionization energy, but the resulting Li^+ has a $1s^2$ configuration, equivalent to the noble gas He. This will be give the greatest difference.

(b) Sr has a small first and second ionization energy, but the resulting Sr^{2+} has a $1s^22s^22p^63s^23p^63d^{10}4s^24p^6$ configuration, equivalent to the noble gas Kr.

17-67. The ranking goes from smallest (top of list) to largest (bottom)

Co^{25+}: Such a highly charged ion must be very, very small

F^+: F is a small atom to begin with. It's cation will be very small

F: F is among the smallest neutral atoms

Br: This is in the same row as K. Br will be smaller

K: This is smaller than its heavier group-mate Rb

17-67. (continued)

> Rb: An alkali metal low in the periodic table will be large
>
> Rb⁻: Adding an electron to a large atom will make it very large

17-69. The difference in electronegativity for the elements in LiCl is 2.18, and that for the elements in HF is 1.78. We are not surprised that LiCl exhibits the high melting and boiling points typical of an ionic compound, while HF exhibits covalent properties.

17-71. As the text indicates, electronegativity should be larger when the electrons of an atom are tightly held. This is true literally in the sense that smaller atoms should have higher electronegativity and figuratively in that atoms where the principal quantum number of the valence shell electrons is smaller -- atoms higher in a group -- will spend more time "in close" to the nucleus. The elements that don't fit the expected trend of decreasing electronegativity down the group are germanium and lead. Germanium appears right after the first transition series and its valence electrons are not well shielded by the newly filled $4d$ shell. This makes the element both smaller and more electronegative than expected. Supporting this analysis, a similar anomaly appears when the electronegativities and atomic radii of aluminum and gallium are compared.

In the case of lead, size arguments (atomic radius of Pb is 1.75 Å; that of Sn is 1.40Å) and poorly shielding shells are probably not good arguments for the unexpectedly high electronegativity of Pb. The effects are probably more subtle, and may be due to the fact that the huge charge of the Pb nucleus is inducing relativistic effects on the core electrons (*J. Chem. Educ.* **1991**, *68*, 110).

17-73. The "mechanism" of bond formation in both cases is electrons going to the lowest energy position in a compound. For ionic bonds, this means they largely transfer from the less to the more electronegative element. For covalent bonds, this means they occupy regions of space between the elements, with significant attachment to both elements. By all means, both mechanisms can contribute to bond strength, as in the formation of strongly polar covalent bonds in the hydrogen halides.

17-75. Element 114 will probably be a super-heavy member of the carbon group -- right under lead. The maximum oxidation state of the element should be 4, but the trend to lower stability of higher oxidation states at the bottom of the representative elements means that the +2 oxidation state may dominate the chemistry of this element.

17-77. This refers to the strong oxidizing power of fluorine, which is capable of making the metals achieve very high oxidation states associated with covalent bonding and molecular character.

CHAPTER 18 - MOLECULAR ORBITALS AND SPECTROSCOPY

18-1. The same principles are used in determining the configuration of homonuclear diatomic ions as were used in the text with neutral species. One must remember, though, that the dication has two fewer electrons than a neutral molecule.

ION	CONFIGURATION	bond order
He_2^{2+}	$(\sigma_{1s})^2$	1
Li_2^{2+}	$(\sigma_{1s})^2(\sigma_{1s}^*)^2$	0
Be_2^{2+}	$(\sigma_{1s})^2(\sigma_{1s}^*)^2(\sigma_{2s})^2$	1
B_2^{2+}	$(\sigma_{1s})^2(\sigma_{1s}^*)^2(\sigma_{2s})^2(\sigma_{2s}^*)^2$	0
C_2^{2+}	$(\sigma_{1s})^2(\sigma_{1s}^*)^2(\sigma_{2s})^2(\sigma_{2s}^*)^2(\pi_{2p})^2$	1
N_2^{2+}	$(\sigma_{1s})^2(\sigma_{1s}^*)^2(\sigma_{2s})^2(\sigma_{2s}^*)^2(\pi_{2p})^4$	2
O_2^{2+}	$(\sigma_{1s})^2(\sigma_{1s}^*)^2(\sigma_{2s})^2(\sigma_{2s}^*)^2(\sigma_{2p})^2(\pi_{2p})^4$	3
F_2^{2+}	$(\sigma_{1s})^2(\sigma_{1s}^*)^2(\sigma_{2s})^2(\sigma_{2s}^*)^2(\sigma_{2p})^2(\pi_{2p})^4(\pi_{2p}^*)^2$	2
Ne_2^{2+}	$(\sigma_{1s})^2(\sigma_{1s}^*)^2(\sigma_{2s})^2(\sigma_{2s}^*)^2(\sigma_{2p})^2(\pi_{2p})^4(\pi_{2p}^*)^4$	1

18-3. We consider only the molecular orbitals derived from $n = 2$ atomic orbitals.

(a) $$F_2: (\sigma_{2s})^2(\sigma_{2s}^*)^2(\sigma_{2p})^2(\pi_{2p})^4(\pi_{2p}^*)^4$$
$$F_2^+: (\sigma_{2s})^2(\sigma_{2s}^*)^2(\sigma_{2p})^2(\pi_{2p})^4(\pi_{2p}^*)^3$$

(b) The bond order for F_2 is **1**. The bond order for F_2^+ is $^3/_2$.

(c) Only $\mathbf{F_2^{2+}}$ is expected to be **paramagnetic**.

(d) The $\mathbf{F_2^{2+}}$ molecule should have a stronger bond.

18-5. The electronic configuration should resemble that of O_2, except for sulfur the $n = 3$ orbitals are the valence orbitals. We will only focus on the molecular orbitals derived from $n = 3$ atomic orbitals.

Configuration: $(\sigma_{3s})^2(\sigma_{3s}^*)^2(\sigma_{3p})^2(\pi_{3p})^4(\pi_{3p}^*)^2$ Bond Order: $^1/_2 (8 - 4) = \mathbf{2}$. Partial occupancy of the π_{3p}^* orbitals will make S_2, like O_2, **paramagnetic**.

18-7. All of these molecules are in the second row of the periodic table with the $n = 2$ shell used to form molecular orbitals.

(a) 14 valence electrons in a neutral homonuclear diatomic molecule -- **F**

(b) 9 valence electrons in a homonuclear diatomic molecule with +1 charge -- **N**

(c) 13 valence electrons in a homonuclear diatomic molecule with -1 charge -- **O**

18-9. (a) $^1/_2$ (8 bonding electrons - 6 anti-bonding electrons) = **1.**

(b) $^1/_2$ (7 bonding electrons - 2 anti-bonding electrons) = **2 $^1/_2$.**

(a) $^1/_2$ (8 bonding electrons - 5 anti-bonding electrons) = **1$^1/_2$**

18-11. (a) Even number of electrons with no partially filled orbitals: **diamagnetic**

(b) Odd number of electrons: **paramagnetic**

(c) Odd number of electrons: **paramagnetic**

18-13.

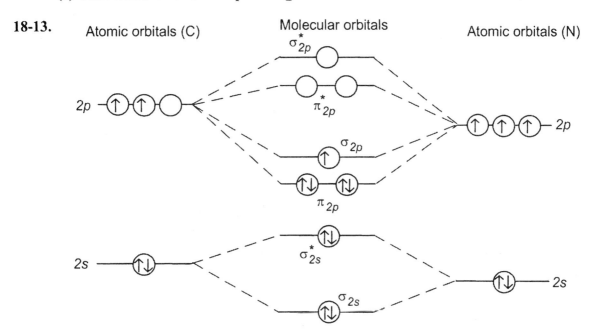

There are seven bonding and two anti-bonding electrons, so the bond order is **2$^1/_2$**. The unpaired electron will make the molecule **paramagnetic**.

18-15.

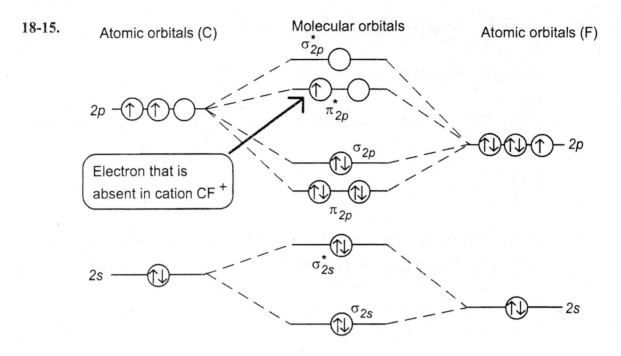

The appropriate molecular orbital diagram for CF shows an unpaired electron in an antibonding orbital. This is the electron removed in forming CF^+. The bond order in CF is $2\,^1/_2$. The bond order in CF^+ is 3. This is manifested in the length of the two bonds. The bond is shorter in the molecular ion CF^+ than in the molecule CF because the bond order in CF^+ is higher.

18-17.

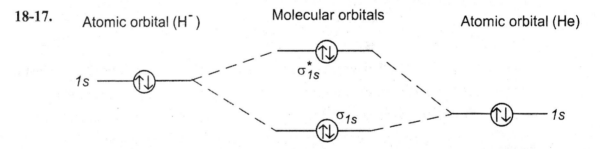

Both the bonding and the antibonding orbitals are fully occupied in HeH^-. The electronic configuration is $(\sigma_{1s})^2(\sigma^*_{1s})^2$, so the bond order is zero and the species should not be stable.

18-19. The Lewis dot structure for NH_2^- is:

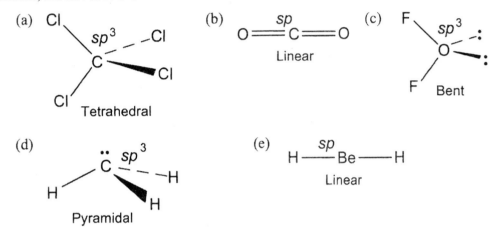

The molecule has $S.N. = 4$ on the central nitrogen, which is therefore sp^3 hybridized. Two of the sites are occupied by lone pairs and two by the H's. The structure is bent.

18-21. All non-hydrogen outer atoms have a complete octet of electrons; the lone pairs, where present, are not drawn on outer atoms.

(a) Cl — C(Cl)(Cl)(Cl), sp^3, Tetrahedral
(b) O=C=O, sp, Linear
(c) F—O—F, sp^3, Bent
(d) C—H(H)(H), sp^3, Pyramidal
(e) H—Be—H, sp, Linear

18-23. Both compounds have an sp^2-hybridized central chlorine atom with $S.N. = 3$. These Lewis structures also show the shapes.

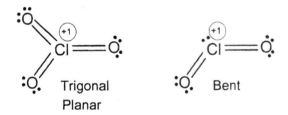

Trigonal Planar | Bent

18-25. Based on the most reasonable Lewis structure, shown below, orthonitrate will have sp^3 hybridization on the nitrogen and a tetrahedral geometry.

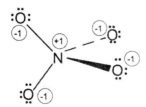

18-27. (a) Ethylene: 5-σ bonds, 1-π bond

(b) Fluorine: 1-σ bond. Note that, though the MO diagram for F_2 has π electrons, there are no *net* π bonds.

(c) Propane has 10 σ bonds and 0 π bonds. It can be written $H_3C-CH_2-CH_3$.

(d) Carbon tetrachloride has 4 σ bonds and 0 π bonds.

18-29.

The geometry about the central C is trigonal with 120° bond angles. The geometry about the other C's is tetrahedral with approximately 109.5° bond angles. There is one π bond, between the oxygen and the central carbon.

18-31.

The geometry and bonding are essentially the same as for acetone in #19-29 above.

18-33. (a) and (b) Each can have three isomers, one with the two halogens on the same carbon, another with the halogens on different carbon atoms but on the same side of the double bond ("*cis*") and the third with the halogens on different carbons atoms and on the opposite side of the double bond from one another ("*trans*").

18-35. The Lewis dot structure for nitrite ion is:

The steric number of the nitrogen is **3** and the molecule is expected to be **bent**. The nitrogen atom will be sp^2 hybridized.

We can draw a single π orbital to accommodate the multiple bonding among the atoms, shown at right. In this case, the single π orbital is split among the two N-O bonds, giving each N-O bond an order of $1\ ^1/_2$.

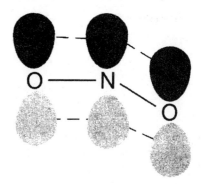

18-37. As indicated in Table 19-3, microwave radiation affects the *rotational* energy a molecule. Knowledge of rotational energy levels gives information about bond lengths and angles, which are the fundamental characteristics of molecular structure.

18-39. Ethylene has a single, doubly occupied π-orbital between the carbon atoms. It will also have a single, unoccupied π^*-orbital. The added electron in $C_2H_4^-$ will go into this π^* orbital, weakening the C-C bond. The bond order will **decrease** from 2 to $1\,{}^1/_2$.

18-41. The color that is complementary to orange is blue; indanthrene is expected to absorb blue light, around **400-475 nm**, strongly.

18-43. There is one localized π bond in cyclohexene, compared to the three delocalized π bonds in benzene. The gap between the energy of the highest π-orbital and the lowest π^*-orbital in cyclohexene should be larger than the gap between the energy of the highest π-orbital and the lowest π^*-orbital in benzene. The promotion of an electron in cyclohexene will therefore require light of higher energy -- which means light of shorter wavelength.

18-45. It takes 440 kJ to break one mole of C-F bonds, which means it takes

$$\frac{440,000 \text{ J mol}^{-1}}{6.022 \times 10^{23} \text{ atoms mol}^{-1}} = 7.31 \times 10^{-19} \text{ J atom}^{-1}.$$ The light for this has wavelength:

$$\lambda = \frac{hc}{\Delta E} = \frac{\left(6.6261 \times 10^{-34}\,\text{J s}\right)\left(2.998 \times 10^8\,\text{m s}^{-1}\right)}{7.31 \times 10^{-19}\,\text{J}} = 2.7 \times 10^{-7} \text{ m}$$

$$= \textbf{270 nm}; \quad \text{This is in the UV region of the spectrum.}$$

18-47.(a) The Lewis dot structure is:

This is essentially the same as for NO_2^- problem (18-35). Ozone has a single delocalized π orbital and net bond order of $^3/_2$ between the oxygens. The steric number of the central O is **3** and the molecule is expected to be **bent**. The central oxygen atom will be sp^2 hybridized.

18-49. (a)

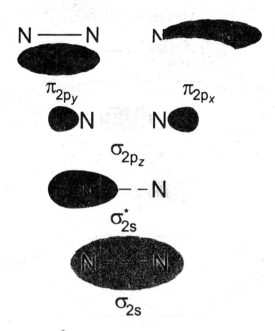

All the orbitals but the σ_{2s}^* orbital are bonding

(b) The electron removed will come from the π_{2p} level. It will be a bonding electron, weakening the bond. This will make make the bond longer n the N_2^+ molecular ion than in the N_2 molecule.

18-51. The molecular orbital diagram for B_2 is given below. Paramagnetism results from the partial occupancy of the π_{2p} orbital with one electron in each of the two orbitals. By Hund's rule, this means the electrons have parallel spins, and therefore the molecule is paramagnetic. If the ordering of the σ_{2p} and the π_{2p} orbitals were reversed, then the two electrons could occupy a single orbital. Their spins would be paired and the molecule would be diamagnetic.

18-51. (continued)

Atomic orbitals (B)   Molecular orbitals   Atomic orbitals (B)

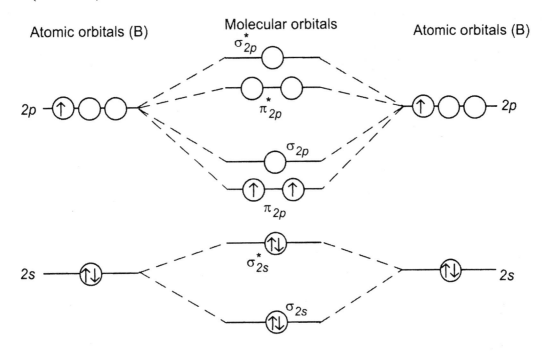

18-53. (a) The highest energy electron in H_2 occupies a H-H bonding orbital, by definition an orbital that is *lower* in energy than the lone H $1s$ orbital. So it is harder to ionize H_2 than H. The highest energy electron in O_2, on the other hand, occupies a π^* orbital, and this antibonding orbital is *higher* in energy than the O $2p$ orbital that is its origin. So it is easier to remove an electron from the highest energy orbital in O_2 than from an O atom. (b) As with oxygen, the highest occupied MO in F_2 is antibonding, at higher energy than in an F atom alone. It will be easier to ionize F_2.

18-55.

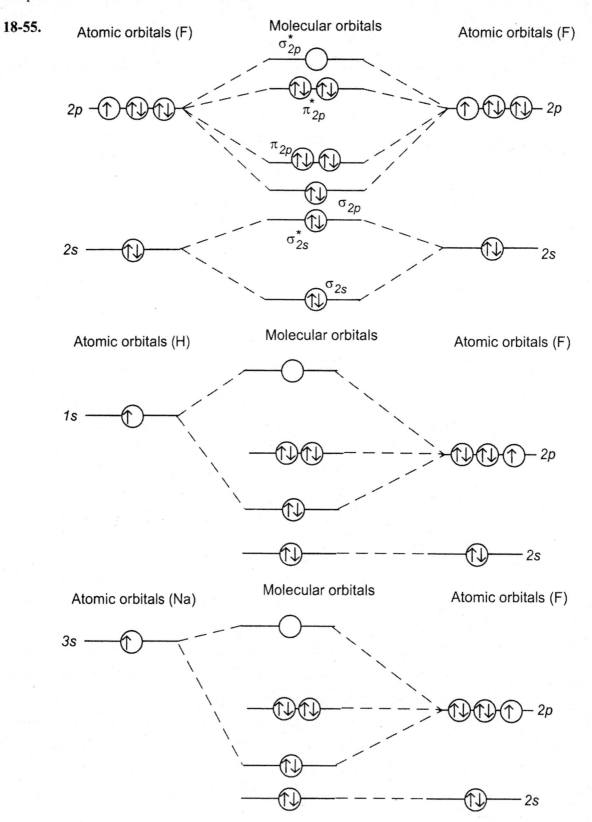

18-55. (continued)

MO theory accounts for the increasing polarity by making the orbitals more and more like the F atomic orbitals, with the electron from the H and then the Na "essentially" transferred to an orbital that is very similar to a simple F $2p$ orbital.

18-57. The two structures have very different Lewis structures:

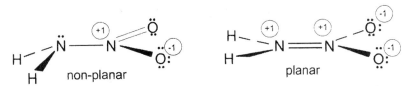

(a) In the non-planar structure, the nitrogen with the hydrogens has $S.N. = 4$ and sp^3 hybridization. It has a lone pair and the N-N bond order is **1**.

(b) In the planar structure both N's have $S.N. = 3$ and sp^2 hybridization. There are no lone pairs on nitrogen, and there *is* a π bond between the nitrogens, for a N-N bond order of **2**.

18-59. Here are the relevant Lewis structures:

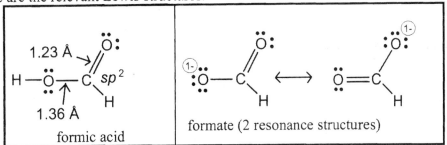

formic acid | formate (2 resonance structures)

(a) For formic acid, there is only a single Lewis structure, with a planar, sp^2 hybridized carbon with SN = 3 and a trigonal planar geometry.

(b) In formic acid the π bonding is localized between the C and one of the oxygen atoms- -the one without the hydrogen. In formate there is π bonding between the C and both of the oxygen atoms. It is a π-system that is delocalized over all three atoms.

(c) In formic acid, there is one C-O single bond to the OH group and one C=O double bond to the lone oxygen. It is not surprising that the single bond is longer and the double bond is shorter. The hybridization, steric number, and geometry about the carbon in formate ion is the same as for formic acid. But in formate ion there are two resonance structures as the double bond can be drawn to *either* of the oxygens. We expect that the C-O bond distance will be intermediate between 1.23 Å and 1.36 Å -- about 1.30 Å.

18-61. (a) The number of valence electrons in C_{60} is $4 \times 60 = 240$. For $C_{12}B_{24}N_{24}$, there are $(4 \times 12) + (3 \times 24) + (5 \times 24) = 240$ valence electrons, also with 60 atoms. An equal number of valence electrons and atoms suggests likely similarities in electronic and molecular structure.

(b) The structure of C_{60} has 12 pentagonal faces. A symmetric distribution of the C, N, and B atoms would have each pentagonal face with one C, two B's and two N's. They could be arranged so that, within a face, there were no B-B or N-N bonds. This structure can be repeated around the sphere to give a completely symmetric structure. One face is shown at right.

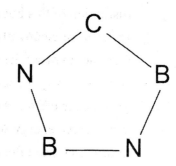

18-63. If a light appears green because of the transmission of just green light, then a prism will leave the light the same color. But if a light appears green to the eye because of a mixing of yellow and blue light, then these will be separated by the prism.

18-65. (a) Retinal has four double bonds, not counting the one in the ring. In the structure on the left, all of these are *trans*. In the structure on the right, three are *trans*, one is *cis*. The photon causes a *trans* to *cis* isomerization about the third (counting out from the ring) double bond.

(b) The double bonds in the CHO group and the ring will act to delocalize further the electrons in the π-orbitals of retinal. This delocalization will *lower* the energy required for promotion of an electron, compared to a CH_3 group at the same position. This means the replacement of CHO and the ring by CH_3 groups will cause the absorption to shift to higher energy, and *lower* wavelength.

18-67. As the text discusses, both nitrogen dioxide and ozone are important components in the photochemistry of the stratosphere and the troposphere. In the stratosphere, ozone acts as a very effective absorber of ultraviolet radiation from the sun, in the process forming O_2 + O. Eventually, many of the O atoms recombine to give ozone back, unless there are other molecules to react with them. NO_2 in particular can react with O atoms to give NO and O_2, effectively removing a source of ozone from the stratosphere. Hence, ozone has beneficial properties in the stratosphere but NO_2 is a pollutant.

In the troposphere, ozone is important for its toxicity, as is NO_2. Hence both are pollutants in the troposphere. But NO_2, because it is photochemically active at visible wavelengths, also acts as an intermediate for the *formation* of ozone in the troposphere, acting in a completely opposite way from its role in the stratosphere.

18-69. Relatively little infrared radiation reaches the earth from the sun, but the earth is an excellent source of this radiation. Part of the cosmic energy budget for the planet, then, is absorption of visible wavelength radiation and re-emission of energy into space by infrared radiation. The greenhouse effect describes the problems associated with the disruption of this budget by molecules that do a very good job at absorbing infrared radiation and converting it into thermal energy. Hence such molecules, which include CO_2, act to make the earth retain energy as thermal energy, possibly inducing global warming. Any process that adds a net amount of CO_2 to the atmosphere adds to the greenhouse effect. An example is the combustion of fossil fuels (coal, oil, and natural gas). Three energy sources that do not act as a net source of CO_2 are nuclear power, hydropower, and the direct conversion of solar energy into either electrical or chemical energy (for example, hydrogen gas). A fourth source of power is the burning of *recently grown* biomass (including derived fuels like ethanol from corn). While the process does give off CO_2, the carbon in, say, corn, was only recently converted from CO_2 by the plant. A cycle

$$CO_2 + \text{sunlight} \rightarrow \text{biomass} \rightarrow CO_2 + \text{energy}$$

does *not* increase the amount of CO_2 in the atmosphere.

CHAPTER 19 - COORDINATION COMPLEXES

19-1. (a) $[Fe(CN)_6]^{3-}$ (b) $[Mn(NH_3)(H_2O)_2(Cl)_3]$ (c) $[Pt(H_2O)Br(en)]^+$

19-3. Methylamine is a monodendate ligand. It binds through the nitrogen atom, the only site of a lone pair on the molecule.

19-5. The first acid ionization reaction is:

$$[Pt(en)_3]^{4+} \text{ (aq)} + H_2O \text{ (l)} \rightarrow [Pt(en)_2(NH_2CH_2CH_2NH)]^{3+} \text{ (aq)} + H_3O^+ \text{ (aq)}$$

We can neglect the second ionization for this calculation. At this concentration of such a weak acid the extent of the reaction is small,

Then, with $[H_3O^+] = [Pt(en)_2(NH_2CH_2CH_2NH)]^{3+} = x$

and $[Pt(en)_3]^{4+} \approx 0.124$ M:

$$x = \sqrt{[Pt(en)_3]^{4+} K_a} = \sqrt{(0.124)(4.5 \times 10^{-6})} = 7.47 \times 10^{-4}$$

$[H_3O^+] = 7.467 \times 10^{-4}$ M;   **pH = 3.13**

19-7. To get the oxidation state of the metal, we only need to consider the charged ligands and the net charge on the complex.

V (**+2**) + 2 Cl$^-$ = zero charge   2 Mo (**+2**) + 8 Cl$^-$ = - 4 charge

Co (**+2**) + 3 Cl$^-$ = -1 charge   Ni (**0**) + no charged ligands = zero charge

19-9. (a) tetrahydroxo = 4 OH$^-$ groups; zincate(II) = Zn^{2+}: overall charge of -2   Formula: **$Na_2[Zn(OH)_4]$**

(b) dichloro = 2 Cl$^-$ groups; cobalt(III) = Co^{3+}: overall charge of +1

bis(ethylenediamine) = 2 en groups nitrate = NO_3^- counterion

Formula: **$[CoCl_2(en)_2]NO_3$**

(c) bromo = Br group *on* Pt; platinum(II) = Pt^{2+}: overall charge of +1

triaqua = three H_2O groups; chloride: Cl$^-$ ion *not* bound to Pt.

Formula: **$[Pt(H_2O)_3Br]Cl$**

(d) dinitro = 2 NO_2^- groups; platinum(IV) = Pt^{4+}: overall charge of +2

tetraammine = 4 NH_3 groups; bromide = Br$^-$ ions *not* bound to Pt.

Formula: **$[Pt(NH_3)_4(NO_2)_2]Br_2$**

19-11. (a) Ammonium diamminetetraisothiocyanatochromate(III)

(b) Pentacarbonyliodotechnetium(I)

(c) Potassium pentacyanomanganate(IV)

(d) Tetraammineaquachlorocobalt(III) bromide

19-13. A thermodynamically stable complex will not react to give a complex of a different formula. It can still be kinetically labile, which means it will exchange bound ligands for other ligands that are the same -- ammonia for ammonia, for example. This can be detected through isotopic labeling.

19-15. In *basic* aqueous solution, the ammonia will come off the copper(II) as NH_3:

$$[Cu(NH_3)_4]^{2+} \ (aq) \rightarrow Cu^{2+} \ (aq) + 4 \ NH_3 \ (aq)$$

$$\Delta G_r^o = (NH_3, aq) + \Delta G_f^o \ (Cu^{2+}, aq) - \Delta G_f^o \ ([Cu(NH_3)_4]^{2+}, aq)$$

$$= (4 \ mol \times -26.50\frac{kJ}{mol}) + (1 \ mol \times 65.49\frac{kJ}{mol}) - (1 \ mol \times -111.07\frac{kJ}{mol})$$

$$= + \ 70.56 \ kJ$$

The reaction is non-spontaneous; **$[Cu(NH_3)_4]^{2+}$ is stable.**

In pH = 0 solution $[H_3O^+]$ = 1 M; ammonia will gain an H^+ ion as it comes off the copper. The reaction then is:

$$4 \ H_3O^+ \ (aq) + [Cu(NH_3)_4]^{2+} \ (aq) \rightarrow Cu^{2+} \ (aq) + 4 \ NH_4^+ \ (aq) + 4 \ H_2O \ (l)$$

The thermodynamics indicate that ΔG_r^o = **-140.68 kJ**, so the **dissociation is spontaneous** in acidic solution.

19-17. The number of ions per formula is determined by the number of anions that will *not* remain bound to the metal when the complex dissolves

$$Cu(NH_3)_2Cl_2 \ < \ KNO_3 \ < \ Na_2[PtCl_6] \ < \ [Co(NH_3)_6]Cl_3$$

ions per formula: 0   2   3   4

19-19. (a)

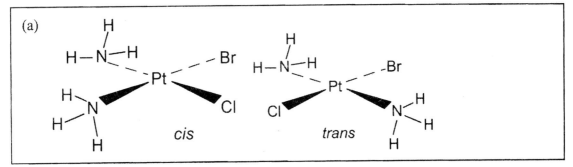

19-19. (continued)

(b)

meridional   facial

The *meridional* isomers each have a mirror plane perpendicular to the NC-Co-CN axis, and hence they are achiral. The *facial* isomer has a mirror plane bisecting the O-Co-O angle, and hence is achiral.

(c)

These enantiomers are the only isomers

19-21.

There is a single *trans* isomer   There are two *cis* isomers, and they are enantiomers
                                                      of one another

19-23.

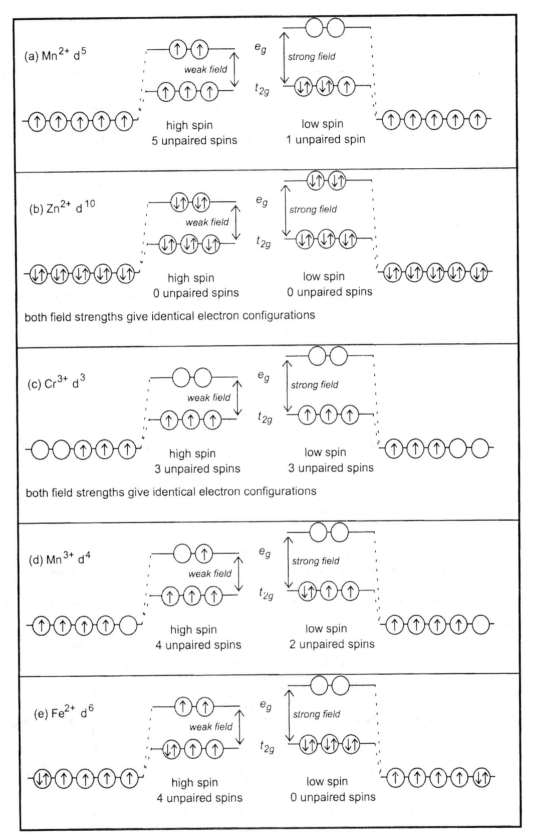

19-25. For all three cases, only the t_{2g} orbital will have any electrons; there is no need for a population of the higher energy e_g orbital. An example is Ti^{3+}, with 1 d-electron and only a $(t_{2g})^1$ configuration. Another example is V^{3+}, with 2 d-electrons and only a $(t_{2g})^2$. A third example is Cr^{3+} with a $(t_{2g})^3$ configuration (this is illustrated in Figure 19-23(c) above).

19-27. Both of these complexes have iron in the +3 oxidation state, with 5 d-electrons on the metal. For the hexacyano complex, there is a stronger crystal field than for the hexaaqua complex. The **hexacyano** complex has a $(t_{2g})^5$ configuration, for one unpaired spin, and the **hexaaqua** complex has a $(t_{2g})^3(e_g)^2$ configuration, for a total of five unpaired spins.

19-29. Complexes with a d^3, $(t_{2g})^3$ and a d^8, $(t_{2g})^6(e_g)^2$ configuration have half-filled or fully-filled t_{2g} valence shells regardless of the strength of the crystal field. These complexes are especially stable with respect to ligand dissociation or association. Similar arguments hold for d^5 ions *in a weak ligand field* (configuration $(t_{2g})^3(e_g)^2$) and d^6 ions in a *strong ligand field* $((t_{2g})^6)$.

19-31. A colorless solution is absorbing essentially no visible light. We conclude that $[Zn(H_2O)_6]^{2+}$ absorbs no significant amount of light in the visible region. The configuration of the d electrons must occupy all of the available levels in a closed shell of d electrons. This is evidence in a complex ion that the $4s$ orbitals lie at higher energy than the $3d$ orbitals.

19-33. The higher oxidation state will decrease the size of the metal ion, increasing the stength of the metal-ligand interaction. This will make a stronger field. The increased charge of the metal will make the electrostatic attraction of negatively charged ligands or strongly polar ligands stronger.

19-35. Red transmitted light corresponds to absorption of green light, $\lambda \approx 450 - 500$ nm. This corresponds to a Δ_o of:

$$\Delta_o \text{ (mol)} = \frac{hc}{\lambda} N_o = \frac{(6.6262 \times 10^{-34} \text{ J s})(2.998 \times 10^8 \text{ m})}{475 \times 10^{-9} \text{ m}}(6.022 \times 10^{23} \text{ mol}^{-1})$$
$$= 250,000 \text{ J mol}^{-1} = \mathbf{250 \text{ kJ mol}^{-1}}$$

19-37.(a) The color complementary to blue-violet is **yellow-orange**.

(b) The maximum absorbance is at the border of yellow and orange, at about **600 nm**.

(c) The introduction of cyano ligands will make the crystal field stronger; this should *increase* the t_{2g} - e_g gap and shift the absorbance maximum to **shorter** wavelengths.

19-39. (a) All of these complexes contain Fe^{3+}, but they differ in the ligand environment. Note the relative crystal field strengths, based on the spectrochemical series: $[Fe(H_2O)_6]^{3+}$ < $[Fe(CN)_6]^{3-}$. The first complex has a weak crystal field, the second a strong crystal field. The hexaaquo complex will have a $(t_{2g})^3(e_g)^2$ configuration, and this configuration, with five unpaired spins and half-filled subshells, should absorb light weakly. The hexacyano complex, with no unpaired spins and a completely vacant e_g level, should absorb light relatively well. Note that these are indications of how readily light is absorbed, not the energy of the transition. The hexafluoro complex $[FeF_6]^{3-}$ should have a ligand field even weaker than that for the hexaaquo complex. Its solutions should also be pale.
(b) This is Hg(II) and it will have a d^{10} configuration. It should be colorless.

19-41. Bond formation by Lewis-acid base interactions is generally favorable, so we expect to form as many bonds as possible, favoring octahedral complexes (6 bond) over tetrahedral (4 bond). Relatively small ligands and large metal atoms mean that there is room for many ligands around the metal. Six is an optimal value; tetrahedral complexes with four ligands are, apparently, uncrowded.

19-43. Both of these iron complexes are also d⁶. But the hexacyano complex is more likely to be in the same strong field configuration as $[Co(NH_3)_3]^{3+}$ and is more likely to be inert.

19-45. The percentage indicates that there are 4.43 g of Co for 100.00 g of the vitamin. Therefore:

$$\text{molar mass B}_{12} = \frac{\text{g B}_{12}}{\text{mol B}_{12}} = \left(\frac{100 \text{ g B}_{12}}{4.43 \text{ g Co}}\right)\left(\frac{58.93 \text{ g Co}}{1 \text{ mol Co}}\right)\left(\frac{1 \text{ mol Co}}{1 \text{ mol B}_{12}}\right)$$
$$= \mathbf{1330 \text{ g mol}^{-1}}$$

19-47. The red color corresponds to absorbtion of green light. This is relatively high energy light in the visible region, and we expect that the oxygen complex is strong field.

19-49. Ligands typically act as Lewis bases. It makes sense that ligands are often ready sources of lone pairs, which is more likely to happen with neutral or anionic species, species some chemists call "electron rich."

19-51. The diruthenium complex has an overall charge of +2 (note the two perchlorate counterions). The complex also has three anionic ligands with three negative charges. Therefore:

charge on Ru's + charge on anionic ligands = total charge

charge on Ru's = total charge - charge on anions
$$= +2 - (-3) = 5$$

The charge *per Ru atom* is half of this value, or **Ru(+2$^1/_2$)**. There is no problem with a fractional oxidation number *provided it is shared among enough atoms to make the fraction disappear.*

19-53. Among the non-metal containing species in this problem, the most important molecule to figuring out this puzzle is HCl. This is a strong acid and cannot have been present as a molecule in the original formula. Hence, there must have been some base that held the hydrogen ion in HCl -- and the logical candidate for this is the ammonia that is produced. Thus, we can imagine that the original complex also had one more ammonium and one more chloro group. The last thing to consider is that the two moles of water were probably present as ligands, not just waters of solvation. We then have:

ORIGINAL COMPOUND: **[NH$_4$]$_2$[IrCl$_5$(H$_2$O)]**

NAME: **Ammonium aquopentachloroiridium(III)**

19-55. The experiments give us the number of water molecules available for removal, which we presume are the water of *solvation* not waters of *ligation*. Therefore:

Compound 1   two H$_2$O available on heating....................[Cr(H$_2$O)$_4$Cl$_2$]Cl•2H$_2$O

Compound 2   one H$_2$O available on heating......................[Cr(H$_2$O)$_5$Cl]Cl$_2$•H$_2$O

Compound 3   no H$_2$O available on heating.................................[Cr(H$_2$O)$_6$]Cl$_3$

All three compounds have identical empirical formulas and molar masses -- 266.447 g mol^{-1}. Therefore, 100.0 g of each will contain 0.37531 mol of compound. The amount of silver chloride that will precipitate will depend on the number of free chloride ions in the formula, not on the bound chloro ligands.

Note: the molar mass of AgCl is 143.323 g mol^{-1}

COMPOUND	Chloride ions per formula	Moles of free Cl$^-$ in 100 g	Moles of AgCl	Mass of AgCl
[Cr(H$_2$O)$_4$Cl$_2$]Cl•2 H$_2$O	1	0.37531	0.37531	**53.79 g**
[Cr(H$_2$O)$_5$Cl]Cl$_2$• H$_2$O	2	0.75062	0.75062	**107.6 g**
[Cr(H$_2$O)$_6$]Cl$_3$	3	1.12593	1.12593	**161.4 g**

19-57. The two structures under consideration are:

Both exist as a single structure with at least one mirror plane perpendicular to the N-Co-N plane. There is no isomerism possible with either complex, unless one considers any folding or puckering of the five membered ring.

19-59.

$[Mn(CN)_6]^{5-}$	$Mn(I)$	d^6	$(t_{2g})^6$
$[Mn(CN)_6]^{4-}$	$Mn(II)$	d^5	$(t_{2g})^5$
$[Mn(CN)_6]^{3-}$	$Mn(III)$	d^4	$(t_{2g})^4$

19-61. The two obvervations are not in conflict, because whether a complex is high or low spin depends on whether Δ_o is large enough. Water is such a weak field ligand that even when it is associated with a "relatively" large Δ_o the complex may still be high spin.

19-63. The empirical formula $[Pt(SCN)_2Br(en)]$ suggests molecular units with the Pt in the 3+ oxidation state, which gives the Pt a d^7 configuration. This odd electron count makes the formula almost certainly incompatible with the observed diamagnetism. In the alternative formulation, half of the Pt atoms occur in $[Pt(SCN)_2(en)_2]^{2+}$ units, which makes them Pt(IV), d^6 and the other half occur in $[Pt(SCN)_2Br_2]^{2-}$ units, which makes them Pt(II), d^8 Both metal sites have an even number of electrons, so the diamagnetism is reasonable.

19-65. The anion in this compound is CuF_6^{2-}, so the oxidation state of copper is +4. This gives the copper a $\mathbf{d^7}$ configuration. The geometry of the coordination around copper will presumably be **octahedral**. Fluoride is a weak field ligand, so we expect the complex to have a high spin configuration for the electrons -- $\mathbf{(t_{2g})^5(e_g)^2}$. The corresponding low spin complex would have a configuration of $(t_{2g})^6(e_g)^1$

19-67. We presume that nitrogen remains uncharged when it is a ligand, as is true of the isoelectronic molecule CO. Then, at low temperature the $[V(N_2)_6]$ complex will have V(0) and a d^5 configuration. It must be paramagnetic. The corresponding CO complex is $V(CO)_6$.

CHAPTER 20 - STRUCTURE AND BONDING IN SOLIDS

20-1. "Second-order" means $n = 2$ for this reflection. Given an experimental angle $2\theta = 54.70^O$, then $\theta = 27.35^O$. Therefore:

$$2d \sin \theta = n\lambda \quad \Rightarrow \quad d = \frac{n\lambda}{2 \sin\theta} = \frac{(2)\left(1.660 \text{ Å}\right)}{2 \sin\left(27.35^\circ\right)} = \textbf{3.613 Å}$$

20-3. "Fourth-order" means $n = 4$.

$$2d \sin \theta = n\lambda$$

$$\theta = \sin^{-1}\left(\frac{n\lambda}{2d}\right) = \sin^{-1}\left(\frac{(4)\left(1.936 \text{ Å}\right)}{2\left(4.950 \text{ Å}\right)}\right) = \sin^{-1}(0.7822) = 51.46^O$$

$$\mathbf{2\theta = 102.92^O}$$

20-5. The maximum possible value for the argument in the $\sin^{-1}$ function is 1.00, because the sine of any angle cannot exceed 1. We insert values of n into the Bragg formula to determine when this maximum value is reached. The equation that we need is:

$$\frac{n\lambda}{2d} = n\,\frac{\left(2.167 \text{ Å}\right)}{2\left(2.570 \text{ Å}\right)} = (0.4216)\,n$$

n	$\dfrac{n\lambda}{2d}$	$\theta = \sin^{-1}\left(\dfrac{n\lambda}{2d}\right)$	2θ
1	0.4216	24.935	49.87^O
2	0.8432	57.48	115.0^O
3	1.265	Not Reasonable	

20-7. A single crystal will give diffraction as *spots*. As an analogy, think of a mirror lying at a fixed angle reflecting a beam of light from a single source. The reflection will be a point of light. If there are many micro-crystals present, the Bragg relationship is satisfied for each and every micro-crystal. But there is no unique crystal orientation. One obtains a *ring* of diffracted radiation. It is all at the angle 2θ, but spread around in a circle.

20-9. (a) This does not have three fold symmetry. Three-fold symmetry requires all corners to be identical. In an isosceles triangle, two angles (corners) are identical but the third is not.

(b) Yes, this has three-fold symmetry. The equilateral triangle has identical corners.

(c) Yes, this has three-fold symmetry. The *face* of a tetrahedron is an equilateral triangle.

(d) Yes, this has three-fold symmetry The *corner* of a cube has three equal edges coming out of it. This is on an axis of three-fold symmetry.

(e) Yes, this has three-fold symmetry. It also has six-fold symmetry.

20-11. (a) Yes, the number 1881 has **two** mirror planes, one vertical and lying in between the 8's and the other horizontal and bisecting the 1's and 8's. But the written image of the number 1881 may or may not have mirror symmetry, depending on whether there are little wiggles on the "ones," which remove mirror symmetry.

(b) Yes. The ideal box has two mirror planes perpendicular to all its faces.

(c) Yes, but the mirror plane is in the plane of the hook.

(d) No, there is no mirror plane. The club is attached to the shaft at one side only, and only one face is designed to be used in hitting the golf ball.

20-13. As noted in Table 20-1, the tetragonal system must possess *one* **4-fold rotation** element.

20-15. For an orthorhombic crystal with $\alpha = \beta = \gamma = 90^0$:

$$V = abc = (8.06 \text{ Å}) (9.12 \text{ Å}) (10.06 \text{ Å}) = 739 \text{ Å}^3$$

Note that $\text{Å} = 10^{-10}$ m, so
$$V = (739 \text{ Å}^3) \left(\frac{10^{-10} \text{ m}}{1 \text{Å}} \right)^3 = 7.39 \times 10^{-28} \text{ m}^3$$

20-17. (a) Volume of C = 45.385 Å³ cell⁻¹ × $\left(\dfrac{1 \text{ cell}}{8 \text{ C atoms}}\right)$ = **5.6731 Å³ per C atom**

(b) The average C atom weighs 12.011 u, or 1.9945×10^{-23} g. We need density in grams cm⁻³. Therefore:

$$\left(\frac{1.9945 \times 10^{-23} \text{ g}}{5.6731 \text{ Å}^3}\right)\left(\frac{1 \text{ Å}}{10^{-8} \text{ cm}}\right)^3 = \textbf{3.5157 g cm}^{\textbf{-3}}$$

20-19. (a) With a cubic unit cell, $\alpha = \beta = \gamma = 90^{\circ}$ and $a = b = c = 5.431$ Å. Therefore, noting that 1 Å = 10^{-8} cm:

$$V = (5.431 \text{ Å})^3 \left(\frac{10^{-8} \text{ cm}}{1 \text{ Å}}\right)^3 = \textbf{1.602} \times \textbf{10}^{\textbf{-22}} \textbf{ cm}^{\textbf{3}}$$

(b) $\qquad\qquad$ Mass = volume × density

$$= (1.6019 \times 10^{-22} \text{ cm}^3) \times (2.328 \text{ g cm}^{-3})$$

$$= \textbf{3.729} \times \textbf{10}^{\textbf{-22}} \textbf{ g}$$

(c) $(3.729 \times 10^{-22} \text{ g unit cell}^{-1}) \left(\dfrac{1 \text{ unit cell}}{8 \text{ Si atoms}}\right) = 4.6616 \times 10^{-23}$ g (Si atom)⁻¹

$$= \textbf{4.662} \times \textbf{10}^{\textbf{-23}} \textbf{ g (Si atom)}^{\textbf{-1}}$$

(d) If the mass of one atom of Si is 28.0855 u, then the mass of one mole of Si is 28.0855 g, and the number of atoms in one mole of Si is equal to the molar mass divided by the mass of one atom: $\qquad N_0 \quad = \dfrac{28.0855}{4.6616 \times 10^{-23}} = \textbf{6.0249} \times \textbf{10}^{\textbf{23}}$

This is 0.003×10^{23} away from the correct number, an error of 0.05%

20-21. We begin by noting the general relation of density, mass, and volume will apply to any sample size large enough to reflect the entire crystal's properties. The smallest such unit is the unit cell.

$$\frac{\text{density } (\beta \text{ - quartz})}{\text{density } (\alpha - \text{quartz})} = \frac{\left(\text{mass } (\beta \text{ - quartz}) \middle/ V \text{ } (\beta \text{ - quartz})\right)}{\left(\text{mass } (\alpha \text{ - quartz}) \middle/ V \text{ } (\alpha \text{ - quartz})\right)}$$

20-21. (continued)

Since the contents of the unit cells are the same, the masses of the unit cells are the same, so we may simplify the above equation.

$$\frac{\text{density } (\beta \text{ - quartz})}{\text{density } (\alpha - \text{quartz})} = \frac{V(\alpha - \text{quartz})}{V(\beta - \text{quartz})}$$

$$\text{density } (\beta\text{-quartz}) = \frac{V(\alpha - \text{quartz})}{V(\beta - \text{quartz})} \text{density } (\alpha \text{ - quartz})$$

$$= \frac{113.01 \text{ Å}^3}{118.15 \text{ Å}^3} 2.648 \text{ g cm}^{-3} = \mathbf{2.533 \text{ g cm}^{-3}}$$

20-23. With a single non-90° angle the unit cell must be monoclinic, so $\alpha = \gamma = 90°$ and $\beta = 97.07°$. Then, in the general volume formula (neglecting terms in $\cos \alpha$ or $\cos \gamma$, which are zero):

$$V = abc \sqrt{1 - \cos^2 \beta} = (21.021 \text{ Å}) (4.014 \text{ Å}) (18.898 \text{ Å}) \sqrt{1 - \cos^2 \left(97.07^o\right)}$$

$$= 1590.08 \text{ Å}^3 = 1.5825 \times 10^{-21} \text{ cm}^3$$

The molar mass of $Pb_4In_3B_{17}S_{18}$ is 1934.2 g mol^{-1}, which means one formula unit weighs $\frac{1934.2 \text{ g mol}^{-1}}{6.0221 \times 10^{-23} \text{ molecules mol}^{-1}} = 3.2119 \times 10^{-21}$ g. Each unit cell contains two formula units, so each cell has a mass of 6.4238×10^{-21} g.

The density is: $\frac{6.4238 \times 10^{-21} \text{ g}}{1.5901 \times 10^{-21} \text{ cm}^3} = \mathbf{4.039 \text{ g cm}^{-3}}$.

20-25. The volume of a unit cell is: $V = abc = (5.863 \text{ Å}) (12.304 \text{ Å}) (9.821 \text{ Å})$

$$= 708.47 \text{ Å}^3 = 7.0847 \times 10^{-22} \text{ cm}^3$$

The mass of the contents of one unit cell is:

$$\text{mass} = (7.08427 \times 10^{-22} \text{ cm}^3) (2.663 \text{ g cm}^{-3}) = 1.8867 \times 10^{-21} \text{ g}$$

In atomic mass units this is:

$$\text{mass} = (1.88627 \times 10^{-21} \text{ g}) (6.0221 \times 10^{23} \text{ u g}^{-1}) = 1136.16 \text{ u}$$

The formula mass for Na_2SO_4 is 142.04 u, so the unit cell contains

$$\frac{1136.160 \text{ u}}{142.04 \text{ u}} = \mathbf{8 \text{ formula units.}}$$

20-27 (a) $V_{\text{unit cell}} = a^3 = (5.6402 \times 10^{-10} \text{ m})^3 = \mathbf{1.7943 \times 10^{-28} \text{ m}^3}$

(b) The crystal itself has a volume of $(1.00 \times 10^{-3} \text{ m})^3 = 1.00 \times 10^{-9} \text{ m}$

$$\text{unit cells in crystal} = \frac{1.00 \times 10^{-9} \text{ m}^3 - \text{crystal}^{-1}}{1.79423 \times 10^{-28} \text{ m}^3 - (\text{unit cell})^{-1}}$$

$$= 5.5734 \times 10^{18} = \mathbf{5.57 \times 10^{18}} \frac{\text{unit cell}}{\text{crystal}}$$

(c) The area of the crystal face is $1.00 \times 10^{-6} \text{ m}^2$.

The area of the unit cell face is $(5.6402 \times 10^{-10} \text{ m})^2 = 3.1812 \times 10^{-19} \text{ m}^2$.

$$\text{unit cells on surface} = \frac{1.00 \times 10^{-6} \text{ m}^2 - (\text{crystal face}^{-1})}{3.1812 \times 10^{-19} \text{ m}^2 - (\text{unit cell})^{-1}}$$

$$= 3.143 \times 10^{-12} \frac{\text{unit cell}}{\text{crystal face}}$$

With 6 faces (and neglecting the fact that we are counting cells on the edges and corners more than once), there are $(3.143 \times 10^{12}) \times 6 = \mathbf{1.89 \times 10^{13}}$ unit cells on all the faces. This represents $\dfrac{1.886 \times 10^{13}}{5.5734 \times 10^{18}} = 3.38 \times 10^{-6} = \mathbf{0.000338\%}$ of all the unit cells

(or one out of every 300,000 unit cells).

20-29. The magnitude of the difference in the electronegativities of the two elements in a binary compound indicates the major form of bonding.

(a) $BaCl_2$ $\Delta EN = 2.27$ Ionic   (b) SiC $\Delta EN = 0.65$ Covalent

(c) CO   Molecular   (d) Co   Metallic

20-31. The only ambiguity in predicting the melting points is removed by the data given in the problem.

$$\text{m.p. (CO)} < \text{m.p. (BaCl}_2) < \text{m.p. (Co)} < \text{m.p. (SiC)}$$

20-33. Since antimony is a semi-metal, copper is a metal, and oxygen is a non-metal, the conductivity should go as: $O_2 < Sb < Cu$

20-35. There are **eight** nearest neighbors--the eight Cl's bound to each Cs. There are **six** Cs atoms that then bridge these Cl's. They are the second-nearest neighbors.

20-37. 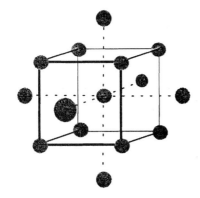 There are **eight** nearest neighbors in this structure. For a sodium atom in the center of the unit cell body, these are the eight sodium atoms at the corners of the cell. Beyond each face of the unit cell lies the second nearest neighbors -- and there are **six** of these, to go with the six faces of a cube.

20-39. The distance from the corner to the center of a body centered tetragonal cell can be determined as follows:

(i) Distance from corner to center of top face is determined because it is equal to one leg of an right triangle with a hypotenuse equal to a : $d_1 = \sqrt{a^2 / 2} = 4.1235$ Å

(ii) Distance from center to the top face: $d_2 = 1/2\, c = 1.59065$ Å

(iii) These two distances form the legs of a right triangle. The distance from the center to the corner is the hypotenuse of this triangle:
$$d_3 = \sqrt{d_1^2 + d_2^2} = \mathbf{4.4197 \text{ Å}}.$$

20-41. Frenkel defects are *displacement* defects. Ions do not leave the crystal. They move where they don't (in a perfect crystal) belong--to an interstitial site. Thus the crystal has as many of each ion as it should. There will be no effect on density.

20-43. Based on the assumption of a 100 g sample (See Chapter 2):

(a)
$$\frac{76.55 \text{ g Fe}}{55.847 \text{ g mol}^{-1}} = 1.3707 \text{ mol Fe}$$
$$\frac{23.45 \text{ g O}}{15.9994 \text{ g mol}^{-1}} = 1.4657 \text{ mol O}$$

20-43. (continued)

We set the stoichiometric coefficient of O to 1.000 and determine Fe by division:
$$\frac{Fe_{1.3707}}{1.4657} O_{1.000} = \mathbf{Fe_{0.9352}O}$$

(b) The non-stoichiometric nature of the material is due to the fact that a certain fraction of the iron is Fe^{3+}. We can denote the formula as "$(Fe^{2+})_{x-y}(Fe^{3+})_y$"O, where $x = 0.9352$ here. The variable y represents the stoichiometric coefficient for Fe^{3+}. Then (as developed in part (b) to example 21-9) we solve for y in:

$$3y + 2(x - y) = 2$$

or: $$y = 2 - 2x$$

where x = the stoichiometric coefficient for Fe.

Here: $$y = 2 - 2(0.9352) = 0.1296$$

This represents a fraction $^{0.1296}/_{0.9352} = \mathbf{0.1356}$ of the iron atoms.

20-45. The liquid phase will have rapid molecular motion (rotational and transitional) for all molecules. The plastic crystalline phase will have some fraction of the molecules tumbling in their places, but no translational motion. The plastic phase will have lower enthalpy (you must heat it to melt it to the liquid phase) and lower entropy than the liquid phase.

20-47. *Scattering* refers to the diffraction of x-rays by a sample. *Interference* occurs when this scattering is coherent--in phase--and this requires crystalline material.

20-49. The conditions for this problem are the same as for Bragg diffraction.

$$2d \sin \theta = n\lambda$$

For first order, $n = 1$, so: $$\sin \theta = \frac{\lambda}{2d}$$

$$\theta = \sin^{-1}\left(\frac{\lambda}{2d}\right) = \sin^{-1}\left(\frac{3.0 \text{ m}}{2(5.0 \text{ m})}\right)$$

$$= 17.46^0 \text{ and } \mathbf{2\theta = 35^0}$$

20-51. Neglecting questions about type-faces:

(a) 10801 has two mirror planes and a two-fold rotation axis.

(b) 86198 has a two-fold rotation axis.

(c) A teacup, exclusive of decoration, has a mirror plane.

20-53. (a) Unit cell volume: $V = a^3 = (5.6402 \text{ Å})^3 = \textbf{179.43 Å}^3$

(b) The primitive unit cell has the volume associated with one lattice point:

$$\frac{179.43 \text{ Å}^3}{4} = \textbf{44.86 Å}^3$$

(c) # beams $= \dfrac{4}{3}\pi\left(\dfrac{2}{2.2896 \text{ Å}}\right)^3 44.86 \text{ Å}^3 = 125.24 = \textbf{125 beams}$ observable

Note that there is no such thing as a fraction of a diffracted beam--hence we round the answer down.

(d) # beams $= \dfrac{4}{3}\pi\left(\dfrac{2}{0.7093 \text{ Å}}\right)^3 44.86 \text{ Å}^3 = \textbf{4212 beams}$ observable

20-55.

Eight C's at the corners = 1 C

Six C's in the faces = 3 C

4 C's in the interior = 4 C

TOTAL = 8 C

20-57. A cubic unit cell has $V_{cell} = a^3$. Such a unit cell also has the atoms in contact along the diagonal of the unit cell. The length of the diagonal of a cube is $d = \sqrt{3a^2} = a\sqrt{3}$, and this is 4 times the radius of the atom, so $r = 1/4\, d = a\sqrt{3}/4$.

The volume of each atom is:

$$V_{atom} = 4/3\,\pi\, r^3$$
$$= 4/3\,\pi\,(a\sqrt{3}/4)^3 = (\pi\sqrt{3}/16)a^3 = 0.340087\, a^3$$

There are two atoms per unit cell, so the volume of the atoms per unit cell is

$2 \times 0.340087a^3 = 0.68\, a^3$. The ratio of this volume to the total volume is just:

$[0.68\, a^3] / a^3 = 0.68 = 68\%$

20-59. As in problem No. 25, in this monoclinic structure we may neglect terms in α and γ because their cosines are 0.

Then:
$$V = abc\sqrt{1 - \cos^2 \beta}$$

$$= (11.04 \text{ Å}) (10.98 \text{ Å}) (10.92 \text{ Å}) \sqrt{1 - \cos^2 (96.73)}$$

$$= 1314 \text{ Å}^3 = 1.314 \times 10^{-21} \text{ cm}^3$$

This is the volume of 48 S atoms. Their mass will be:

$$48 \text{ S atoms} \times \left(\frac{32.066 \text{ u}}{1 \text{ S atom}} \right) \left(\frac{1 \text{ g}}{6.0221 \times 10^{23} \text{ u}} \right) = 2.5559 \times 10^{-21} \text{ g}$$

The density is then:

$$d = \frac{2.5559 \times 10^{-21} \text{ g}}{1.314 \times 10^{-21} \text{ cm}^3} = \textbf{1.945 g cm}^{-3}$$

20-61. It possesses no three fold symmetry because one of the arrows folds *under* the ribbon it is attached to. A suitable substitution has all three arrows folded under or all three arrows folded over.

20-63. (a) With $\alpha = \gamma = 90^\circ$ and $\beta \neq 90^\circ$, the system is **monoclinic**.

(b) $(HgNO_2)_4$ is the contents of the unit cell. It has a mass of 986.62 u, or 1.6379×10^{-21} g. We can then compare the data.

Report	Mass per cell	V_{cell}	Density = Mass/$_V$
1	1.6379×10^{-21} g	271.02 Å^3 $= 2.7102 \times 10^{-22} \text{ cm}^3$	6.0435 g cm^{-3}
2	1.6379×10^{-21} g	273.74 Å^3 $= 2.7374 \times 10^{-22} \text{ cm}^3$	5.983 g cm^{-3}

(c) The unit cell contents must contain a whole-number multiple of the formula unit. Hence, each Hg(I) must be matched by one NO_2^-.

20-63. (continued)

(d) The second report contains less precise data for the axis lengths, suggesting a precision to ±0.001 Å. The first report is more precise, judging by significant figures. Clearly, however, the variation in the data--by 1% in the volume--is greater than that suggested by the significant figures.

20-65. The Lewis structure, shown at right, has only four electrons $\ddot{:}\ddot{C}l\!\!-\!\!Be\!\!-\!\!\ddot{C}l\ddot{:}$ around the Be, so it is an octet deficient molecule. We expect strong Lewis acid-base interactions in the solid state between the lone pairs on Cl and the Lewis acid site Be.

20-67. (a) The wavelength of the neutrons is determined by the de Broglie relation:

$$\lambda = \frac{h}{mv} = \frac{\left(6.6261\ 10^{-34}\ \text{J s}\right)}{\left(1.6749 \times 10^{-27}\ \text{kg}\right)\left(2.639 \times 10^3\ \text{m s}^{-1}\right)}$$

$$= 1.499 \times 10^{-10}\ \text{m} = 1.499\ \text{Å}$$

(b) "Scattered in second order" means $n = 2$ in the Bragg equation. We have $2\theta = 36.26^o$, so $\theta = 18.13^o$. Then: $d = \dfrac{n\lambda}{2\sin\theta} = \dfrac{(2)\left(1.499\ \text{Å}\right)}{2\sin\left(18.13^o\right)} = \mathbf{4.817\ \text{Å}}$.

(c) If the crystal has a rock salt structure, then there are Na^+ ions at the unit cell corners with H^- ions lying half-way between them. Thus, the Na-H distance is one-half the unit cell edge length, or **2.408 Å**.

(d) The Na-H distance is the sum of the ionic radii. Therefore, the H^- radius is 2.408 Å - 0.98 Å = **1.43 Å**.

20-69. (a) In a hypothetical 100 g sample, there will be 28.31 g O (= 1.7694 mol O) and 71.69 g Ti (= 1.4973 mol Ti), so we get: $Ti_{\frac{1.4973}{1.7694}}O = Ti_{\mathbf{0.8462}}O$

(b) The information about the vacancy of the Ti sites tells us that the formula of the compound is "$Ti_{0.765}O_x$". We know that $0.765{:}x$ is the same as $0.750{:}1$. Solving this proportion gives $x = 1.02$. 0.020 of the O sites are not occupied.

20-71. Mineralogical orpiment had the opportunity to grow over geological time spans -- perhaps millions of years -- and this may be absolutely required for growth of large single crystals. Also, in nature the material may have been subject to countless cycles of heating and cooling that enabled it to anneal to a low energy form. Neither of these conditions are easily matched in the laboratory.

CHAPTER 21 - SILICON AND SOLID-STATE MATERIALS

21-1. As we have seen, most particularly in Chapter 17, there is a simple relationship between light energy and wavelength:

$$E = \frac{hc}{\lambda} = \frac{\left(6.626 \times 10^{-34}\,\text{J s}\right)\left(2.998 \times 10^{8}\,\text{m s}^{-1}\right)}{920 \times 10^{-9}\,\text{m}} = \textbf{2.16} \times \textbf{10}^{\textbf{-19}}\,\textbf{J}$$

This is the *minimum* energy that will cause an electron to move from the valence to the conduction band of the semiconductor. Light of higher energy will also move electrons accross the band gap.

21-3. We will have: E (mole) = E (band-gap width) $\times N_O$

$$= (8.7 \times 10^{-19}\,\text{J})(6.0221 \times 10^{23}\,\text{mol}^{-1})$$
$$= 524{,}000\,\text{J mol}^{-1}$$

$$n_O = (4.80 \times 10^{15}\,\text{cm}^{-3}K^{-3/2})(300\,\text{K})^{3/2}\,e^{-\left[524{,}000\,\text{J mol}^{-1}/(2 \times 8.3145\text{J mol K}^{-1} \times 300\,\text{K})\right]}$$

$$= \textbf{6.1} \times \textbf{10}^{\textbf{-27}}\,\textbf{cm}^{\textbf{-3}}$$

This is considerably less -- about 50 orders of magnitude less -- than the number of *atoms* per cubic centimeter. There will be no electrons in diamond's conduction band at 300 K.

21-5. The starting point of this class of semiconductors is the number of electrons in a pure group IV element -- four per atom. If we dope Si with P, we are adding an element with five valence electrons in place of one with four. There will be extra electrons and the conduction will be of the *n*-type to indicate the negative charge carriers. If we dope InSb, a III-V semiconductor, with Zn, then we are adding an element with fewer than 4 electrons. There will be electrons missing from the structure relative to pure InSb. Conduction will be of the *p*-type, to indicate the positive charge carriers, known as holes.

21-7. This is the reverse of the calculation in No. 1 above:

$$\lambda = \frac{hc}{E} = \frac{\left(6.626 \times 10^{-34}\,\text{J s}\right)\left(2.998 \times 10^{8}\,\text{m s}^{-1}\right)}{2.9 \times 10^{-19}\,\text{J}} = 6.85 \times 10^{-7}\,\text{m} = \textbf{680 nm}$$

Light of this or shorter wavelengths will have enough energy to promote electrons from the valence to the conduction band.

21-9. White room temperature ZnO absorbs no visible light. When heated, it appears yellow because it absorbs purple light (complementary to yellow). We can explain this by envisioning light of longer and longer wavelengths becoming capable of promoting electrons across a band gap that is **decreasing** with temperature.

21-11.

If we went to P and then S as the central atom, we would be introducing elements with additional valence electrons *when neutral*. They would form structures of the same overall shape and electron distribution, but with two fewer total charges on the "O₃ZOZO₃" unit. Charge neutrality would be maintained by the placement of positive formal charges on the "Z" atoms -- +1 on P, +2 on S. Taking this process one step further, with Cl there would be the same structure and electron distribution on the *neutral* molecule Cl_2O_7.

21-13. The structure of a silicate is largely set by the Si : O ratio. The ratio is determined by the number of lone O's only. Hydroxide anions are operating as counterions in the structure, as in talc. In all of these structures O and Si will be in their most geochemically stable oxidation states, O(-2) and Si(+4).

The other oxidation states are determined either by prior knowledge (e.g., Na is always Na(1+) in compounds) or by the need for overall charge neutrality.

	Si : O ratio	Structure type	Other oxidation numbers
(a)	1 : 4	discrete tetrahedral SiO_4^{4-} anions	Ca(+2), Fe(+3)
(b)	1 : 2 1/2	infinite sheets	Na(+1), Zr(+2)
(c)	1 : 3 1/2	pairs of tetrahedra	Ca(+2), Zn(+2)
(d)	1 : 2 1/2	infinite sheets	Mg(+2), H(+1)

21-15. For these compounds, it is the ratio of the (Al + Si) : O that determines the structure. The oxidation number of Al is +3, of Si is +4, and of O is -2 in all.

	(Al +Si) : O ratio	Structure type	Other oxidation numbers
(a)	1 : 2	infinite network	Li(+1)
(b)	1 : 2 1/2	infinite sheets	K(+1), H(+1)
(c)	1 : 3	closed rings or infinite single chains	Mg(+2)

21-17. (a) Quartz is a form of SiO_2. The reaction then is:

$$CaCO_3\ (s) + SiO_2\ (s) \rightarrow CaSiO_3\ (s) + CO_2\ (g)$$

(b) For this reaction we will have:

$$\Delta H^o = \Delta H^o_f(CaSiO_3\ (s)) + \Delta H^o_f(CO_2\ (g)) - \Delta H^o_f(CaCO_3\ (s)) - \Delta H^o_f(SiO_2\ (s))$$

$$= (1\ mol \times -1634.94\frac{kJ}{mol}) + (1\ mol \times -393.51\ \frac{kJ}{mol})$$

$$- (1\ mol \times -1206.92\frac{kJ}{mol}) - (1\ mol \times -910.94\frac{kJ}{mol}) = +\textbf{89.41 kJ}$$

$$\Delta S^o = S^o(CaSiO_3\ (s)) + S^o(CO_2\ (g)) - S^o(CaCO_3\ (s)) - S^o(SiO_2\ (s))$$

$$= (1\ mol \times 81.92\frac{J}{mol - K}) + (1\ mol \times 213.63\ \frac{J}{mol - K})$$

$$- (1\ mol \times 92.9\frac{J}{mol - K}) - (1\ mol \times 41.84\ \frac{J}{mol - K}) = +\textbf{160.8 J K}^{-1}$$

(c) As in Chapter 11, for equilibrium we evaluate the point where ΔG is zero:

$$T = \frac{\Delta H}{\Delta S} = \frac{89,410\ J}{160.8\ J\ K^{-1}} = 556\ K$$

Because ΔH and ΔS are both greater than zero, the reaction will be spontaneous at temperatures above this point: **T > 556 K**

21-19. Ceramics have tremendous strength in resisting deformation, but they are very vulnerable to catastrophic cracking. Metals are ductile, which makes them flexible enough to withstand repeated stresses but also vulnerable to bending under intense stress.

21-21. Ceramics are shaped either by setting a precursor in a certain form followed by heating or by grinding. Metals can be shaped by casting molten metal, but they can also be bent to a final shape after they are produced.

21-23. The SiO_2 and $MgSiO_3$ are generated as a grainy network within the product. Still, though the product is heterogeneous, each microcrystal is pure SiO_2 or $MgSiO_3$, so we write those substance as products of the reaction.

$$Mg_3Si_4O_{10}(OH)_2\ (s) \rightarrow 3\ MgSiO_3\ (s) + SiO_2\ (s) + H_2O\ (g)$$

21-25. The molar mass of this sample of glass, which corresponds to the "soda-lime glass" discussed in the text, is 478.57 g mol^{-1}. Thus, 2.50 kg (=2500 g) of glass contains 2500 g/478.57 g mol^{-1} = 5.224 mol of glass.

Carbon dioxide is produced in the manufacture of glass because of the use of sodium carbonate and calcium carbonate as the ultimate sources of the sodium and calcium ions in the glass. In this case, the overall process is:

$$6 \ SiO_2 + Na_2CO_3 + CaCO_3 \rightarrow Na_2O \bullet CaO \bullet (SiO_2)_6 + 2 \ CO_2.$$

Thus we have a 2 : 1 ratio of CO_2 to glass, and for the manufacture of 5.224 mol of glass we will generate 2 × 5.224 mol = 10.448 mol of CO_2. This will occupy a volume of

$$V = \frac{nRT}{P} = \frac{10.448 \ \text{mol} \times 0.08206 \ (\text{L - atm / mol - K}) \times 273.15 \ \text{K}}{1 \ \text{atm}} = \textbf{234 L of } CO_2$$

21-27. We can make our old reliable assumption of a 100.00 g working sample for this analysis. We must then determine the masses of each of the elements from the percentage composition of the glass and the composition of the oxides.

Oxide	Fraction of glass	Mass fraction of M in oxide	Mass of M in 100 g of glass	Moles of M in 100 g of glass
SiO_2	0.724	$\frac{28.086}{60.085} = 0.4674$	33.840 g	1.2049 mol
Na_2O	0.181	$\frac{45.980}{61.979} = 0.7419$	13.428 g	0.5841 mol
CaO	0.081	$\frac{40.078}{56.077} = 0.7147$	5.789 g	0.1444 mol
Al_2O_3	0.010	$\frac{53.964}{101.962} = 0.5293$	0.529 g	0.0196 mol
MgO	0.002	$\frac{24.305}{40.304} = 0.6030$	0.121 g	0.0050 mol
BaO	0.002	$\frac{137.33}{153.33} = 0.8956$	0.179 g	0.0013 mol

The total mass of all the metal in this sample is 53.886 g, which means the mass of oxygen is 100.000 - 58.886 = 46.114 g. This corresponds to 2.882 mol of O.

We conclude the problem by relating the number of moles of the metal atoms to one mole of O. To do this, we divide the mole amounts in the 100 g sample by 2.882.

We then obtain, per mol of O:

Si: **0.418 mol**   Na: **0.203 mol**   Ca: **0.050 mol**

Al: **0.0068 mol**   Mg: **0.002 mol**   Ba: **0.0005 mol**

21-29. For this reaction we will have:

$$\Delta H^o = \Delta H^o_f((CaO)_3 SiO_2 \ (s)) - 3 \ \Delta H^o_f(CaO \ (s)) - \Delta H^o_f(SiO_2 \ (s))$$

$$= (1 \ mol \times -2929.2 \frac{kJ}{mol}) - (3 \ mol \times -635.09 \ \frac{kJ}{mol}) - (1 \ mol \times -910.94 \frac{kJ}{mol})$$

$$= - \ \textbf{113.0 kJ}$$

21-31. (a) $SiO_2 \ (s) + 3 \ C \ (s) \ \rightarrow \ SiC \ (s) + 2 \ CO \ (g)$

(b) $\Delta H^o = \Delta H^o_f(SiC \ (s)) + 2 \ \Delta H^o_f(CO \ (g)) - 3 \ \Delta H^o_f(C \ (s)) - \Delta H^o_f(SiO_2 \ (s))$

$$= (1 \ mol \times -65.3 \frac{kJ}{mol}) + (2 \ mol \times -110.52 \ \frac{kJ}{mol})$$

$$- (3 \ mol \times 0.00 \frac{kJ}{mol}) - (1 \ mol \times -910.94 \frac{kJ}{mol}) = + \ \textbf{624.6 kJ}$$

(c) Because silicon carbide is an extended three dimensional structure formed from elements of similar electronegativity, it will have a high melting point, great hardness, and poor conductivity.

21-33. The reaction is: $\quad SiC \ (s) + 2 \ O_2 \ (g) \rightarrow SiO_2 \ (s) + CO_2 \ (g)$

$$\Delta G^o = \Delta G^o_f(SiO_2 \ (s)) + \Delta G^o_f(CO_2 \ (g)) - 2 \ \Delta G^o_f(O_2 \ (g)) - \Delta G^o_f(SiC \ (s))$$

$$= (1 \ mol \times -856.67 \frac{kJ}{mol}) + (1 \ mol \times -394.36 \ \frac{kJ}{mol})$$

$$- (2 \ mol \times 0.00 \frac{kJ}{mol}) - (1 \ mol \times -62.8 \frac{kJ}{mol}) = \textbf{-1188.2 kJ}$$

Because this process has a negative standard free energy change, we expect that silicon carbide is **unstable** with respect combustion with oxygen.

21-35. The sum of the charges on the elements must add to zero. We already have:

2 Ba (+2) + 1 Y (+3) + 7 O(-2) = 7- charges

The copper ions must counter these charges with +7 charges overall. The charge per copper -- and their oxidation number -- is $+^7/_3$.

21-37. The silicon atoms in solid silicon are sp^3 hybridized while the carbon atoms in graphite are sp^2 hybridized. If silicon adopted a graphite structure, then it would be more conducting than in the normal silicon structure. However, conduction in the graphite structure occurs through an almost infinite π-bonding system formed by the p orbitals. π-bonding is strong with carbon $2p$ orbitals but it is much weaker with silicon (and other atoms with a $3p$ valence shell). Silicon in the graphite structure would probably be a poorer conductor than carbon in the graphite structure.

21-39. It is easy to focus on the actions of the people in this problem. After all, it is they, not the empty seat, that move. But the *unique* thing is the empty seat that appears to move one position to the right every 5 minutes, with a velocity of 0.2 positions min^{-1}. The same situation applies for the holes in a *p*-type semiconductor. A hole is, well, nothing. The particles that move are the electrons. But there are so many electrons that watching them move and interact with the environment (an external electric field, for example) is difficult. It is easier to get an exact description of the electronic properties of a *p*-type semiconductor by watching the positions and interactions of the holes.

21-41. (a) This problem asks for a standard state calculation for the reaction:

$$SiO_2 \ (quartz) \rightarrow SiO_2 \ (cristobalite)$$

$\Delta H_r^o = \Delta H_f^o(SiO_2, \text{cristobalite}) - \Delta H_f^o(SiO_2, \text{quartz})$

$= (1 \text{ mol} \times -909.48 \frac{kJ}{mol}) - (1 \text{ mol} \times -910.94 \frac{kJ}{mol}) = \textbf{+1.46 kJ}$

$\Delta S_r^o = S^o(SiO_2, \text{cristobalite}) - S^o(SiO_2, \text{quartz})$

$= (1 \text{ mol} \times 42.68 \frac{J}{mol-K}) - (1 \text{ mol} \times -856.67 \frac{J}{mol-K}) = \textbf{+0.84 J K}^{-1}$

$\Delta G_r^o = \Delta G_f^o(SiO_2, \text{cristobalite}) - \Delta G_f^o(SiO_2, \text{quartz})$

$(1 \text{ mol} \times -855.43 \frac{kJ}{mol}) - (1 \text{ mol} \times -856.67 \frac{kJ}{mol}) = \textbf{+1.24 kJ}$

(b) The quartz form is more stable under standard state conditions.

(c) With a positive entropy of reaction, there may be a temperature where cristobalite becomes more stable than quartz.

21-43. This silicate has a Si : O ratio in the silicate network of $12 : 30 = 1 : 2^{1}/_2$. We expect it to be built of **infinite sheets** of $Si_2O_5^{2-}$ units. The oxidation numbers of the "invariant" elements are: Si(+4), H(+1), Mg(+2), O(-2) and Cl(-1). These elements have a total charge of -26. This must be balanced by positive charges on the 13 transition metal ions. Average oxidation states of **Mn(+2)** and **Fe(+2)** are expected.

21-45. (a) The reaction is a simple dehydration reaction:

$$Mg_3Si_4O_{10}(OH)_2 \ (s) + Mg_2SiO_4 \ (s) \rightarrow 5 \ MgSiO_3 \ (s) + H_2O \ (g)$$

(b) The only substance in this reaction that is not a pure condensed phase is water vapor. Therefore, $K = P_{H_2O}$. Increasing the pressure of the system at equilibrium will cause P_{H_2O} to rise above K and the reaction will shift **right to left** as a result.

21-47. The difference between the reactant anorthite and the product kaolinite is the exchange of one Ca^{2+} ion in anorthite for two hydrogen ions and a water molecule in kaolinite.

21-47. (continued)

This reaction is promoted by carbonic acid:

$$CaAl_2Si_2O_8 \ (s) + H_2CO_3 \ (aq) + H_2O \ (l) \rightarrow CaCO_3 \ (s) + Al_2Si_2O_5(OH)_4 \ (s)$$

As the pH is lowered, the relative amount of carbonic acid compared to carbonate and hydrogen carbonate will increase. This will **favor** the weathering of anorthite.

21-49. Air-rich kilns will have a relatively large amount of O_2 present, compared to the amount needed for fuel combustion. That oxygen will maintain the iron in a high oxidation state, so an **air rich kiln will favor red colors**. Conversely, the presence of a smoky atmosphere suggests soot formed by incomplete oxidation of a fossil fuel. A smoky kiln will have a relatively small amount of oxygen present, so low oxidation states of iron will be favored. **A smoky kiln will favor black colors**.

21-51. The composition of leaded glass in Table 22-2 includes 56% of "SiO_2" and 29% of "PbO" *by mass*. To obtain relative chemical amounts, we must convert these to moles. For convenience, let us assume a 100 g sample size containing 56 g of SiO_2 and 29 g of PbO.

$$\text{mol Si} = (56 \text{ g } SiO_2) \times \left(\frac{28.086 \text{ g Si}}{60.085 \text{ g } SiO_2} \right) \times \left(\frac{1 \text{ mol Si}}{28.086 \text{ g Si}} \right) = 0.932 \text{ mol Si}$$

$$\text{mol Pb} = (29 \text{ g } PbO) \times \left(\frac{207.2 \text{ g Pb}}{223.2 \text{ g } PbO} \right) \times \left(\frac{1 \text{ mol Pb}}{207.2 \text{ g Pb}} \right) = 0.130 \text{ mol Pb}$$

Therefore, the Pb : Si ratio is 0.130 : 0.932 = **1: 7.17**.

21-53. Since the formulas are so similar, we will start by determining the relative chemical amounts of $MgCO_3$ and $CaCO_3$. In exactly 100 grams of dolomite, there will be 45.7 g of $MgCO_3$ and 54.3 g of $CaCO_3$. Then:

$$MgCO_3\text{: } 45.7 \text{ g } MgCO_3 \times \left(\frac{1 \text{ mol } MgCO_3}{84.31 \text{ g } MgCO_3} \right) = 0.542 \text{ mol } MgCO_3$$

$$CaCO_3\text{: } 54.3 \text{ g } CaCO_3 \times \left(\frac{1 \text{ mol } CaCO_3}{100.087 \text{ g } CaCO_3} \right) = 0.542 \text{ mol } CaCO_3$$

There are equal chemical amounts of both compounds. So we can write the empirical formula of dolomite as $MgCO_3 \bullet CaCO_3$, or **$CaMg(CO_3)_2$**.

Chapter 21

21-55. Beryllia is an amphoteric oxide, which can behave as an acid or a base (see Chapter 4-6). Like another amphoteric oxide, alumina (Al_2O_3), beryllia can react with either aqueous acid or base, or even with water:

With acid: $BeO\ (s) + 2\ H^+\ (aq) \rightarrow H_2O\ (l) + Be^{2+}\ (aq)$

With base: $BeO\ (s) + 2\ OH^-\ (aq) + H_2O\ (l) \rightarrow Be(OH)_4^-\ (aq)$

With water: $BeO\ (s) + H_2O\ (l) \rightarrow Be(OH)_2\ (s)$

This last reaction is probably the source of volatile beryllium that the book mentions.

21-57. $$Si_3N_4\ (s) + 12\ HF\ (aq) \rightarrow 3\ SiF_4\ (g) + 4\ NH_3\ (g)$$
This, and many other reactions involving the formation of Si-F bonds, is driven by the strength of the Si-F linkage.

21-59. Hmmmmm. Sure sounds like glass to me. There had better be some big advantages to using this stuff on a wide scale. These include:

(a) Cost

(b) The incredible importance of being able to see through a container in some cases.

(c) The fact that the *starting* materials aren't biodegradable, either.

and, finally:

(d) The ability for ceramists to incorporate other elements to make the glass much harder and shatter-resistant in particular situations.

CHAPTER 22 - CHEMICAL PROCESSES

22-1. In a batch process, the reactants are all put into a single reactor and the reaction is allowed to proceed. The product or products are removed only after the completion of the process. Thus, a batch process has a discrete beginning, middle, and end. A continuous process is characterized by the continual addition of one or more of the reactants to some reactor, which contains embedded in it the heaters, pressurizers, catalysts, etc., needed to produce the product or products. The product or products are removed continuously also, though a separate purification step may still be required.

22-3. Geochemical processes are not driven by any "requirements," while biochemical processes must meet the needs of the organism they support. Thus, although both geochemical and biochemical processes must obey the laws of thermodynamics, only biochemical processes often exhibit high selectivity associated with a specific goal. Since organisms are also under time pressures, biochemical processes must often be very fast. Geochemical processes can be very slow, though exceptions are known, such as the eruption of a volcano.

The process diagram for the conversion of igneous calcium carbonate might look something like:

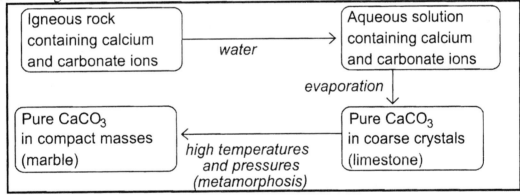

22-5. (a) Pure sulfur is obtained directly from mining the S *(s)* formed by reduction of calcium sulfate (gypsum) by methane in the presence of carbon dioxide.

(b) Elemental carbon is the major constituent of coal, itself produced by the geochemical and biochemical conversion of biomass.

(c) Molybdenum is a less-common metal, and as such is rarely obtained as in its own ores. It is often obtained as a constituent of other ores, especially in igneous rocks. Certain exceptional deposits of MoS_2 do, however, exist.

22-7. Lead sulfate will deposit in the presence of lead (II) ions -- the likely product of the oxidation of lead -- and this will coat the lead that makes up the chamber.

Chapter 22

22-9. The following observations about the direct process suggest certain improvements, given the reaction stoichiometry and thermodynamics.

$$SO_2 (g) + {}^1/_2 O_2 (g) \rightarrow SO_3 (g) \qquad\qquad \Delta H^o_f = -98.9 \text{ kJ}$$

OBSERVATION		POSSIBLE IMPROVEMENT
There are fewer particles of gas products than gaseous reactants	$\Rightarrow$	Increase the reaction pressure
Reaction is exothermic	$\Rightarrow$	Cool the reaction / lower the temperature
SO_3 (g) is more soluble than SO_2 (g) in H_2SO_4 (aq)	$\Rightarrow$	Include H_2SO_4 (aq) in the unit if it is done in a batch process; pass gases over water before recycling in a continuous process. This will remove SO_3 selectively

22-11. The reaction must convert SO_2 (g) into SO_3 (g) or H_2SO_4 (aq), either of which contains sulfur (VI). A balanced aqueous reaction with H_2O_2 is:

$$SO_2 (g) + H_2O_2 (aq) \rightarrow HSO_4^- (aq) + H^+ (aq)$$

In this case, the free energy calculation is as follows:

$$\Delta G^0_r = \Delta G^o_f (HSO_4^- (aq)) + \Delta G^o_f (H^+ (aq)) - \Delta G^o_f (SO_2 (g)) - \Delta G^o_f (H_2O_2 (aq))$$

$$= (1 \text{ mol} \times -755.91 \frac{\text{kJ}}{\text{mol}}) + (1 \text{ mol} \times 0.00 \frac{\text{kJ}}{\text{mol}})$$

$$- (1 \text{ mol} \times -300.19 \frac{\text{kJ}}{\text{mol}}) - (1 \text{ mol} \times -134.03 \frac{\text{kJ}}{\text{mol}}) = \textbf{-321.7 kJ}$$

Note that the sulfuric acid will completely ionize to hydrogen ion and hydrogen sulfate. This reaction would be spontaneous; and aqueous H_2O_2 would work to oxidize SO_2 to SO_3, at least thermodynamically.

22-13. The answer for this comes from the many reactions discussed in section 22-2.

Extraction of ore:	$ZnS (s) + {}^3/_2 O_2 (g) \rightarrow ZnO (s) + SO_2 (g)$
Direct oxidation of SO_2:	$SO_2 (g) + {}^1/_2 O_2 (g) \rightarrow SO_3 (g)$
Dissolution in water:	$SO_3 (g) + H_2O (l) \rightarrow H_2SO_4 (aq)$
Overall:	$H_2O (l) + ZnS (s) + 2 O_2 (g) \rightarrow ZnO (s) + H_2SO_4 (aq)$

22-15. This is both an oxidation-reduction and an acid base reaction. In balancing it, we treat elemental phosphorus as having a simple formula "P". One may use half reactions for greatest insight, even though the reaction is done directly. Note that this is a reaction in aqueous acid.

290

22-15. (continued)

Oxidation: $4 H_2O \ (l) + P \ (s) \rightarrow H_3PO_4 \ (aq) + 5 H^+ \ (aq) + 5 \ e^-$

Reduction: $2 \ e^- + 2 H^+ \ (aq) + H_2SO_4 \ (aq) \rightarrow SO_2 \ (g) + 2 H_2O \ (l)$

The oxidation reaction is multiplied by 2 and the reduction reaction is multiplied by 5 to cancel the electrons (H^+ cancels in the process, too.)

Overall: $2 P \ (s) + 5 H_2SO_4 \ (aq) \rightarrow 2 H_3PO_4 \ (aq) + 5 SO_2 \ (g) + H_2O \ (l)$

22-17. The best industrial starting materials for almost any inorganic acid are often sulfuric acid and an ore. In the case of these minerals and acids:

$$CaF_2 \ (s) + H_2SO_4 \ (l) \rightarrow 2 HF \ (g) + CaSO_4 \ (s)$$
$$2 NaCl \ (s) + H_2SO_4 \ (l) \rightarrow 2 HCl \ (g) + Na_2SO_4 \ (s)$$

22-19. The Kraft process is used in converting wood pulp into paper. An interesting aspect is the use of organic matter to reduce sulfate to sulfide. In that step "[C]" stands for many different compounds, all of which serve as reductants of sulfate, yielding CO_2 also. Together, the important individual steps are:

Synthesis of sodium sulfate:

$$H_2SO_4 \ (aq) + 2 NaOH \ (aq) \rightarrow 2 H_2O \ (l) + Na_2SO_4 \ (s)$$

Reduction of sodium sulfate:

$$Na_2SO_4 \ (s) + 2 \ [C] \rightarrow Na_2S \ (s) + 2 CO_2 \ (g)$$

22-21. Disulfuric acid is made up of sulfuric acid and sulfur trioxide. When it reacts with water, we can envision it splitting into its components followed by hydration of SO_3 to give H_2SO_4: $H_2S_2O_7 \ (s) + H_2O \ (l) \rightarrow 2 H_2SO_4 \ (l)$

22-23. Note that the phrase "amount of fixed nitrogen" refers to elemental N, which is appropriate because the emitted nitrogen is in a variety of substances. Also note that, 2×10^8 kg is equal to 2×10^{11} g and, as discussed in problem 16-21, 1 year is equal to 3.16×10^7 sec. In this case:

$$\left(\frac{2 \times 10^{11} \text{ g N}}{1 \text{ year}} \right) \times \left(\frac{1 \text{ mol N}}{14.00674 \text{ g N}} \right) \times \left(\frac{6.0221 \times 10^{23} \text{ N atoms}}{1 \text{ mol N}} \right) \times \left(\frac{1 \text{ year}}{3.16 \times 10^7 \text{ s}} \right)$$

$$= 3 \times 10^{26} \text{ N atoms s}^{-1}$$

22-25. *Synthesis of lime:* $CaCO_3 \ (s) \rightarrow CaO \ (s) + CO_2 \ (g)$

Synthesis of calcium carbide:

$$CaO \ (s) + 3 C \ (s) \rightarrow CaC_2 \ (s) + CO \ (g)$$

Fixation of nitrogen into calcium cyanamide:

$$CaC_2 \ (s) + N_2 \ (g) \rightarrow CaCN_2 \ (s) + C \ (s)$$

22-25. (continued)

Hydrolysis of calcium cyanamide:

$$CaCN_2 \ (s) + 4 \ H_2O \ (l) \rightarrow Ca(OH)_2 + CO_2 \ (g) + 2 \ NH_3 \ (g)$$

Overall: $CaCO_3 \ (s) + 2 \ C \ (s) + N_2 \ (g) + 4 \ H_2O \ (l)$

$$\rightarrow Ca(OH)_2 \ (s) + 2 \ CO_2 \ (g) + CO \ (g) + 2 \ NH_3 \ (g)$$

The data in the appendix give *for this reaction:* ΔH = **+374.4 kJ.** The reaction has two moles of ammonia; for each molecule of ammonia, the enthalpy change is ΔH = **+187.2 kJ mol^{-1}**.

22-27. Since the Haber-Bosch process has fewer gas molecules in the product than there are in the reactants, it will go more to the product if the pressure is raised. On the other hand, this creates engineering and safety issues. Concerning temperature, the text shows that the reaction is exothermic. It will go more to the product if the temperature is kept lower, regardless of the pressure. But, finally, the reaction will go more quickly at **higher** temperatures.

22-29. All of the steps in the Ostwald process are exothermic. So higher conversions are favored by **lower** temperatures.

22-31. These reactions involve the same half reaction for the nitric acid:

$$1 \ e^- + H^+ \ (aq) + HNO_3 \ (aq) \rightarrow NO_2 \ (g) + H_2O \ (l)$$

The reactions for the metal both follow the same pattern: $M \ (s) \rightarrow M^{3+} + 3 \ e^-$

These are combined in the right amounts to cancel the electrons:

$$3 \ H^+ \ (aq) + 3 \ HNO_3 \ (aq) + Cr \ (s) \rightarrow Cr^{3+} \ (aq) + 3 \ NO_2 \ (g) + 3 \ H_2O \ (l)$$
$$3 \ H^+ \ (aq) + 3 \ HNO_3 \ (aq) + Fe \ (s) \rightarrow Fe^{3+} \ (aq) + 3 \ NO_2 \ (g) + 3 \ H_2O \ (l)$$

22-33.

22-35. Specifying pH = 14 means that the reaction will occur with a standard 1 M concentration of hydroxide ion, and that ammonia will not be protonated. The disproportionation reaction would be:

$$3 \ N_2H_4 \ (aq) \rightarrow N_2 \ (g) + 4 \ NH_3 \ (aq)$$

$\Delta \mathcal{E}^o = \mathcal{E}^o(\ N_2H_4 \mid N_2) - \mathcal{E}^o(\ NH_3 \mid N_2H_4) = 0.1 \ V - (-1.16 \ V) = + 1.3 \ V$

A positive potential indicates the disproportionation reaction will be spontaneous. Hydrazine is thermodynamically **unstable** in aqueous base.

22-37. This kind of reaction is known as a "con-proportionation" because two oxidation states of nitrogen are present in the reactants and only one oxidation state of nitrogen is present among the products.

Oxidation: $\qquad\qquad\qquad\qquad N_2H_4\ (l) \rightarrow N_2\ (g) + 4\ H^+\ (aq) + 4\ e^-$

Reduction: $\qquad 10\ e^- + 2\ HNO_3\ (l) + 10\ H^+\ (aq) \rightarrow N_2\ (g) + 6\ H_2O\ (l)$

Combined: $\qquad\qquad 4\ HNO_3\ (l) + 5\ N_2H_4\ (l) \rightarrow 7\ N_2\ (g) + 12\ H_2O\ (l)$

$\Delta H_r^o = 12\ \Delta H_f^o\ (H_2O\ (l)) + 7\ \Delta H_f^o\ (N_2\ (g)) - 5\ \Delta H_f^o\ (N_2H_4\ (l)) - 4\ \Delta H_f^o\ (HNO_3\ (l))$

$\qquad = (12\ \text{mol} \times -285.83\ \dfrac{kJ}{mol}) + (7\ \text{mol} \times 0.00\ \dfrac{kJ}{mol})$

$\qquad - (5\ \text{mol} \times +50.63\ \dfrac{kJ}{mol}) - (4\ \text{mol} \times -174.10\ \dfrac{kJ}{mol}) = \textbf{-2986.71 kJ}$

22-39. This process is not described in the chapter, but we can come up with a plan by working backwards ("retrosynthetically" in the language of modern synthetic chemistry) from our target to our starting material through pathways we know:

$$N_2O_4 \quad \Rightarrow \quad NO_2 \quad \Rightarrow \quad NO\ \text{and}\ O_2 \quad \Rightarrow \quad NH_3\ \text{and}\ O_2 \Rightarrow N_2\ \text{and}\ H_2$$

We can then reconstruct the synthesis in the forward direction:

Ammonia synthesis: $\qquad\qquad\qquad N_2\ (g) + 3\ H_2\ (g) \rightarrow 2\ NH_3\ (g)$
(Haber-Bosch process)

Nitrogen oxide synthesis: $\qquad\quad 4\ NH_3\ (g) + 5\ O_2\ (g) \rightarrow 4\ NO\ (g) + 6\ H_2O\ (g)$
(Pt catalyzed)

Nitrogen dioxide synthesis: $\qquad\quad 2\ NO\ (g) + O_2\ (g) \rightarrow 2\ NO_2\ (g)$

Condensation of nitrogen dioxide: $\qquad 2\ NO_2\ (g) \rightarrow N_2O_4\ (s)$

22-41. TNT by itself does not have all the oxygen required for complete combustion of its carbon to carbon dioxide. Getting full combustion requires an oxidizing agent. Ammonium nitrate is an excellent oxidant and powerful explosive in its own right. Its decomposition ($NH_4NO_3 \rightarrow N_2 + 2\ H_2O + {}^3/_2\ O_2$) also liberates oxygen gas to serve to complete the oxidation of the CO produced from TNT to CO_2.

22-43. (a) A plausible reaction converts all the nitrogen to N_2, the H and some of the O to H_2O, and the excess O to O_2: $\qquad NH_4N(NO_2)_2 \rightarrow 2\ N_2 + 2\ H_2O + O_2$

(b) $\Delta H_r^o = 2\ \Delta H_f^o\ (N_2) + 2\ \Delta H_f^o\ (H_2O) + \Delta H_f^o\ (O_2) - \Delta H_f^o\ (NH_4N(NO_2)_2)$

$\qquad = 2\ \text{mol} \times -285.83\ \dfrac{kJ}{mol} + 2\ \text{mol} \times 0\ \dfrac{kJ}{mol} + 1\ \text{mol} \times 0\ \dfrac{kJ}{mol} - 1\ \text{mol} \times -36\ \dfrac{kJ}{mol}$

$\qquad = \textbf{- 536 kJ}$

22-45. (a) Aspirin is at a lower free energy than coal, hydrogen, and air. Coal, hydrogen, and the oxygen in air are all high-free energy species. As is mentioned in the text, they can serve as useful raw materials for the synthesis of intermediate free energy substances, like aspirin.

(b) Virtually any mixture, especially of salts in water, is a lower free energy than the separated components. Sea water is at lower free energy in this case.

(c) Carbon dioxide and water are at the lowest free energy of any carbon, hydrogen, and oxygen containing species. They are at lower free energy than vitamin A and oxygen.

22-47. This question, in both its parts, was answered in the "Chemistry in Progress" module at the end of Chapter 12 in the text. Plants use the energy of the sun, which they collect as light energy, to drive the processes of their life cycles. This coupling of a non-spontaneous process to a spontaneous one is the key to many processes. Included are the processes by which animals use the liberation of chemical energy available in their food (ultimately, plants) to drive their own, otherwise non-spontaneous biochemical pathways.

22-49. Over history, the following sources have been important sources of S for sulfuric acid. Note that one source, elemental sulfur, has been prominent in two different eras. In chronological order:

1a. Elemental sulfur, easily extracted.

2. Sulfide ores, as by-product of metal processing (early 19th century)

1b. Elemental sulfur, extracted from earth by the Frasch process (1890s)

3. Recovery of sulfur from fossil fuels as hydrogen sulfide (1970s)

4. Scrubbing of SO_2 from gases with hydrogen sulfide (Claus process , 1970s)

22-51.

Process	ΔH^o (kJ)	ΔS^o (J K^{-1})	ΔG^o (kJ)
$SO_2\ (g) + \frac{1}{2}\ O_2\ (g) \rightarrow SO_3\ (g)$	-98.89	- 93.98	-70.89
$SO_2\ (g) + NO_2\ (g) \rightarrow SO_3\ (g) + NO\ (g)$	-41.82	-20.76	-35.63

Oxidation of SO_2 by O_2 has a larger driving force at room temperature than oxidation by NO_2. Therefore, the NO_2 reaction will function best if it is a catalytic pathway linked to the reoxidation of NO to NO_2 by O_2. Note, however, that the NO_2 reaction will become thermodynamically superior to the direct reaction of SO_2 by O_2 at higher temperatures because the NO_2 reaction is less endothermic.

22-53. An electrolysis reaction that makes $S_2O_8^{2-}$ from SO_4^{2-} will use the reaction as an anode connected to a hydrogen half-cell as the cathode. Then, for the overall reaction

$$SO_4^{2-}\ (aq) + 2\ H^+\ (aq) \rightarrow S_2O_8^{2-} + H_2\ (g)$$

we have $\Delta E^o = E^o(H^+ \mid H_2) - E^o(S_2O_8^{2-} \mid SO_4^{2-}) = 0.00\ V - (+2.0\ V) = -2.0\ V$

The number of electrons transferred in this case, n, is 2.

In the Nernst equation:

$$\Delta E = -2.0\ V - \frac{0.0592\ V}{2} \log\left(\frac{\left[S_2O_8^{2-}\right](P_{H2})}{\left[SO_4^{2-}\right]^2\left[H^+\right]^2}\right)$$

$$= -2.0\ V - \frac{0.0592\ V}{2} \log\left(\frac{(0.5)(0.10)}{(1.0)^2(1.0)^2}\right)$$

$$= -2.0\ V - (-0.04\ V) = -1.96\ V$$

The electrolysis unit must provide an electrical potential sufficient to overcome this thermodynamic limitation, or a potential of at least **+1.96 V**.

22-55. *Precipitation*: The product must be a sparingly soluble sulfate if it is to precipitate in the presence of the high concentration of hydrogen ions generated

$$Pb^{2+}\ (aq) + H_2SO_4\ (aq) \rightarrow PbSO_4\ (s) + 2\ H^+\ (aq) \quad K = 1.5 \times 10^{-6}$$

Acid base reaction: The "pickling" of steel, for example to dissolve iron oxide:

$$Fe_2O_3\ (s) + H_2SO_4\ (aq) \rightarrow 2\ Fe^{3+}\ (aq) + 6\ SO_4^{2-}\ (aq) + 6\ H_2O\ (l)$$

Oxidation-reduction reaction: The strong acid properties of sulfuric acid will cause many metals to oxidize, as in Charles' process for manufacturing hydrogen gas for balloons in the eighteenth century (problem 5-52):

$$Fe\ (s)\ +3\ H_2SO_4\ (aq) \rightarrow Fe^{2+}\ (aq) + 6\ SO_4^{2-}\ (aq) + H_2\ (g)$$

22-57. (a) We determine the free energy change at 1000 K by assuming that the enthalpy and entropy changes of the reaction are independent of temperature.

$$\Delta H_r^o = \Delta H_f^o\ (CO_2\ (g)) + 2\ \Delta H_f^o\ (H_2O\ (g)) + 4\ \Delta H_f^o(SO_2\ (g))$$
$$- 4\ \Delta H_f^o(SO_3\ (g)) - \Delta H_f^o(CH_4\ (g))$$

$$= (1\ mol \times -393.51\ \frac{kJ}{mol}) + (2\ mol \times -241.82\ \frac{kJ}{mol}) + (4\ mol \times -293.83\ \frac{kJ}{mol})$$

$$- (4\ mol \times -395.72\ \frac{kJ}{mol}) - (1\ mol \times -74.81\ \frac{kJ}{mol}) = -406.78\ kJ = -406,780\ J$$

$$\Delta S_r^o = S^o\ (CO_2\ (g)) + 2\ S^o\ (H_2O\ (g)) + 4\ S^o(SO_2\ (g)) - 4\ S^o(SO_3\ (g)) - S^o(CH_4\ (g))$$

$$= (1\ mol \times 213.63\ \frac{J}{mol\text{-}K}) + (2\ mol \times 188.72\ \frac{J}{mol\text{-}K}) + (4\ mol \times 248.11\ \frac{J}{mol\text{-}K})$$

$$- (4\ mol \times 256.65\ \frac{J}{mol\text{-}K}) - (1\ mol \times 186.15\ \frac{J}{mol\text{-}K}) = 370.76\frac{J}{K}$$

22-57. (continued)

At 1000 K : $\Delta G^\circ = -406{,}780$ J $- (1000$ K $\times 370.76\frac{J}{K}) = -777{,}540$ J $= \mathbf{-777.54\ kJ}$

(b) For the critic's reaction:

$$\Delta H_r^o = \Delta H_f^o\ (CO_2\ (g)) + \Delta H_f^o\ (H_2O\ (g)) + \Delta H_f^o(H_2S\ (g))$$
$$- \Delta H_f^o(SO_3\ (g)) - \Delta H_f^o(CH_4\ (g))$$

$$= (1\ mol \times -393.51\ \frac{kJ}{mol}) + (1\ mol \times -241.82\ \frac{kJ}{mol}) + (1\ mol \times -20.63\ \frac{kJ}{mol})$$
$$- (1\ mol \times -395.72\ \frac{kJ}{mol}) - (1\ mol \times -74.81\ \frac{kJ}{mol}) = -185.43\ kJ = -185{,}430\ J$$

$$\Delta S_r^o = S^o\ (CO_2\ (g)) + S^o\ (H_2O\ (g)) + S^o(H_2S\ (g)) - S^o(SO_3\ (g)) - S^o(CH_4\ (g))$$

$$= (1\ mol \times 213.63\ \frac{J}{mol\text{-}K}) + (1\ mol \times 188.72\ \frac{J}{mol\text{-}K})$$
$$+ (1\ mol \times 205.68\ \frac{J}{mol\text{-}K}) - (1\ mol \times 256.65\ \frac{J}{mol\text{-}K})$$
$$- (1\ mol \times 186.15\ \frac{J}{mol\text{-}K}) = 165.23\ \frac{J}{K}$$

At 1000 K : $\Delta G^\circ = -185{,}430$ J $- (1000$ K $\times 165.23\frac{J}{K}) = -350{,}700$ J $= \mathbf{-350.70\ kJ}$

(c) It seems there is more driving force for your proposed reaction, until you realize that the free energy change *per mole of SO$_3$* is only -194 kJ for your reaction, while it is -351 kJ for the critic's reaction. Thus, in the presence of methane as reducing agent, H_2S is a more favored substance than SO_2. Still, in terms of the conversion of SO_3, your product -- SO_2 -- may be an intermediate on the path to more highly reduced H_2S. Perhaps the right conditions (catalysts, etc.) can stop the reaction at SO_2.

22-59. The basic repeat unit of a polymer indicates its empirical formula. In this drawing (lone pairs are omitted) two of the repeat units are indicated in the brackets.

The bracketed atoms are "HPO$_3$" units, the empirical formula for metaphosphoric acid.

22-61. To rewrite the formula for an analysis of P_2O_5 content, we must pull that formula unit out of the rest of the formula. For superphosphate, the overall formula has 7 moles of $CaSO_4 \bullet 2\ H_2O$ (gypsum) 3 moles of $Ca(H_2PO_4) \bullet H_2O$:

$$(CaSO_4 \bullet 2\ H_2O)_7(Ca(H_2PO_4)_2 \bullet H_2O)_3 = Ca_{10}P_6S_7O_{69}H_{46}$$
$$Ca_{10}P_6S_7O_{69}H_{46} = (P_2O_5)_3(Ca_{10}S_7O_{54}H_{46})$$

22-61. (continued)

The molar mass of "$(P_2O_5)_3$" is 425.835 g mol^{-1}. The molar mass of "$Ca_{10}P_6S_7O_{69}H_{46}$" is 1961.41 g mol^{-1}.

The mass fraction of "P_2O_5" in superphosphate is $\dfrac{425.835}{1961.41} = 0.2171$, so regular superphosphate is "**21.7% P_2O_5**." Triple superphosphate has a phosphorus content about three times (actually 2.6 times) this amount, whence its name.

22-63. The Haber-Bosch process is very energy intensive, both in the actual hydrogen and nitrogen reaction and in the need to make high-free-energy hydrogen gas, which is usually done with coal and water as the starting materials.

22-65. This is a reaction where two different nitrogen oxidation states become one. For urea, the oxidation state of N is -3 (treating carbon as C (+4) in comparison with the related carbonic acid) and in nitrogen dioxide the oxidation state of N is +4. The balanced reaction is: $4 (NH_2)_2CO + 6 NO_2 \rightarrow 7 N_2 + 4 CO_2 + 8 H_2O$

For the reaction of NO, the nitrogen begins in the +2 oxidation state and we have:
$$2 (NH_2)_2CO + 6 NO \rightarrow 5 N_2 + 2 CO_2 + 4 H_2O$$

22-67. The suggestion of an NH_2Cl intermediate indicates the following possible reaction scheme:

chlorination of ammonia:
$$NH_3 \ (aq) + OCl^- \ (aq) \rightarrow NH_2Cl \ (aq) + OH^- \ (aq)$$

coupling to hydrazine:
$$NH_3 \ (aq) + NH_2Cl \ (aq) + OH^- \ (aq) \rightarrow N_2H_4 \ (aq) + Cl^- \ (aq) + H_2O(l)$$

Hydroxide is also an intermediate in this case. This reaction is related to the very dangerous reaction that can result when household ammonia and chlorine bleach are mixed.

22-69. (a) $\Delta G_r^o = 4 \Delta G_f^o(Fe_3O_4 \ (s)) + 2 \Delta G_f^o(H_2O \ (l)) + \Delta G_f^o(N_2 \ (g))$

$\qquad\qquad - 6 \Delta G_f^o(Fe_2O_3 \ (s)) - \Delta G_f^o(N_2H_4 \ (aq))$

$\qquad = (4 \ \text{mol} \times -1015.5 \ \dfrac{kJ}{mol}) + (2 \ \text{mol} \times -237.18 \ \dfrac{kJ}{mol}) + (1 \ \text{mol} \times 0.00 \ \dfrac{kJ}{mol})$

$\qquad\qquad - (6 \ \text{mol} \times -742.2 \ \dfrac{kJ}{mol}) - (1 \ \text{mol} \times +128.1 \ \dfrac{kJ}{mol}) = \textbf{-211.26 kJ}$

(b) Including concentrated acid would result in two possible problems. First, the acid might actually dissolve some of the metal oxides. Second, the hydrazine would now be present as hydrazinium ion, $N_2H_5^+$.

22-69. (continued)

The simple reaction stoichiometry would then become:

$6 \ Fe_2O_3 \ (s) + N_2H_5^+ \rightarrow 4 \ Fe_3O_4 \ (s) + N_2 \ (g) + 2 \ H_2O \ (l) + H^+ \ (aq)$

This reaction is less favored in the presence of high concentrations of hydrogen ion, so hydrazine is a poorer reducing agent under strong acid conditions.

22-71. (a) $H_2SO_4 + 2 \ NaNO_3 \rightarrow Na_2SO_4 + 2 \ HNO_3$

(b) We expect the product to have the composition of "fuming nitric acid", about 98% by weight.

CHAPTER 23 - CHEMISTRY OF THE HALOGENS

23-1. The determination of the equilibrium constant requires the standard free energy change for this reaction, which is calculated as follows:

$$\Delta G^o = 2\,\Delta G^o_f(OCl^-\ (aq)) + \Delta G^o_f(H_2O\ (l)) - 2\,\Delta G^o_f(OH^-\ (aq)) - \Delta G^o_f(Cl_2O\ (g))$$

$$= (2\text{ mol} \times -36.8\frac{kJ}{mol}) + (1\text{ mol} \times -237.18\ \frac{kJ}{mol}) - (2\text{ mol} \times -157.24\frac{kJ}{mol})$$

$$- (1\text{ mol} \times +97.9\ \frac{kJ}{mol}) = -94.2\text{ kJ} = -94{,}200\text{ J}$$

This means that, per mole of reaction, $\Delta G^o = -94{,}200$ J mol^{-1}

$$K = e^{-\left(\Delta G^o/RT\right)} = e^{-\left(-94{,}200\text{ J mol}^{-1}/8.3145\text{J mol}^{-1}\text{ K}^{-1}\times 298.15\text{ K}\right)} = e^{+38.00} = \mathbf{3.2 \times 10^{16}}$$

23-3. The Solvay and the Leblanc processes both convert sodium chloride and calcium carbonate into sodium carbonate and other materials. For the Solvay process, the "other materials" are the simple by-product calcium chloride required by the stoichiometry. The Solvay process does employ carbon dioxide and ammonia to form important intermediates, but these are recovered and recycled. It is much easier to produce pure product from the Solvay process, which goes through the selective formation of sodium hydrogen carbonate. The Leblanc process achieves the transformation through the use of expensive sulfuric acid and coke (carbon). These additional reactants also contribute to the by-products calcium sulfide, hydrochloric acid, and sodium hydroxide, all of which must be separated from the sodium carbonate product.

23-5. For this thermochemical analysis, we will treat the reaction in the following manner:

$$2\ NaCl\ (s) + H_2SO_4\ (l) + 2\ C\ (s) + CaCO_3\ (s)$$
$$\rightarrow 2\ HCl\ (g) + CaS\ (s) + 2\ CO_2\ (g) + Na_2CO_3\ (s)$$

With so many reactants and products, it is best to tabulate the enthalpy changes:

Substance	Stoichiometric coefficient n	$\Delta H^o_f\left(\dfrac{kJ}{mol}\right)$	$n \times \Delta H^o_f$
2 HCl	+2 mol	-92.31	-184.62 kJ
CaS	+1 mol	-482.4	-482.4 kJ
2 CO$_2$	+2 mol	-393.51	-787.02 kJ
Na$_2$CO$_3$	+1 mol	-1130.68	-1130.68 kJ
NaCl	-2 mol	-411.15	+822.30 kJ
H$_2$SO$_4$	-1 mol	-813.99	+813.99 kJ
C	-1 mol	0.00	0.00 kJ
CaCO$_3$	-1 mol	-1206.92	+1206.92 kJ
		TOTAL:	**+258.5 kJ**

The reaction is endothermic; it will give more product at higher temperatures.

23-7. Chlorous acid ($pK_a = 1.96$) is much stronger than hypochlorous acid ($pK_a = 7.53$). Only chlorous acid is competent to convert sodium fluoride directly to hydrofluoric acid ($pK_a = 3.18$).

23-9. (a) $2 \, NaCl \, (l) \rightarrow 2 \, Na \, (s) + Cl_2 \, (g)$

(b) The data in the Appendix represent values at 298.15 K. More important, the data in the appendix represent the *minimum* values for the potential (*i.e.*, assuming that $\Delta G = - w_{elec}$). The applied potential must generally be higher. This "overpotential" is needed for kinetic reasons (to make the reaction go faster) and to account for the loss of energy in other ways than in driving the reaction.

23-11. Both calcium and carbonate ions are possible impurities in NaOH made from $Ca(OH)_2$ and Na_2CO_3. Since electrolytic sodium hydroxide is made with only water and sodium chloride, neither calcium nor carbonate will be an impurity.

23-13. The moles of Br_2 available in the sea water is half the number of moles of Br^-:

$$\left(0.0036 \frac{mol \, Br^-}{L}\right) \times \left(1000 \, m^3\right) \times \left(\frac{10 \, dm}{m}\right)^3 \times \left(\frac{1 \, L}{dm^3}\right) \times \left(\frac{1 \, mol \, Br_2}{2 \, mol \, Br^-}\right) = 1800 \, mol \, Br_2$$

There is a 1 : 1 ratio of Cl_2 : Br_2 in this process, so we need 1800 mol Cl_2.

At STP this is: 1800 mol $Cl_2 \times (22.4 \, L \, mol^{-1}) = $ **40,000 L** of gaseous chlorine

23-15. (a) Chlorine has a +1 oxidation state in Cl_2O

(b) The oxyacid formed is the one with $Cl(+1)$ -- **hypochlorous acid**.

23-17. For the disproportionation reaction, the first reaction will be the "cathode", the second the "anode." Then the standard cell potential will be;

$$\Delta \mathcal{E}^o = \mathcal{E}^o \, (ClO_2 \mid ClO_2^-) - \mathcal{E}^o \, (ClO_3^- \mid ClO_2)$$
$$= 0.954 \, V - (-0.25 \, V) = +\textbf{1.204 V}$$

A positive cell potential indicates the reaction is spontaneous. **ClO_2 is unstable** with respect to disproportionation in aqueous base.

23-19. As was discussed with carbon in the text, the small size and high electronegativity of fluorine mean its bond to Si will be stronger than that of Cl.

23-21. (a) *Oxidation of oxide to oxygen:*
$$2 \, SrO \, (s) + 2 \, F_2 \, (g) \rightarrow 2 \, SrF_2 \, (s) + O_2 \, (g)$$

(b) *Oxidation of oxygen to O(+2)*
$$O_2 \, (g) + 2 \, F_2 \, (g) \rightarrow 2 \, OF_2 \, (g)$$

(c) *Oxidation of U(+4) to U(+6), its maximum oxidation state*
$$UF_4 \, (s) + F_2 \, (g) \rightarrow UF_6 \, (s)$$

23-21. (continued)

(d) *Oxidation of chloride to chlorine*

$$2\,NaCl\,(s) + F_2\,(g) \to 2\,NaF\,(s) + Cl_2\,(g)$$

23-23. All we need here is the molar ratio of F to Na, beginning from the mass ratio:

$$\left(\frac{2.48\ g\ F}{1.00\ g\ Na}\right) \times \left(\frac{22.990\ g\ Na}{1\ mol\ Na}\right) \times \left(\frac{1\ mol\ F}{18.998\ g\ F}\right) = \frac{3.00\ mol\ F}{1\ mol\ Na}$$

Therefore, the stoichiometry requires three F's for every Na, something possible only with the acid fluoride **NaF•2 HF**

23-25. The empirical formula of a binary compound can be determined by a ratio analysis, as in problem #21. Here we may assume that we have a sample of exactly 100 g, which will contain 38.35 g Cl and 61.65 g F:

$$\left(\frac{61.65\ g\ F}{38.35\ g\ Cl}\right) \times \left(\frac{35.4527\ g\ Cl}{1\ mol\ Cl}\right) \times \left(\frac{1\ mol\ F}{18.998\ g\ F}\right) = \frac{3.00\ mol\ F}{1\ mol\ Cl}$$

The empirical formula is then ClF_3, which has a formula mass of 92.45.

The gas density data provide a means to directly determine the molar mass of the compound: $4.13\ g\ L^{-1} \times 22.4\ L\ mol^{-1} = 92.5\ g\ mol^{-1}$.

This value is equal to the formula mass. The molecular formula is **ClF_3**.

23-27. (a) OF_2: Bent, with an F-O-F angle slightly less than 109.5^0. SN (oxygen) = 4.

(b) BF_3: Trigonal planar, with F-B-F angles equal to 120^0. SN(boron) = 3.

(c) BrF_3: "T"-shaped, with all the atoms lying in a plane and F-Br-F angles between the central and outer F's of slightly less than 90^0. SN(bromine) = 5.

(d) BrF_5: Square pyramid, with F-Br-F angles slightly less than 90^0. SN(bromine) = 6.

(e) Pentagonal bipyramid, with F-I-F angles in the equatorial plane of $360/5 = 72^0$. SN(I) = 7.

(f) Octahedral, with all F-Se-F angles equal to 90^0. SN(Se) = 6.

23-29. (a)

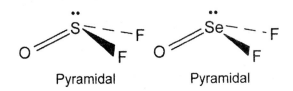

There will be a steric number of 4 in each case about the central atoms.

(b) A Lewis base is defined as a molecule that is capable of donating an electron pair. In this case, thionyl difluoride acts as a **Lewis base.**

Chapter 23

23-31. The autoionization in this case involves the transfer of F^-:

$$2\ BrF_3 \rightleftharpoons BrF_4^- + BrF_2^+$$

This is a Lewis acid-base reaction. An atom and an electron pair on one of the BrF_3 molecules is transferred to the other BrF_3 molecule. The BrF_3 and BrF_2^+ species are Lewis acids.

23-33. The boiling points will increase as the ionic character of the bonding increases.

$$b.p.(PtF_6) < b.p.(PtF_4) < b.p.(CaF_2).$$

23-35. The explosive decomposition of tetrafluoroethylene proceeds according to the equation:

$$C_2F_4\ (g) \rightarrow C\ (s) + CF_4\ (g)$$

For this,
$$\Delta H_r^o = \Delta H_f^o(C\ (s)) + \Delta H_f^o(CF_4\ (g)) - \Delta H_f^o(C_2F_4\ (g))$$

$$= (1\ mol \times 0.00\frac{kJ}{mol}) + (1\ mol \times -925\frac{kJ}{mol}) - (1\ mol \times -651\frac{kJ}{mol})$$

$$= -274\ kJ.\ \text{This is the heat released for 1 mol of } C_2F_4.$$

The molar mass of C_2F_4 is 100.01 g mol^{-1} and 1.00 kg is equal to 1000 g. So the enthalpy change for the decomposition of 1.00 kg of C_2F_4 is:

$$\Delta H = 1000\ g\ C_2F_4 \times \left(\frac{1\ mol\ C_2F_4}{100.01\ g\ C_2F_4}\right) \times \left(\frac{-274\ kJ}{1\ mol\ C_2F_4}\right)$$

$$= -2740\ kJ.$$

Note that the compounds are not in their standard states; this value is an estimat only. This is the opposite to the heat absorbed by the surroundings, **+2740 kJ.**

23-37. The enthalpy change for this reaction includes the quantity we seek, so we can rearrange the expression for ΔH^o to get $\Delta H_f^o(XeF_6)$ (all substances are gases):

$$\Delta H^o = \Delta H_f^o(Xe) + 6\ \Delta H_f^o(HF) - \Delta H_f^o(XeF_6) - 3\ \Delta H_f^o(H_2)$$

Note that $\Delta H_f^o(Xe)$ and $\Delta H_f^o(H_2)$ are zero, so:

$$\Delta H^o = 6\ \Delta H_f^o(HF) - \Delta H_f^o(XeF_6)$$

$$\Delta H^o - 6\ \Delta H_f^o(HF) = -\Delta H_f^o(XeF_6)$$

$$6\ \Delta H_f^o(HF) - \Delta H^o = \Delta H_f^o(XeF_6)$$

When we write this to solve for $\Delta H_f^o(XeF_6)$, we get:

$$1\ mol\ XeF_6 \times \Delta H_f^o(XeF_6) = 6\ \Delta H_f^o(HF) - \Delta H^o$$

$$= (6\ mol \times -271.1\frac{kJ}{mol}) - (-1282\ kJ) = -345\ kJ$$

$$\Delta H_f^o(XeF_6) = \mathbf{-345\ \frac{kJ}{mol}}$$

23-37. (continued)

(b) For thermochemical purposes, we may calculate the enthalpy of formation for XeF_6 by comparing the bond enthalpies for the Xe-F bonds formed and the F-F bonds destroyed.

We can consider the formation of Xe-F as a two step cycle. In the first step, we "invest" the enthalpy of 3 F-F bonds and in the second step we "gain" the enthalpy of six Xe-F bonds.

$$3 \text{ F-F} \rightarrow 6 \text{ F} \qquad \Delta H_1 = 3 \times \text{F-F enthalpy} = 3 \text{ mol} \times 158 \frac{\text{kJ}}{\text{mol}} = +474 \text{ kJ}$$

$$6 \text{ F} + \text{Xe} \rightarrow XeF_6 \qquad \qquad \Delta H_2 = -6 \times \text{Xe-F enthalpy}$$
$$3 \text{ F-F} + \text{Xe} \rightarrow XeF_6 \qquad \qquad \Delta H = \Delta H_1 + \Delta H_2 = \Delta H^o_f (XeF_6)$$

This last equation can be rearranged to obtain ΔH_2:

$$\Delta H_2 = \Delta H^o_f (XeF_6) - \Delta H_1 = (1 \text{ mol} \times -345 \frac{\text{kJ}}{\text{mol}}) - 474 \text{ kJ} = -819 \text{ kJ}$$

Then:
$$\Delta H_2 = -6 \times \text{Xe-F enthalpy} = -819 \text{ kJ}$$
$$\text{Xe-F bond enthalpy} = -(^{819}/_6) = \mathbf{136 \text{ kJ}}$$

23-39. $\Delta H^o_r = \Delta H^o_f (XeOF_4 \ (l)) + 2 \Delta H^o_f (HF \ (g)) - \Delta H^o_f (XeF_6 \ (g)) - \Delta H^o_f (H_2O \ (l))$

$$= (1 \text{ mol} \times 148 \frac{\text{kJ}}{\text{mol}}) + (2 \text{ mol} \times -271.1 \frac{\text{kJ}}{\text{mol}})$$

$$- (1 \text{ mol} \times -298 \frac{\text{kJ}}{\text{mol}}) - (1 \text{ mol} \times -285.83 \frac{\text{kJ}}{\text{mol}}) = \mathbf{+190 \text{ kJ}}$$

23-41. <u>Oxidation:</u> This is the opposite of the reduction of permanganate to manganese(II) ion, which was treated in problem 13-9b:

$$Mn^{2+} \ (aq) + 4 \ H_2O \ (l) \rightarrow 5 \ e^- + 8 \ H^+ \ (aq) + MnO_4^- \ (aq)$$

<u>Reduction:</u> $\qquad\qquad\qquad\qquad\qquad XeO_6^{4-} \rightarrow Xe$

Add H_2O to balance O: $\qquad\qquad XeO_6^{4-} \rightarrow Xe + 6 \ H_2O$

Add H^+ to balance H: $\qquad 12 \ H^+ + XeO_6^{4-} \rightarrow Xe + 6 \ H_2O$

Add e^- to balance charge: $\qquad 8 \ e^- + 12 \ H^+ + XeO_6^{4-} \rightarrow Xe + 6 \ H_2O$

Add appropriate phase descriptions:

$$8 \ e^- + 12 \ H^+ \ (aq) + XeO_6^{4-} \ (aq) \rightarrow Xe \ (g) + 6 \ H_2O \ (l)$$

<u>Overall:</u> The least common multiple of 5 and 8 is 40, so we must multiply the oxidation reaction by 8 and the reduction reaction by 5. When common substances are removed from both sides of the reaction, we get:

$$5 \ XeO_6^{4-} \ (aq) + 8 \ Mn^{2+} \ (aq) + 2 \ H_2O \ (l) \rightarrow 4 \ H^+ \ (aq) + 8 \ MnO_4^- \ (aq) + 5 \ Xe \ (g)$$

23-43. Bromine will also be an oxidizing bleach, but because bromine is a weaker oxidant than chlorine ($\mathcal{E}^o$ (Cl_2 | Cl^-) = +1.358 V and $\mathcal{E}^o$ (Br_2 | Br^-) = +1.065 V), it will be a weaker bleach than chlorine.

23-45. $\Delta H^o = \Delta H^o_f$ (Cl_2 (g)) + ΔH^o_f (H_2O (g)) - $^1/_2$ ΔH^o_f(O_2 (g)) - 2 ΔH^o_f(HCl (g))

$$= (1 \text{ mol} \times 0.00\frac{kJ}{mol}) + (1 \text{ mol} \times -241.82\frac{kJ}{mol})$$

$$- (^1/_2 \text{ mol} \times 0.00\frac{kJ}{mol}) - (2 \text{ mol} \times -92.31\frac{kJ}{mol}) = -57.2 \text{ kJ} = -57,200 \text{ J}$$

$\Delta S^o = S^o$ (Cl_2 (g)) + S^o (H_2O (g)) - 1/2 S^o(O_2 (g)) - 2 S^o(HCl (g))

$$= (1 \text{ mol} \times 222.96\frac{J}{mol\text{-}K}) + (1 \text{ mol} \times 188.72\frac{J}{mol\text{-}K})$$

$$- (^1/_2 \text{ mol} \times 205.03\frac{J}{mol\text{-}K}) - (2 \text{ mol} \times 186.80\frac{J}{mol\text{-}K}) = -64.44\frac{J}{K}$$

The reaction is run at 450 ^{o}C, which is equal to 723 K. Assuming that ΔH and ΔS do not change from their standard state values:

$$\Delta G \approx \Delta H^o - T\,\Delta S^o = -57,200 \text{ J} - (723 \text{ K} \times -64.44\frac{J}{K}) = -10,610 \text{ J}$$

This is equal to a free energy change per mole of reaction of -10,610 J mol^{-1}
The equilibrium constant is then

$$K = e^{-(\Delta G/RT)} = e^{-\left(-10,610 \text{ J mol}^{-1}/8.3145 \text{ J mol}^{-1}\text{ K}^{-1} \times 723 \text{ K}\right)} = e^{+1.765}$$
$$= \mathbf{5.84}$$

(b) The H-Br bond is weaker than the H-Cl bond by 66 kJ mol^{-1}. The Br-Br bond is weaker than the Cl-Cl bond by 50 kJ mol^{-1}. Although we are making a weaker bond in Br_2, it is so much easier to break H-Br bonds that we expect the reaction to have a **larger** equilibrium constant with bromine.

23-47. The chlor-alkali reaction is given in the text as:

2 H_2O (l) + 2 Cl^- (aq) $\rightarrow$ 2 OH^- (aq) + H_2 (g) + Cl_2 (g)

$$\Delta G^o = +422 \text{ kJ} = +422,000 \text{ J}.$$

For this reaction, the number of electrons transferred $n = 2$. The standard cell potential for this reaction is:

$$\Delta\mathcal{E}^o = -\frac{\Delta G^o}{n\,\mathcal{F}} = -\frac{422,000 \text{ J}}{\left(2 \text{ mol e}^-\right) \times (96,485 \text{C (mol e}^-)^{-1})} = -2.19\frac{J}{C} = \mathbf{-2.19 \text{ V.}}$$

A potential of at least 2.19 V must be applied to the cell to cause the reaction to proceed from left to right.

23-49. The article must be referring to the weight of the product. There are 2 moles of sodium hydroxide (caustic soda) made for every 1 mole of Cl_2.

One mole of Cl_2 has a mass of 70.9 g.

Two moles of NaOH have a mass of 2 mol × 40.00 g mol^{-1} = 80.00 g.

The theoretical mass ratio of NaOH : Cl_2 is 80.0 : 70.9 = 1.13. Thus, we expect about 13% more NaOH than Cl_2, by weight. The "as usual" must also refer to a typically lower actual yield of NaOH.

23-51. The blue black color is a sure sign of iodine present. It shows up because the spaghetti left behind some starch in the pot, and iodine and starge form a blue black complex. The boiling drove off the iodine in the first case (before the cooking) because the I_2 present turned into I_2 (g).

23-53. We can work this out "retrosynthetically" by going backwards from F_2 to CaF_2.

$$F_2 \quad \Rightarrow \quad HF \quad \Rightarrow \quad CaF_2$$

We can then reconstruct the synthesis in the forward direction:

Extraction of HF from CaF_2 (fluorspar):

$$CaF_2\ (s) + H_2SO_4\ (aq) \rightarrow CaSO_4\ (s) + 2\ HF\ (g)$$

Electrolysis of HF: $\qquad\qquad\qquad$ $2\ HF\ (g) \rightarrow F_2\ (g) + H_2\ (g)$

23-55. The fluoride ion is a Lewis base because it has lone pair electrons to donate to other atoms and molecules. In accepting two fluoride ions TiF_4 acts as a Lewis acid.

23-57. There are several non-standard units in this problem. Let's clarify them first.

The market value of gold is $350 (troy oz)$^{-1}$ × ($^{1\ \text{troy oz}}/_{31.3\ g}$) = $11.18 g^{-1}.

A 1000 g piece of quartz has 1000 g × 0.001% = 1000 g × 0.00001 = 0.01 g Au and essentially 1000 g of SiO_2. The value of the gold in the quartz is then $11.18 g^{-1} × 0.01 g = $0.11.

The reaction has to destroy all of the SiO_2 in the quartz. The mass of HF required is:

$$1000\ \text{g SiO}_2 \times \left(\frac{1\ \text{mol SiO}_2}{60.065\ \text{g SiO}_2}\right) \times \left(\frac{4\ \text{mol HF}}{1\ \text{mol SiO}_2}\right) \times \left(\frac{20.006\ \text{g HF}}{\text{mol HF}}\right) = 1332\ \text{g HF}$$

The HF is sold as a commercial solution. The density of the solution is 1.17 g cm^{-3}, or 1,170 g L^{-1}. Then the amount of the solution needed is:

$$1332\ \text{g HF} \times \left(\frac{100\ \text{g solution}}{50\ \text{g HF}}\right) \times \left(\frac{1\ \text{L solution}}{1,170\ \text{g solution}}\right) = 2.28\ \text{L solution}$$

This will cost 2.28 L × $0.25 L^{-1} = $0.57. This is more than the value of the gold that could be recovered.

23-59.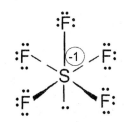

The bond order for all the bonds is 1. The geometry will be bent around both oxygens, with O-O-F angles slightly less than 109.5°. It is analogous to the compound of oxygen and hydrogen called hydrogen peroxide.

23-61. The geometry is a square pyramid, based on an SN = 6 for the sulfur with one lone pair. The F-S-F angles will be less than 90°, because of the extra volume of the lone pair on the sulfur.

23-63. Teflon is a polymer of tetrafluoroethylene. It has no chlorine. The empirical formula of teflon is CF_2, with a formula mass of 50.007 g mol^{-1}. The theoretical mass fraction of fluorine is:

$$\left(\frac{2 \text{ mol F}}{1 \text{ mol } CF_2}\right) \times \left(\frac{18.998 \text{ g F}}{1 \text{ mol F}}\right) \times \left(\frac{1 \text{ mol } CF_2}{50.007 \text{ g } CF_2}\right) = \frac{0.7598 \text{ g F}}{1 \text{ g } CF_2}.$$

This corresponds to a mass percentage of 76.0%. There was far too little fluorine found in the original sample, possibly because of the challenge of completely degrading such a famously inert substance as Teflon.

23-65. The standard molar free energy of formation of XeF_4 *is* the standard free energy change for this reaction, because ΔG_f^o for both Xe and F_2 will be zero. Thus,

$$\Delta G_r^o = \Delta G_f^o(XeF_4 \; (g)) = 1 \text{ mol} \times -134 \frac{\text{kJ}}{\text{mol}} = -134 \text{ kJ} = -134,000 \text{ J}.$$

This is equal to a free energy change per mole of reaction of -134,000 J mol^{-1}

Then:

$$K = e^{-\left(\Delta G^o / RT\right)} = e^{-\left(-134,000 \text{ J mol}^{-1} / 8.3145 \text{ J mol}^{-1} \text{ K}^{-1} \times 298 \text{ K}\right)} = e^{+54.1}$$
$$= 3.07 \times 10^{23}$$

CHAPTER 24 - FROM PETROLEUM TO PHARMACEUTICALS

24-1. An octane number of 100 means a gasoline burns with the same amount of knocking as the reference compound *iso*-octane. If a compound burns with *less* knocking than *iso*-octane, then it will have an octane number exceeding 100.

24-3. We know that for each mole of the alkane we obtain two moles of carbon dioxide. This means that each mole of the alkane contains two moles of carbon. So we have the alkane containing two carbon atoms. The formulas of alkanes are always C_nH_{2n+2}. Since $n = 2$, we have $2n + 2 = 6$, and the formula must be C_2H_6. The alkane is **ethane**.

$$2 \ C_2H_6 \ (g) + 7 \ O_2 \ (g) \rightarrow 4 \ CO_2 \ (g) + 6 \ H_2O \ (g)$$

24-5. (a) The formation of cyclopropane can be thought of as occurring through the atomization of C and H and then the recombination of these atoms into C_3H_6.

ATOMIZATION STEPS

$$3 \ H_2 \ (g) \rightarrow 6 \ H \ (g) \quad 6 \ \Delta H_f^o(H, g) = 6 \ \text{mol} \times +217.96 \ \frac{kJ}{mol} \quad = +1307.76 \quad kJ$$

$$3 \ C \ (s) \rightarrow 3 \ C \ (g) \quad 3 \ \Delta H_f^o(C, g) = 3 \ \text{mol} \times +716.682 \ \frac{kJ}{mol} \quad = +2150.046 \quad kJ$$

BOND FORMING STEPS

$$3 \ C \ (g) + 3 \ H \ (g) \rightarrow 6 \ \text{C-H} \quad -6 \ \text{B.E.(C-H)} = -6 \ \text{mol} \times 413 \ \frac{kJ}{mol} \quad = -2478 \quad kJ$$

$$3 \ C \ (g) \rightarrow 3 \ \text{C-C} \quad -3 \ \text{B.E.(C-C)} = -3 \ \text{mol} \times 348 \ \frac{kJ}{mol} \quad = -1044 \quad kJ$$

$$\text{TOTAL } \Delta H^o \quad = \quad \textbf{-64} \quad kJ$$

(b) The combustion reaction will have an enthalpy change related to the standard enthalpy of formation of the reactants and products (all substances are gases):

$$\Delta H^o = 3 \times \Delta H_f^o(CO_2) + 3 \times \Delta H_f^o(H_2O) - 3 \times \Delta H_f^o(O_2) - \Delta H_f^o(C_3H_6)$$

$$1 \times \ \Delta H_f^o(C_3H_6) = 3 \times \Delta H_f^o(CO_2) + 3 \times \Delta H_f^o(H_2O) - 3 \times \Delta H_f^o(O_2) - \Delta H^o$$

$$= 3 \ \text{mol} \times (-393.51 \frac{kJ}{mol}) + 3 \ \text{mol} \times (-241.82 \frac{kJ}{mol})$$

$$- 3 \ \text{mol} \times (0.0 \frac{kJ}{mol}) - (-1959 \ kJ) = \textbf{+53 kJ}$$

$$\Delta H_f^o(C_3H_6) = \textbf{+53 kJ mol}^{\textbf{-1}}$$

(c) Cyclopropane lies at an enthalpy **+117** kJ mol^{-1} higher than expected on the basis of "normal" bond enthalpies. This represents the compound's strain energy.

24-7. Decane has 10 carbon atoms, so the products of cracking into molecules with equal numbers of carbon atoms will give two chains of 5 carbons each. The alkane product will have the formula $C_nH_{2n+2} = C_5H_{12}$. The alkene product will have the formula $C_nH_{2n} = C_5H_{10}$. The chemical equation for this is: $C_{10}H_{22} \rightarrow C_5H_{10} + C_5H_{12}$
(b) Assuming the product alkene is a straight chain, there are two possible isomers: 1-pentene ($H_2C=CHCH_2CH_2CH_3$) and 2-pentene ($H_3CCH=CHCH_2CH_3$). 2-pentene will also exist as *cis* and *trans* isomers.

24-9. (a)

(b)

(c)

(d)

(e)

(f)

(g)

(h)

24-11.

24-13. (a) 1,2-hexadiene

(b) 1,3,5-hexatriene

(b) 2-methyl-1-hexene

(d) 3-hexyne

24-15. Hybridization for carbon can be assigned based on the number of bound atoms:

$$2 \text{ bound atoms} \ldots\ldots\ldots sp \text{ hybrid}$$
$$3 \text{ bound atoms} \ldots\ldots\ldots sp^2 \text{ hybrid}$$
$$4 \text{ bound atoms} \ldots\ldots\ldots sp^3 \text{ hybrid}$$

(a) The first and third carbon atoms (counting from the left) are sp^2 hybridized. The second carbon is sp hybidrized. The other carbons are sp^3 hybridized

(b) All the carbons are sp^2 hybridized.

(c) The first and second carbon atoms (counting from the left) are sp^2 hybridized. The other carbons, including the methyl group, are sp^3 hybridized

(d) The third and fourth carbon atoms are sp hybridized. The other carbons are sp^3 hybridized

24-17.

24-19. (a)   *Addition of bromine:* $CH_2{=}CH_2 + Br_2 \rightarrow CH_2BrCH_2Br$

          Removal of HBr: $CH_2BrCH_2Br \rightarrow CH_2{=}CHBr + HBr$

    (b)   *Addition of water:* $CH_3CH_2CH{=}CH_2 + H_2O \rightarrow CH_3CH_2CH_2CH_2OH$

      (requires acid catalyst)

    (c) *Oxidation of propene*: $CH_3CH{=}CH_2 + 1/2\ O_2 \rightarrow CH_3COCH_3$

24-21.

$$R\text{-}COOH + HO\text{-}CR_3'' \rightarrow RCOOCR_3''$$

The use of R denotes a variable group; most any carboxylic acid will under some reaction like this. The R'' designation on the alcohol indicates that it is a *different* group, anything but hydrogen.

24-23. In large scale problems like this, it is often best to work entirely in kilograms instead of grams. The molar mass of ethylene is 28.0538 g mol^{-1}, or 0.0280538 kg mol^{-1}, and the molar mass of ethylene dichloride ($C_2H_4Cl_2$) is 98.959 g mol^{-1}, or 0.098959 kg mol^{-1}. The synthesis of one mole of ethylene dichloride requires one mole of ethylene.

The mass of ethylene required to yield 6.26×10^9 kg of ethylene dichloride is:

$$6.26 \times 10^9 \text{ kg } C_2H_4Cl_2 \times \left(\frac{1 \text{ mol } C_2H_4Cl_2}{0.098959 \text{ kg } C_2H_4Cl_2} \right) \times \left(\frac{1 \text{ mol } C_2H_4}{1 \text{ mol } C_2H_4Cl_2} \right) \times \left(\frac{0.0280538 \text{ kg } C_2H_4}{1 \text{ mol } C_2H_4} \right)$$

$$= 1.775 \times 10^9 \text{ kg } C_2H_4.$$

This represents $\dfrac{1.775 \times 10^9 \text{ kg } C_2H_4}{15.87 \times 10^9 \text{ kg } C_2H_4} = 0.1118 = \textbf{11.2\%}$ of all production

The synthesis of ethylene dichloride involves the addition of chlorine to ethylene. Therefore, the mass of product that does *not* come from ethylene must come from chlorine.

mass Cl_2 + mass C_2H_4 = mass $C_2H_4Cl_2$

$$\text{mass } Cl_2 = \text{mass } C_2H_4Cl_2 - \text{mass } C_2H_4$$

$$= 6.26 \times 10^9 \text{ kg } - 1.775 \times 10^9 \text{ kg } C_2H_4$$

$$= \textbf{4.49} \times \textbf{10}^9 \textbf{ kg } \textbf{Cl}_2$$

24-25. Ethanol is an alcohol, which means it has an OH bond. This is very important to the phase properties of this molecule. It produces a strong H-bond in the liquid and condensed phase. In addition, the hydrogen bonding matches well with the hydrogen bonding that is so important in water, increasing ethanol's solubility

24-27.

(a)

acid: (structure shown)

CH₃—O ... CH, CH₃, CH, CH₂, CH₂, CH₂, CH₂, CH, CH, CH, C, C, CH₃, CH₃, O, OH, CH₃

alcohol: CH₃—CH(OH)—CH₃ 2-propanol

(b)

H₃C—O ... CH, CH₃, CH, CH₂, CH₂, CH₂, CH₂, CH, CH, C, CH₃, CH₃, CH₃, CH₃

24-29. (a) Aspirin has a molecular formula of $C_9H_8O_4$. The molar mass of this is 180.16 g mol⁻¹.

(b) For 325 mg of aspirin, there will be 0.325 g of aspirin, containing

$$\frac{0.325 \text{ g aspirin}}{180.16 \text{ g mol}^{-1} \text{ aspirin}} = 1.80 \times 10^{-3} \text{ mol aspirin}$$

24-31.

Remove methyl group and convert to ketone

Convert to ketone

Dehydrogenate to ketone

Cut off side chain; add OH group

Add OH group

Move double bond

(steroid structure with labels CH₃, CH₂, CH, C, H₂C, HO, etc.)

24-33. It is tempting to mention the mass of the molecules, but gravity has nothing to do with boiling point. A steady increase in the boiling point means there are stronger intermolecular forces. This is explained because the van der Waals forces increase with the larger size of the molecules.

24-35. Combustion of heptane:C_7H_{16} (g)+ 11 O_2 (g)→ 7 CO_2 (g) + 8 H_2O (g)

Combustion of isooctane:C_8H_{18} (g) + $11\frac{1}{2}$ O_2 (g) → 7 CO_2 (g) + 9 H_2O (g)

In determining the standard enthalpy change for the combustion reactions, we may ignore the oxygen, which has a $\Delta H_f^o = 0$.

The standard enthalpy for combustion of heptane (all substances are gases):

$$\Delta H^o = 7 \, \Delta H_f^o(CO_2) + 8 \, \Delta H_f^o(H_2O) - \Delta H_f^o(C_7H_{16})$$

$$= (7 \text{ mol} \times -393.51 \frac{kJ}{mol}) + (8 \text{ mol} \times -241.82 \frac{kJ}{mol}) - (1 \text{ mol} \times -187.82 \frac{kJ}{mol})$$

=- 4501.31 kJ

The standard enthalpy for combustion of heptane (all substances are gases):

$$\Delta H^o = 8 \, \Delta H_f^o(CO_2) + 9 \, \Delta H_f^o(H_2O) - \Delta H_f^o(C_7H_{16})$$

$$= (8 \text{ mol} \times -393.51 \frac{kJ}{mol}) + (9 \text{ mol} \times -241.82 \frac{kJ}{mol}) - (1 \text{ mol} \times -224.13 \frac{kJ}{mol})$$

= - 5100.33 kJ

(b) Note that this question only asks which give more heat per gallon. We need to determine the relative heat per volume, and this will not be keyed to any unit system. So we might as well ask the relative heat delivered per liter. Moreover, the relative densities will be independent of the units used, so if the relative mass of one gallon of isooctane : one gallon heptane = 5.77 lb : 5.71 lb = 1.0105 : 1, then this will be the ratio of the relative mass of one liter of each.

$$\frac{\Delta H \, L^{-1} \text{ isooctane}}{\Delta H \, L^{-1} \text{ heptane}} = \frac{\Delta H \text{ mol}^{-1} \text{ isooctane}}{\Delta H \text{ mol}^{-1} \text{ heptane}} \times \frac{g \text{ mol}^{-1} \text{ heptane}}{g \text{ mol}^{-1} \text{ isooctane}} \times \frac{g \, L^{-1} \text{ isooctane}}{g \, L^{-1} \text{ heptane}}$$

$$= \frac{5100.33 \text{ kJ}}{4501.31 \text{ kJ}} \times \frac{100.204 \text{ g mol}^{-1}}{114.231 \text{ g mol}^{-1}} \times 1.0105$$

$$= 1.0043$$

There is **0.4%** more heat given off in the combustion of 1 gallon of isooctane than in the combustion of one gallon of heptane.

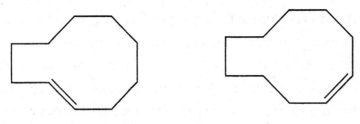

trans-cyclododene   *cis*-cyclododene

24-37.

24-39. The alkene content can be increased by catalytic cracking of longer chain hydrocarbons into shorter chain alkenes and alkanes (see problem 25-5b). Both alkenes and aromatics can also be made from alkanes by dehydrogenation, removing H_2 from one or more C-C bond to give C=C double bonds.

24-41. *Dehydration* is the removal of *water*, as in the conversion of an alcohol to an alkene: $RCH_2CH_2OH \rightarrow H_2O + RCH=CH_2$.

Dehydrogenation is the removal of *hydrogen*, as in the conversion of heptane into heptene: $C_5H_{11}CH_2CH_3 \rightarrow H_2 + C_5H_{11}CH=CH_2$.

24-43. (a)

(b) The reaction of glycerol and stearic acid to give tristearin is an esterification reaction, whether three ester linkages are formed from the acid groups and three alcohol groups. Tristearin is a tri-ester.

24-45. Both pharmaceuticals and pesticides must be biologically active, preferably in as specific a manner as possible. In many cases, the strategy for delivery of the compound to the target is similar: utilizing the organism's own pathways for taking in compounds as a means for getting it to accept a molecule that doesn't, in a basic sense, "belong" in its system. Hence, mimicking the organism's own chemicals is important in delivery of both kinds of chemicals.

Also, the development of both requires a careful examination of the purpose of the compound and also any possible side-effects. Pharmaceuticals are under particular pressure to do one and only one thing in the organism they are given to. This objective may be different for different drugs, and one must carefully consider how to get the pharmaceutical to the target tissue, and what will happen to the pharmaceutical in the body's metabolism. This is in contrast to the objective of a pesticide. It has a very simple objective: killing the target by any means possible. Therefore, pesticides can have a variety of effects, and this makes many of them genuinely broad spectrum in their target. However, side effects outside of the target organism may be more important than the short term goal of killing a pest. Hence, many pesticides are toxic to other organisms, and this severely restricts their usefulness.

CHAPTER 25 - POLYMERS: NATURAL AND SYNTHETIC

25-1. Addition polymerization is the linkage of many monomer units without loss of any atoms. For 1,1-dichloroethylene, this reaction is:

$$n \, Cl_2C=CCH_2 \rightarrow (CCl_2CH_2)_n$$

25-3. The starting monomer is the empirical formula of the polymer, if we assume growth by addition polymerization. In this case the monomer would be **CH_2O**, which is the molecular formula of formaldehyde.

25-5. (a) Condensation polymerization is the linkage of many monomer units *with* loss of some atoms. Polyglycine is built up of peptide linkages, themselves formed from an amino group and a carboxylic acid group with the splitting off of **water**:

$$NH_2-CH_2-COOH \; + \; NH_2-CH_2-COOH$$

$$\longrightarrow \; H_2O \; + \; NH_2-CH_2-CO-NH-CH_2-COOH$$

(b) The repeat unit is (NH-CH_2-$C(O)$), which has a formula of C_2H_3NO. Note that this has a formula of glycine ($C_2H_5NO_2$) - water (H_2O). The structure of two repeat units is:

25-7. The initiator reacts with one monomer unit to form the "living polymer" based on continuous regeneration of the same anion for further polymerization. For sodium amid and acrylonitrile, this will be:

$$NaNH_2 + H_2C=CHCN \rightarrow H_2NCH_2CH^-(CN) \, Na^+$$

The same anionic functional group is generated by butyl lithium and acrylonitrile.

25-9. The first thing that happens is an acid-base reaction, where hydrogen ions from the COOH groups transfer to the NH_2 group. The structure will be:

25-11.

25-13. The hydrophilic groups will have an OH, NH_2, or an SH group. These are serine, threonine, lysine, arginine, histidine, aspartic acid, glutamic acid, asparagines, glutamine, and cysteine. Of course, the amino and carboxylic acid groups of all amino acids will interact well with water, but this question is looking for an extra interaction in the side chain.

25-15. Because a polypeptide has a direction to it -- determined by the carboxylic acid end and the amino end of the molecule -- the three positions in a linear peptide are unique (this is not true for circular polypeptides, however). Therefore, the sequence "A-B-C," where A, B, and C are three different amino acids, is different from the sequence "C-B-A."

In such a case, we can determine the total number of possible sequences by considering the probability of any single one of them. There is a $1/3$ rd chance of finding a given amino acid at any one position. So the probability of finding three specific amino acids at three specific positions is $1/3 \times 1/3 \times 1/3 = 1/27$.

If any one sequence has a probability of $1/27$ of being present, there must be 27 different sequences. This reasoning is similar to that used in problem 12-67, which dealt with flips of a coin.

25-17. The pentapeptide in question has the following structure.

The strongest and most important interactions between a molecule and a solvent will involve the functional groups on the outside of the molecule. For a polypeptide, these groups are the amino acid side chains, outlined above. In this example they contain carbon and hydrogen only. These will interact more strongly with a hydrocarbon solvent, in this case **octane**.

25-19. In its ring form, β-galactose will have the form shown below. The five asymmetric carbon centers are marked with asterisks.

25-21. A polymer built of phenylalanine units will be made by condensation polymerization. Recall from problem No. 5(b) that the empirical formula for such a polymer is the empirical formula of the monomer minus water.

In this case: polymer formula = monomer formula - water

= phenylalanine - water

= $C_9H_{11}NO_2$ - H_2O = **C_9H_9NO**

The molar mass of the repeat unit is 147.2 g mol^{-1}. The number of repeat units in the chain is $17,500 \div 147.2 = $ **119**.

25-23.

25-25. (a) Starch and cellulose are both condensation polymers of glucose. As in No. 15:

polymer formula = monomer formula - water

= glucose - water

= $C_6H_{12}O_6$ - H_2O = **$C_6H_{10}O_5$**

25-27. Nylon-66 fiber is made from the condensation polymerization of adipic acid and hexamethylenediamine:

Note that the repeat unit of the polymer, indicated in the product, has a formula of $C_{12}H_{22}O_2N_2$. The molar mass of this repeat unit is 226.32 g. Each repeat unit is built of one adipic acid and one hexamethylenediamine. Note also that 1.00×10^3 kg = 1.00×10^6 g.

$$\text{Moles of repeat unit} = \frac{1.00 \times 10^6 \text{ g repeat unit}}{226.32 \text{ g mol}^{-1} \text{ repeat unit}} = 4418.5 \text{ mol repeat unit}$$

Chapter 25

25-27. (continued)

This is also the number of moles of adipic acid (molar mass 146.14 g mol^{-1}) and hexamethylenediamine (molar mass 116.21 g mol^{-1}) needed to make the nylon. Therefore:

$$4418.5 \text{ mol adipic acid} \times 146.14 \text{ g mol}^{-1} = \textbf{646,000 g adipic acid}$$
$$4418.5 \text{ mol hexamethylenediamine} \times 116.21 \text{ g mol}^{-1} =$$

$$\textbf{513,000 g} \text{ hexamethylenediamine}$$

25-29. A branch point in a polymer occurs when a single carbon atom is bound to more than two chains, usually when it has three or more other carbon atoms bound to it. In diamond, each carbon is bound to four other carbons, forming four "chains" branching off each carbon atom.

25-31. Because polyethylene is an addition polymer, the mass of ethylene required for to make polyethylene is equal to the mass of the product (assuming 100% yields). In this case, then, 4.37 billion kilograms of ethylene is needed for all the low-density polyethylene. Note that this mass is equal to 4.37×10^9 kg or 4.37×10^{12} g of ethylene. Then:

$$4.37 \times 10^{12} \text{ g } C_2H_4 \times \left(\frac{1 \text{ mol } C_2H_4}{28.05376 \text{ g } C_2H_4} \right) \times \left(\frac{22.4 \text{ L } C_2H_4}{1 \text{ mol } C_2H_4} \right) = 3.49 \times 10^{12} \text{ L}$$

There are 1000 L in a cubic meter, so this is equal to 3.49×10^9 m^3, the equivalent of a 1 meter high bubble of ethylene gas over an area 59 km on a side.

25-33. The butyl lithium is needed to create the first side of an anionic living polymer based on acrylonitrile. Once butyl lithium adds to an acrylonitrile monomer, then polymerization can proceed. The butyl lithium is not released; it is a permanent "cap" on the polymer. Hence, though it may form a very small part of the mass of the product polymer, the butyl lithium **is consumed** in the reaction, so it is not a catalyst.

25-35. For these polymers, the direction of growth may be as important as the stoichiometry. An arrow is used to indicate this in the following diagrams of a polymer based on butadiene ("B") and vinylchloride ("V"):

RANDOM:

$$\rightarrow V \rightarrow B \rightarrow B \rightarrow V \rightarrow V \rightarrow B \rightarrow V \rightarrow B \rightarrow$$

BLOCK:

$$all\ V \rightarrow V \rightarrow V \rightarrow V \rightarrow V \rightarrow B \rightarrow B \rightarrow B \rightarrow B \rightarrow all\ B$$

GRAFT:

$$B \rightarrow B \rightarrow B \rightarrow B \rightarrow B \rightarrow all\ B$$

$$all\ V \rightarrow V \rightarrow V \rightarrow V \rightarrow V \rightarrow V \rightarrow V \rightarrow V \rightarrow V \rightarrow all\ V$$

$$B \rightarrow B \rightarrow all\ B$$

25-37. This is the same as problem #9 earlier, except there is now a $1/2$ chance of finding a given amino acid at a given site. The probability of any single polypeptide being formed is then:

$$(1/2) \times (1/2) \times (1/2) \times (1/2) \times \ldots \times (1/2) = (1/2)^{22} = 1/4,194,304.$$

By the same rationale given earlier, there are 4,194,304 different polypeptides that can form.

25-39. The mass of Fe in one mole of hemoglobin is:

$$65,000 \text{ g hemoglobin} \times \frac{0.00344 \text{ g Fe}}{1 \text{ g hemoglobin}} = 223.6 \text{ g Fe}$$

This corresponds to $223.6 \text{ g} / 55.847 \text{ g mol}^{-1} = 4.00 \text{ mol Fe}$.

There are four moles of iron in one mole of hemoglobin, so there are four atoms of iron in one molecule of hemoglobin.

25-41. L-sucrose is the enantiomer of D-sucrose, so the structure of L-sucrose is:

25-43. The name ethylene dichloride refers to the chemical "$C_2H_4Cl_2$" (see problem No. 24-21). Its conversion to vinyl chloride involves the loss of one mole of HCl per mole of vinyl chloride. We can then think of the synthesis of PVC by the *overall* equation:

$$n \, C_2H_4Cl_2 \rightarrow (C_2H_3Cl)_n + n \, HCl$$

We need to determine the mass of PVC that could theoretically be obtained from the indicated amount of ethylene dichloride. For this problem, we might as well remain in the mass units given -- pounds.

The relative mass of equal chemical amounts of ethylene dichloride and PVC will be equal to the ratio of the molar mass of their empirical units.

$$\text{theoretical yield of PVC} = \text{mass of ethylene dichloride} \times \frac{\text{mass } C_2H_3Cl}{\text{mass } C_2H_4Cl_2}$$

$$= 950 \text{ lb } C_2H_4Cl_2 \times \frac{62.498 \text{ lb } C_2H_3Cl}{98.0592 \text{ lb } C_2H_4Cl_2}$$

$$= 600 \text{ lb } C_2H_3Cl = \textbf{600 lb PVC}$$

The highest percentage yield would then be $^{550}/_{600} = \textbf{91.7\%.}$
The lowest percentage yield would then be $^{500}/_{600} = \textbf{86.7\%.}$

25-45. The major similarity in the formation of the major varieties of polypropylene and the major varieties of polyisoprene is that the stereochemistry of the growing chain is most important. But whereas in polypropylene this is the stereochemistry about a tetrahedral sp^3 hybridized carbon, in polyisoprene this is the stereochemistry about a carbon-carbon double bond. The relevant structures are (note that a wiggly line indicates variable stereochemistry).

25-45. (continued)

isotactic

syndiotactic

atactic

all *cis*

all *trans*

mixed *cis* and *trans*

APPENDIX A - SCIENTIFIC NOTATION AND EXPERIMENTAL ERROR

A-1. (a) 5.82×10^{-4} (b) 1.402×10^3

        (c) 7.93   (d) -6.59300×10^3

        (e) 2.530×10^{-3} (f) 1.47

A-3. (a) 0.000537   (b) 9,390,000

        (c) -0.00247   (d) 0.006020

        (e) 20,000

A-5. All four are given from biggest to smallest.

1.90×10^2	>	9.7×10^{-2}	>	-4.10×10^{-2}	>	-4.10×10^{-2}
190	>	0.097	>	-0.00410	>	-490

A-7. 746,000,000 kg

A-9. (a) The **135.64 g** value seems much too large. It should be discarded.

        (b) The average of the six valid measurements is **111.34 g**.

        (c) The standard deviation σ is **0.22 g**.

A-11. The measurements in problem 9 are reported to a precision of one part out of 10^5. The measurements in problem 10, to one part out of 10^3. The measurements in problem 9 are more precise.

A-13. (a) Five   (b) Three   (c) Two or three

        (d) Three   (e) Four

A-15. We will retain the notation system used in each part.

        (a) 14 L   (b) -0.0034 °C   (c) 34016

        (d) 3.4×10^2 miles   (e) 6.2×10^{-27} J

A-17. 2,997,215.55

A-19. We report the "raw" value first, then the value to the proper number of significant figures.

        (a) $-167.254 = -167.25$

        (b) $75.50 = 76$

A-19. (continued)

(c) A clearer layout of this is:

$$3.2156 \times 10^{15} - 0.04631 \times 10^{15} = 2.16929 \times 10^{15} = 3.1693 \times 10^{15}$$

(d) As in (c):

$$0.241 \times 10^{-25} - 7.83 \times 10^{-25} = -7.589 \times 10^{-25} = -7.59 \times 10^{-25}$$

A-21. (a) $\qquad -8.4008 = -8.40$

(b) $\qquad 0.14735 = 0.147$

(c) $\quad 3.2446 \times 10^{-12} = 3.24 \times 10^{-12}$

(d) $\qquad 4.464 \times 10^{13} = 4.5 \times 10^{13}$

A-23. The formula for the area of a triangle is:

$$Area = \frac{1}{2} \times base \times altitude$$

$$= \frac{1}{2} \times (42.07 \text{ cm}) \times (16.0 \text{ cm})$$

$$= 336.56 \text{ cm}^2 = 337 \text{ cm}^2$$

APPENDIX B - S.I. UNITS AND THE CONVERSION OF UNITS

B-1. (a) 65.2 nanograms $\times \left(\dfrac{10^{-9} \text{ grams}}{\text{nanograms}} \right) \times \left(\dfrac{1 \text{ kg}}{1000 \text{ grams}} \right) = \mathbf{6.52 \times 10^{-11}}$ **kg**

(b) 88 picoseconds $\times \left(\dfrac{10^{-12} \text{ seconds}}{1 \text{ picosecond}} \right) = \mathbf{8.8 \times 10^{-11}}$ **s**

(c) 5.4 terawatts $\times \left(\dfrac{10^{12} \text{ watts}}{1 \text{ terawatt}} \right) \times \left(\dfrac{1 \text{ kg m}^2 \text{ s}^{-3}}{1 \text{ watt}} \right) = \mathbf{5.4 \times 10^{12}}$ **kg m^2 s^{-3}**

(d) 17 kilovolts $\times \left(\dfrac{1000 \text{ volts}}{1 \text{ kilovolt}} \right) \times \left(\dfrac{1 \text{ kg m}^2 \text{ s}^{-3} \text{ A}^{-1}}{1 \text{ volt}} \right) = \mathbf{1.7 \times 10^{4}}$ **kg m^2 s^{-3} A^{-1}**

B-3. $1 \text{ Wb} = 1 \text{ Vs} = 1 \ (\text{kg m}^2 \text{ s}^{-3} \text{ A}^{-1}) \ (\text{s}) = \mathbf{1 \text{ kg m}^2 \text{ s}^{-2} \text{ A}^{-1}}$

B-5. We should first note what each value is on a common scale, the liter:

$100 \text{ cm}^3 = 0.100 \text{ L}$ $500 \text{ mL} = 0.500 \text{ L}$

$100 \text{ dm}^3 = 100 \text{ L}$ $1.5 \text{ gall} = 5.678 \text{ L}$

$100 \text{ cm}^3 < 500 \text{ mL} < 1.5 \text{ gall (US)} < 100 \text{ dm}^3 < 150 \text{ L}$

B-7. The general formula is: $^{\circ}C = \dfrac{5 \ ^{\circ}C}{9 \ ^{\circ}F} \left(t_{\circ F} - 32 \ ^{\circ}F \right)$

(a) 4983 $^{\circ}$C

(b) 37.0 $^{\circ}$C

(c) The boiling point of water at 1 atm pressure is 212 $^{\circ}$F.
Then $^{\circ}C = 111 \ ^{\circ}C$

(d) -40. $^{\circ}$C (This is the only temperature where the two scales are exactly equal.)

B-9. (a) 5256 K   (b) 310 K

(c) 384 K   (d) 233 K

B-11.

Physical Quantity	Name of Unit	Value in Familiar Units
volume	m^3	264.1722 gall (US)
length	m	39.37008...in
mass	kg	2.2046226 lb
energy	J	0.2390057... cal

B-13. (a) $\left(55.0\ \dfrac{\text{miles}}{\text{hr}}\right) \times \left(\dfrac{1609.344\ \text{m}}{\text{mile}}\right) \times \left(\dfrac{1\ \text{hour}}{3600\ \text{s}}\right) = \mathbf{24.6\ m\ s^{-1}}$

(b) $\left(1.51\ \dfrac{\text{g}}{\text{cm}^3}\right) \times \left(\dfrac{1\ \text{kg}}{1000\ \text{g}}\right) \times \left(\dfrac{100\ \text{cm}}{1\ \text{m}}\right)^3 = \mathbf{1.51 \times 10^3\ kg\ m^{-3}}$

(c) $\left(1.6 \times 10^{-19}\ \text{C Å}\right) \times \left(\dfrac{1\ \text{A}}{1\ \text{C-s}^{-1}}\right) \times \left(\dfrac{1.0 \times 10^{-10}\ \text{m}}{1\ \text{Å}}\right) = \mathbf{1.6 \times 10^{-29}\ A\ s\ m}$

(d) $\left(0.15\ \dfrac{\text{mol}}{\text{L}}\right) \times \left(\dfrac{1\ \text{L}}{1\ \text{dm}^3}\right) \times \left(\dfrac{10\ \text{dm}}{\text{m}}\right)^3 = \mathbf{1.5 \times 10^2\ mol\ m^{-3}}$

(e)

$\left(5.7 \times 10^3\ \dfrac{\text{L atm}}{\text{min}}\right) \times \left(\dfrac{1\ \text{dm}^3}{1\ \text{L}}\right) \times \left(\dfrac{10\ \text{m}}{10\ \text{dm}}\right)^3$

$\times \left(\dfrac{101{,}325\ \text{Pa}}{1\ \text{atm}}\right) \times \left(\dfrac{1\ \text{kg m}^2\ \text{s}^{-2}}{1\ \text{Pa}}\right) \times \left(\dfrac{1\ \text{min}}{60\ \text{s}}\right)$

$= 9.6 \times 10^3\ \text{kg m}^2\ \text{s}^{-3} = \mathbf{9.6 \times 10^3\ W}$

B-15. Lb in^{-3} and $\mu\ \text{L}^{-1}$ are obviously mass ÷ volume units. For the others:

$$\text{Pa m}^2\ \text{s}^{-2} \times \left(\dfrac{1\ \text{kg m}^{-1}\ \text{s}^{-2}}{1\ \text{Pa}}\right) = \text{kg m s}^{-4}$$

$$\text{J m}^{-1}\ \text{s}^{-2} \times \left(\dfrac{1\ \text{kg m}^2\ \text{s}^{-2}}{1\ \text{J}}\right) = 1\ \text{kg m s}^{-4}$$

These are not density units.

B-17. $(1\ \text{kw-h}) \times \left(\dfrac{1000\ \text{w}}{\text{kw}}\right) \times \left(\dfrac{1\ \text{kg m}^2\ \text{s}^{-3}}{1\ \text{w}}\right) \times \left(\dfrac{3600\ \text{s}}{\text{hr}}\right)$

$= 3.6 \times 10^6\ \text{kg m}^2\ \text{s}^{-2} = 3.6 \times 10^6\ \text{J}$

$15.3\ \text{kw-h} = \mathbf{5.51 \times 10^7\ J}$

B-19. $(404\ \text{in}^3) \times \left(\dfrac{2.54\ \text{cm}}{1\ \text{in}}\right)^3 \times \left(\dfrac{1\ \text{L}}{1000\ \text{cm}^3}\right) = \mathbf{6.62\ L}$

APPENDIX C - MATHEMATICS FOR GENERAL CHEMISTRY

C-1 (a) 4.551   (b) 1.53×10^7 (c) 2.57×10^8 (d) -48.7264

C-3. The number x with the common logarithm 0.4793 is the number x in the equation:

$$\log_{10} x = 0.4793$$
$$x = 10^{+0.4793} = \mathbf{3.015}$$

C-5. Many calculators will not accept exponents beyond 99. But we can solve this equation in parts, using the rules for manipulation of logarithms:

$$\log_{10}(3.00 \times 10^{121}) = \log_{10}(3.00) + \log_{10}(10^{121})$$

C-7. In a logarithmic function, the "power of 10" determines the number *before* the decimal. The pre-exponential determines the number *after* the decimal.

Therefore: $\log_{10}(5.64 \times 10^7) = 7 + .751 = \mathbf{7.751}$

$$\log_{10}(5.64 \times 10^{-3}) = -3 + .751 = \mathbf{-2.249}$$

C-9. (a) slope $= \dfrac{\text{change in distance}}{\text{change in time}} = \dfrac{150 \text{ miles - 75 miles}}{3.0 \text{ h - 1.5 h}} = 50 \dfrac{\text{miles}}{\text{hour}}$

C-11. (a) Slope = 4   Intercept = -7

(b)
$$7x - 2y = 5$$
$$-2y = -7x + 5$$
$$y = +\frac{7}{2}x - \frac{5}{2}$$

Slope $= \dfrac{7}{2}$ Intercept $= -\dfrac{5}{2}$

(c)
$$3y + 6x - 4 = 0$$
$$3y = -6x + 4$$
$$y = -2x + \frac{4}{3}$$

Slope = -2   Intercept $= \dfrac{4}{3}$

C-13.

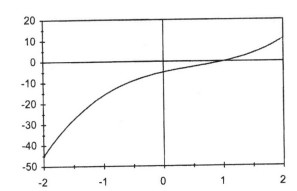

This is not a straight line!

C-15. (a) $7x + 5 = 0$

$7x = -5$

$x = -\dfrac{5}{7}$

(b) $-4x + 3 = 0$

$-4x = -3$

$x = \dfrac{3}{4}$

(c) $-3x = -2$

$x = \dfrac{2}{3}$

C-17. (a) We can take the first equation and rearrange to $x = \dfrac{1}{2} + \dfrac{5}{2}y$. This is substituted into the second equation to obtain $y = 1$. Going back to the first equation gives us $x = 3$.

(b) We note that both the second and the third equations contain the term "$+ y + z$". So we can subtract the second from the third equation to get $x = 1$.

The first equation then is "$2 - y + z = 6$" or "$-y + z = 4$" and the second equation is "$2 + y + z = 3$" or "$y + z = -1$" If we add these two equations, the y's will cancel by subtraction and we will get $2z = 5$, or $z = 5/2$. Finally we go back to any of the equations and calculate $y = -3/2$

(c) Subtracting the third from the second equation gives $x = 1$. The first equation then becomes "$2 - y + 2z = 6$" or "$-y + 2z = 4$". Similarly, the second equation can be converted to "$+y + z = 1$" Adding these two equations cancels the y by subtraction and gives $3z = 5$, or $z = 5/3$. Finally we get to $y = -2/3$.

C-19. (a) $a = 4; b = 7; c = -5$

$$x = \dfrac{-7 \pm \sqrt{(7)^2 - 4(4)(-5)}}{8}$$

$$= \dfrac{-7 \pm \sqrt{129}}{8} = \dfrac{-7 \pm 11.36}{8} = +0.5447 \text{ and } -2.295$$

C-19. (continued)

(b)
$$2x^2 = -3 - 6x$$
$$2x^2 + 6x + 3 = 0 \qquad\qquad a = 2;\ b = 6;\ c = 3$$

ROOTS: x = **-0.6340 and -2.366**

(c)
$$2x + \frac{3}{x} = 6$$

$$2x^2 + 3 = 6x$$

$$2x^2 - 6x + 3 = 0 \qquad\qquad a = 2,\ b = -6;\ c = 3.$$

ROOTS: x = **+0.6340 and +2.366**

C-21. (a) We will try an assumption that x << 2.00:

$$x\,(2.00)^2 \approx 2.6 \times 10^{-6}$$

$x \approx 1.65 \times 10^{-7}$. This is sufficient to justify the assumption

There are three roots to every third order polynomial. We have found one by approximation of small x. Others can be detected by a wide-ranging graph of the function:

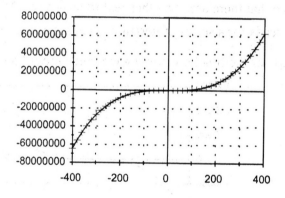

There is clearly only one real root. The other two roots must be complex.

(b) We will start with an assumption of x small.

$$x\,(3.00)\,(2.00) \approx 0.230$$

$$x = 0.046$$

This is **not** small enough to justify the assumption. So we will use it as a key for a solution by successive approximation.

$$x(3.00 - 7(0.046))(2.00 + 2\,(0.046)) \approx 0.230$$

$$x\,(2.678)\,(2.092) = 0.230$$

$$x = 0.043$$

C-21. (continued)

A third iteration using 0.-43 as the key gives x = 0.041. Further iterations refine this value to **0.0407.**

We must check for other roots here, too. A narrow range graph of the function looks like:

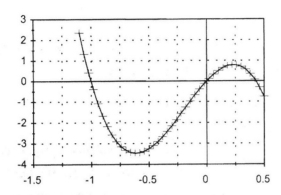

We can see that there are two other real roots -- around 0.40 and -1.00. Using these values as seeds for a series of iterations yields the roots x = +0.399 and x = -1.011.

(c) A graph of the function over a wide range looks like this:

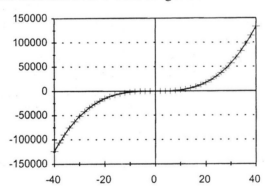

C-21. (continued)

We can "zoom in" to get a better view of the intercept point:

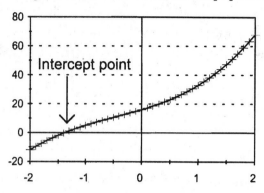

And zooming in even more gives:

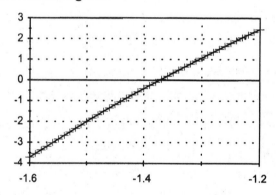

From this we can see that a solution to the equation occurs at $x \approx -1.37$. More precise graphing yields a root at x = **-1.3723**.

C-23. The requirement that the argument of a logarithm -- base 10 or otherwise -- be greater than 0 restricts the domain of the function to $x > 1$ so that
$\ln x > 0$. A wide graph of the function yields the graph on the left.

C-23. (continued)

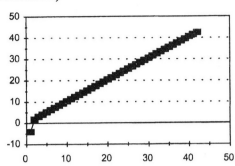 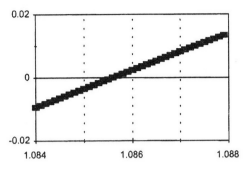

We can see that the function does have an intercept -- at about 1.2. Zooming in, we find from the graph on the right that the intercept is closest to $x = 1.086$.

C-25. (a) The total mass of the system is 24.60 g + 17.2 g + 0.0020 g = 41.802 g

Note that the last two digits in the total mass are not significant, but they *should* be retained in calculations of ratios, to avoid round-off errors..

$$\text{mass fraction of water} = \frac{24.60 \text{ g}}{41.802 \text{ g}} = 0.588$$

$$\text{mass fraction of NaCl} = \frac{17.2 \text{ g}}{41.802 \text{ g}} = 0.411$$

$$\text{mass fraction of mercury} = \frac{0.0020 \text{ g}}{41.802 \text{ g}} = 4.8 \times 10^{-5}$$

(b) the total volume is 32.614147 mL. The volume fractions are 0.756 water, 0.244 NaCl, and 4.5×10^{-6} mercury.

(c) ppmv = volume fraction $\times 10^{6}$ = 4.5 ppmv

(d) ppbm mass fraction $\times 10^{9}$ = 48,000 ppbm